LONDON MATHEMATICAL SOCIETY LECTURE NOTE SERIES

Managing Editor: Professor J.W.S. Cassels, Department of Pure Mathematics and Mathematical Statistics, University of Cambridge, 16 Mill Lane, Cambridge CB2 1SB, England

The books in the series listed below are available from booksellers, or, in case of difficulty, from Cambridge University Press.

34 Representation theory of Lie groups, M.F. ATIYAH *et al*
36 Homological group theory, C.T.C. WALL (ed)
39 Affine sets and affine groups, D.G. NORTHCOTT
40 Introduction to H_p spaces, P.J. KOOSIS
43 Graphs, codes and designs, P.J. CAMERON & J.H. VAN LINT
45 Recursion theory: its generalisations and applications, F.R. DRAKE & S.S. WAINER (eds)
46 p-adic analysis: a short course on recent work, N. KOBLITZ
49 Finite geometries and designs, P. CAMERON, J.W.P. HIRSCHFELD & D.R. HUGHES (eds)
50 Commutator calculus and groups of homotopy classes, H.J. BAUES
51 Synthetic differential geometry, A. KOCK
57 Techniques of geometric topology, R.A. FENN
59 Applicable differential geometry, M. CRAMPIN & F.A.E. PIRANI
62 Economics for mathematicians, J.W.S. CASSELS
66 Several complex variables and complex manifolds II, M.J. FIELD
69 Representation theory, I.M. GELFAND *et al*
74 Symmetric designs: an algebraic approach, E.S. LANDER
76 Spectral theory of linear differential operators and comparison algebras, H.O. CORDES
77 Isolated singular points on complete intersections, E.J.N. LOOIJENGA
78 A primer on Riemann surfaces, A.F. BEARDON
79 Probability, statistics and analysis, J.F.C. KINGMAN & G.E.H. REUTER (eds)
80 Introduction to the representation theory of compact and locally compact groups, A. ROBERT
81 Skew fields, P.K. DRAXL
82 Surveys in combinatorics, E.K. LLOYD (ed)
83 Homogeneous structures on Riemannian manifolds, F. TRICERRI & L. VANHECKE
86 Topological topics, I.M. JAMES (ed)
87 Surveys in set theory, A.R.D. MATHIAS (ed)
88 FPF ring theory, C. FAITH & S. PAGE
89 An F-space sampler, N.J. KALTON, N.T. PECK & J.W. ROBERTS
90 Polytopes and symmetry, S.A. ROBERTSON
91 Classgroups of group rings, M.J. TAYLOR
92 Representation of rings over skew fields, A.H. SCHOFIELD
93 Aspects of topology, I.M. JAMES & E.H. KRONHEIMER (eds)
94 Representations of general linear groups, G.D. JAMES
95 Low-dimensional topology 1982, R.A. FENN (ed)
96 Diophantine equations over function fields, R.C. MASON
97 Varieties of constructive mathematics, D.S. BRIDGES & F. RICHMAN
98 Localization in Noetherian rings, A.V. JATEGAONKAR
99 Methods of differential geometry in algebraic topology, M. KAROUBI & C. LERUSTE
100 Stopping time techniques for analysts and probabilists, L. EGGHE

101 Groups and geometry, ROGER C. LYNDON
103 Surveys in combinatorics 1985, I. ANDERSON (ed)
104 Elliptic structures on 3-manifolds, C.B. THOMAS
105 A local spectral theory for closed operators, I. ERDELYI & WANG SHENGWANG
106 Syzygies, E.G. EVANS & P. GRIFFITH
107 Compactification of Siegel moduli schemes, C-L. CHAI
108 Some topics in graph theory, H.P. YAP
109 Diophantine Analysis, J. LOXTON & A. VAN DER POORTEN (eds)
110 An introduction to surreal numbers, H. GONSHOR
111 Analytical and geometric aspects of hyperbolic space, D.B.A.EPSTEIN (ed)
112 Low-dimensional topology and Kleinian groups, D.B.A. EPSTEIN (ed)
113 Lectures on the asymptotic theory of ideals, D. REES
114 Lectures on Bochner-Riesz means, K.M. DAVIS & Y-C. CHANG
115 An introduction to independence for analysts, H.G. DALES & W.H. WOODIN
116 Representations of algebras, P.J. WEBB (ed)
117 Homotopy theory, E. REES & J.D.S. JONES (eds)
118 Skew linear groups, M. SHIRVANI & B. WEHRFRITZ
119 Triangulated categories in the representation theory of finite-dimensional algebras, D. HAPPEL
121 Proceedings of *Groups - St Andrews 1985*, E. ROBERTSON & C. CAMPBELL (eds)
122 Non-classical continuum mechanics, R.J. KNOPS & A.A. LACEY (eds)
124 Lie groupoids and Lie algebroids in differential geometry, K. MACKENZIE
125 Commutator theory for congruence modular varieties, R. FREESE & R. MCKENZIE
126 Van der Corput's method for exponential sums, S.W. GRAHAM & G. KOLESNIK
127 New directions in dynamical systems, T.J. BEDFORD & J.W. SWIFT (eds)
128 Descriptive set theory and the structure of sets of uniqueness, A.S. KECHRIS & A. LOUVEAU
129 The subgroup structure of the finite classical groups, P.B. KLEIDMAN & M.W.LIEBECK
130 Model theory and modules, M. PREST
131 Algebraic, extremal & metric combinatorics, M-M. DEZA, P. FRANKL & I.G. ROSENBERG (eds)
132 Whitehead groups of finite groups, ROBERT OLIVER
133 Linear algebraic monoids, MOHAN S. PUTCHA
134 Number theory and dynamical systems, M. DODSON & J. VICKERS (eds)
135 Operator algebras and applications, 1, D. EVANS & M. TAKESAKI (eds)
136 Operator algebras and applications, 2, D. EVANS & M. TAKESAKI (eds)
137 Analysis at Urbana, I, E. BERKSON, T. PECK, & J. UHL (eds)
138 Analysis at Urbana, II, E. BERKSON, T. PECK, & J. UHL (eds)
139 Advances in homotopy theory, S. SALAMON, B. STEER & W. SUTHERLAND (eds)
140 Geometric aspects of Banach spaces, E.M. PEINADOR and A. RODES (eds)
141 Surveys in combinatorics 1989, J. SIEMONS (ed)
142 The geometry of jet bundles, D.J. SAUNDERS
143 The ergodic theory of discrete groups, PETER J. NICHOLLS
144 Introduction to uniform spaces, I.M. JAMES
145 Homological questions in local algebra, JAN R. STROOKER
146 Cohen-Macaulay modules over Cohen-Macaulay rings, Y. YOSHINO
147 Continuous and discrete modules, S.H. MOHAMED & B.J. MÜLLER
148 Helices and vector bundles, A.N. RUDAKOV et al
152 Oligomorphic permutation groups, P. CAMERON
154 Number theory and cryptography, J. LOXTON (ed)
156 Twistors in mathematics and physics, T.N. BAILEY & R.J. BASTON (eds)

London Mathematical Society Lecture Note Series. 145

Homological Questions in Local Algebra

Jan R. Strooker
Professor of Algebra, University of Utrecht

CAMBRIDGE UNIVERSITY PRESS
Cambridge
New York Port Chester Melbourne Sydney

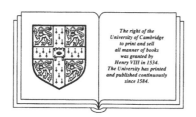

CAMBRIDGE UNIVERSITY PRESS
Cambridge, New York, Melbourne, Madrid, Cape Town, Singapore, São Paulo

Cambridge University Press
The Edinburgh Building, Cambridge CB2 8RU, UK

Published in the United States of America by Cambridge University Press, New York

www.cambridge.org
Information on this title: www.cambridge.org/9780521315265

First published 1990
Re-issued in this digitally printed version 2008

A catalogue record for this publication is available from the British Library

ISBN 978-0-521-31526-5 paperback

CONTENTS

PREFACE

It all began with Serre's beautiful 1957/58 course at the Collège de France on Algèbre locale - Multiplicités [Se]. Here he introduced a local algebraic version of the geometric notion of intersection multiplicity as follows. Let A be a d-dimensional regular local ring and M and N finitely generated A-modules such that $M \otimes_A N$ has finite length. Then he defined their intersection multiplicity $\chi(M,N)$ as $\Sigma_{j=1}^{d}(-1)^j \ell(\mathrm{Tor}_j^A(M,N))$ and the course culminated in the three statements

(M_0) $\dim M + \dim N \leq d$;

(M_1) In case of an inequality above, $\chi(M,N) = 0$;

(M_2) In case there is equality, $\chi(M,N) > 0$.

Serre proved (M_0) in general and (M_1) and (M_2) except for ramified local rings in mixed characteristic. He then cagily added: "Il est naturel de conjecturer que ces résultats sont vrais pour tous les anneaux réguliers.", thus raising the first of the homological questions of the type this book addresses. Almost 30 years later, (M_1) was proved by P. Roberts [Ro 85], [Ro 87b] and by Gillet-Soulé [GS] using different methods, but (M_2) remains open in the ramified case.

That the (M_i) are not true for an arbitrary noetherian local ring is seen in the example of a 3-dimensional local domain $A = k[[X,Y,U,V]]/(XY-UV)$, k a field, and its 2-dimensional modules

$M = A/(X,U)$, $N = A/(Y,V)$. Since $M \otimes_A N$ has length 1, statement (M_0) is violated.

It was then surmised that the operative fact for regular rings - which in those days had just been established by Auslander-Buchsbaum and by Serre himself - is that every module has finite projective dimension. To exclude examples as above, the (M_i) were therefore conjectured for an arbitrary noetherian local ring provided M or perhaps M and N possessed finite projective dimension. This was the point of view taken by M. Auslander, which stimulated a lot of research. It therefore created quite a stir when a module M of finite length and projective dimension was constructed over the 3-dimensional ring A above, such that for our module N the intersection multiplicity $\chi(M,N) = -1$, though $\dim M + \dim N = 2 < 3$, contradicting a generalization of (M_i), $i = 1,2$ with only M of finite projection dimension [DHM]. Over this (and any) complete intersection ring A however, (M_1) is true if both M and N have finite projective dimension, [Ro 85], [Ro 87b], [GS].

In this book I shall concentrate on such a generalization of (M_0) or rather a weaker form as explained in section 8.5. This Intersection Conjecture was proved by Peskine-Szpiro in their pioneering joint thesis [PS 73] in positive characteristic and for rings essentially of finite type over a field of characteristic 0, a restriction which was soon removed by Hochster [Ho 75a]. The statement is now the Intersection Theorem 8.4.4, since P. Roberts recently managed to prove the mixed characteristic case [Ro 87a], [Ro 89]. Fittingly perhaps, his main tool is the theory of local Chern characters and their Riemann-Roch Theorem, which is part of the Fulton-MacPherson treatment of intersection multiplicities in algebraic geometry [Fu], bringing the development back full circle to the underlying geometry. Of this part of Roberts' proof I can only give a brief indication in section 13.1.

The Intersection Theorem with related results and conjectures is a focal point in the book. I also treat the Direct Summand Conjecture and the Monomial Conjecture which are equivalent and prove them in equal characteristic, Theorem 10.3.5, along the general lines of [Ho 73b]. Hochster has shown in a long and rich paper [Ho 83] that the latter, ostensibly not overly homological, yet implies the Intersection Theorem and its consequences. But since the Monomial Conjecture is still open in mixed characteristic, I have not taken this route.

There are two main strategies for attacking these homological questions. The first deploys a range of subtle homological arguments, as in [PS 73], [Ro 76], [Ro 87a]. Most spectacular in the second approach is the brute force construction by Hochster of Big Cohen-Macaulay modules [Ho 75a]. Both methods have in common that they work in positive characteristic by exploiting the Frobenius endomorphism, and then obtain the result in equal characteristic zero by general methods, as explained in Chapter 12. There is however a notable exception, Roberts' proof of New Intersection over the complex field [Ro 80c].

In this book I combine the best of both worlds, plugging our Big Cohen-Macaulay module into the Acyclicity Lemma to obtain New Intersection, as first shown by Foxby [Fo 77b]. I present however a refinement of Hochster's original construction of Big Cohen-Macaulay modules which goes back to [BS] and [Ba].

I do not list all the homological conjectures, nor trace their often surprising interconnections. This has been done in [PS 73] and repeatedly and authoritatively by Hochster [Ho 73a], [Ho 75a], [Ho 75b], [Ho 78], [Ho 79], [Ho 82], [Ho 87]. To some extent I tried my hand at this in [St 85], which may serve as a synopsis of the main thrust of this book.

In writing a monograph of this kind, a problem is where to

start. I have decided to take standard homological algebra and the theory of homological dimension for granted, but in Chapter 3 rather carefully treat injective envelopes and Matlis duality. While commutative algebra is assumed at the level of, say, [AM], completion is discussed in Chapter 2 and regular sequences and parameters in Chapters 5 and 8 respectively. In this way I hope that the book will not only be of interest to commutative algebraists, but that mathematicians in other areas and also persevering graduate students, will be able to partake of these fascinating homological questions.

Now for acknowledgements. First I should like to salute Mel Hochster, whose unflagging enthusiasm for the homological conjectures drew me into the subject. This book originated in a course and seminar which I taught more years ago than I care to remember. Several participants have greatly assisted in the writing of the book. Lex Vermeulen and Ton Vorst contributed drafts and outlines of sections in the early stages, and so did Dick Buijs. Jaap Bartijn deserves a special word of thanks. Quite a bit of material in the book derives from our collaboration on [BS] or from his thesis [Ba]. Moreover, we have often discussed the contents of the book and he has written drafts for certain chapters. Dick Buijs has throughout been a great help in reading innumerable versions of chapters. His pertinent comments have reduced the number of errors and improved readability. This also holds for Anne-Marie Simon who performed the same task during the later stages.

For a friendly gesture of a different nature I am indebted to Lou van den Dries. He offered to write a chapter where Big Cohen-Macaulay modules are shown to exist in equal characteristic 0 once they are known to exist in all finite characteristics. The argument really pertains in greater generality, and Van den Dries has given us his view of the matter in Chapter 12, which differs in certain respects from previous treatments. This chapter also offers a valuable introduction, at fairly elementary level, to Artin

Approximation and henselization. Since this material is almost entirely independent from the rest of the book, and was written several years ago, Utrecht University put out a tract and distributed a limited number of copies in 1983. Van den Dries would like to thank Craig Huneke for spotting a gap in the proof of Approximation in that early rendering.

"De Pauwhof" in Wassenaar deserves thanks for several sojourns of sustained work. The publisher, lastly incarnated in David Tranah, has shown courtesy and patience throughout. Finally, it is a pleasure to thank Karin Berlang for cheerfully and competently processing a hundred visions and revisions before the taking of a toast and tea.

Utrecht, Jan R. Strooker

December 1989

1 HOMOLOGICAL PRELIMINARIES

In writing a book on an advanced topic one cannot start from scratch, so one needs to make a choice of basic subjects which the reader is supposed to be familiar with. Here we assume a working knowledge of homological algebra and the theory of various homological dimensions, as expounded for instance in standard texts like [CE], [No 60], [HSa] or [Rot].

In this brief chapter we merely explain a few technical devices which are less uniformly known. They concern the exactness of certain complexes of modules. In the first section we deal with abstract Acyclicity Lemma's for single complexes in abelian categories, which will i.a. be used to deduce the parent version due to Peskine-Szpiro, 6.1.1. We also restate uniqueness of derived functors in a form suitable to our purposes. In 1.2 we consider tensor products of complexes of modules, preparing for certain results in Chapter 7. Most readers will hardly feel tempted to study this material for its own sake, but may like to refer back when it is needed in these later chapters.

1.1 ACYCLICITY LEMMA'S

Let $\mathscr{C}$ and $\mathscr{D}$ be abelian categories. A sequence F^i, $i = 0,1,2,\ldots$, of covariant additive functors from $\mathscr{C}$ to $\mathscr{D}$ is called right connected exact if to each short exact sequence $0 \to C' \to C \to C'' \to 0$ in $\mathscr{C}$ there corresponds a functorial long exact sequence

$$0 \to F^0C' \to F^0C \to F^0C'' \to F^1C' \to \ldots \to F^iC' \to F^iC \to F^iC'' \to F^{i+1}C' \to \ldots \text{ in } \mathscr{D}.$$

For each object C in $\mathscr{C}$ we put $f^-C = \inf\{i \mid F^iC \neq 0\}$. Notice that f^-C is either ∞ or a nonnegative integer.

1.1.1 PROPOSITION. Let F^i be a right connected exact sequence of functors from $\mathscr{C}$ to $\mathscr{D}$. If $C_\bullet = 0 \to C_s \overset{d_s}{\to} C_{s-1} \to \ldots \overset{d_1}{\to} C_0$ is a complex in $\mathscr{C}$ such that for $i = 1,\ldots,s$ both $f^-C_i \geq i$ and either $H_i(C_\bullet) = 0$ or $f^-H_i(C_\bullet) = 0$, then $C_\bullet$ is exact.

PROOF. Writing H_i for the homology $H_i(C_\bullet)$, we shall successively show that $H_s = 0$, $H_{s-1} = 0$, ... , $H_1 = 0$. If $H_s \neq 0$, then $F^0H_s \neq 0$, and the monomorphism $H_s \to C_s$ yields $F^0C_s \neq 0$ which contradicts the fact that $f^-C_s \geq s > 0$.

For each i we shall write $T_i = \mathrm{coker}(d_{i+1})$ and $K_i = \ker(d_i)$. Then $T_s = C_s$ and $f^-T_s \geq s$, so that, for $0 < r < s$, our induction hypothesis may read: $H_i = 0$ and $f^-T_i \geq i$ for $0 < r < i \leq s$.

To take the induction step, consider the two sequences $0 \to T_{r+1} \to C_r \to T_r \to 0$ and $0 \to T_{r+1} \to K_r \to H_r \to 0$ which are exact because $H_{r+1} = 0$. The first one gives rise to an exact sequence $F^jC_r \to F^jT_r \to F^{j+1}T_{r+1}$ for each $j \geq 0$. Since $f^-T_{r+1} \geq r+1$ and $f^-C_r \geq r$, we find $f^-T_r \geq r$. The second one induces an exact sequence $0 \to F^0T_{r+1} \to F^0K_r \to F^0H_r \to F^1T_{r+1} \to \ldots$ in $\mathscr{D}$. Since $r \geq 1$, we know that $F^0T_{r+1} = F^1T_{r+1} = 0$. If $H_r \neq 0$, then $F^0H_r \neq 0$ and thus $F^0K_r \neq 0$. But then $F^0C_r \neq 0$, which contradicts our assumptions.

One of the advantages of such an abstract formulation in terms of abelian categories, first noticed by A.-M. Simon [Si], [St 90], is that we right away have a dual statement. Terminology and notation should be self-explanatory.

1.1.2 PROPOSITION. Let F_i be a left connected exact sequence of functors between $\mathcal{C}$ and $\mathcal{D}$. If $C^\bullet = C^0 \to C^1 \to \ldots \to C^s \to 0$ is a complex in $\mathcal{C}$ such that, for $i = 1,\ldots,s$, both $f\_C^i \geq i$ and either $H^i(C^\bullet) = 0$ or $f\_H^i(C^\bullet) = 0$, then $C^\bullet$ is exact.

It is left to the reader to formulate companion versions for contravariant functors. To prepare for our next proposition, suppose that $F = (F^i)$ and $G = (G^i)$ are both right connected exact sequences of functors from $\mathcal{C}$ to $\mathcal{D}$. A morphism σ: $F \to G$ is a collection of morphisms σ^i: $F^i \to G^i$, $i \geq 0$, such that for every short exact sequence $0 \to C' \to C \to C'' \to 0$ in $\mathcal{C}$ the square

$$\begin{array}{ccc} F^iC'' & \longrightarrow & F^{i+1}C' \\ \downarrow{\scriptstyle \sigma^i(C'')} & & \downarrow{\scriptstyle \sigma^{i+1}(C')} \\ G^iC'' & \longrightarrow & G^{i+1}C' \end{array}$$

commutes in $\mathcal{D}$. The result we shall need does not occur in this form in the standard texts, but is an immediate consequence of their treatment of satellites and derived functors. A more general version is proved in [St 78, Th. 3.4.3].

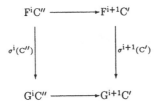

1.1.3 PROPOSITION. Let $F = (F^i)$ and $G = (G^i)$ both be right connected exact sequences of covariant functors between abelian categories $\mathscr{C}$ and $\mathscr{D}$. Suppose there is an isomorphism σ^0: $F^0 \to G^0$. This extends to an isomorphism σ: $F \to G$ provided $F^iI = G^iI = 0$ for every $i \geq 1$ and every injective object I in $\mathscr{C}$.

We shall also have occasion to use two dual versions, where the higher functors are required to vanish on projectives.

1.2 A FEW ISOMORPHISMS OF COMPLEXES

We shall need several facts on complexes, which are of a fairly simple nature, because key results concern complexes of vector spaces. Ad hoc proofs could certainly be provided. On the other hand, this material is rightfully part of the theory of derived categories, to which there exists several introductions in [Ha], [De], [Bor] and [Iv].

We shall strike a compromise, since fortunately the basic techniques which we shall use are treated carefully and in detail in the treatise [Bo 80]. In this section therefore we adhere to notation and terminology of Bourbaki and often refer to this work for the proofs. Our treatment is in the spirit of [Fo 77a] and [Fo 79].

We take a fixed (commutative) ring A and work with complexes $C^\bullet$ and $C_\bullet$ of A-modules. Here $C^\bullet$ will always stand for an ascending complex $\ldots \to C^{i-1} \overset{d^{i-1}}{\to} C^i \overset{d^i}{\to} C^{i+1} \to \ldots$ with $d^i \circ d^{i-1} = 0$ and $C_\bullet$ for a descending complex with d_i: $C_i \to C_{i-1}$ and $d_i \circ d_{i+1} = 0$. It is sometimes convenient to switch from one to the other by writing $C_i = C^{-i}$ and $d_i = d^{-i}$.

A morphism (of degree 0) u: $C^\bullet \to D^\bullet$ is called a **homologism** between these complexes if it induces an isomorphism $H(u)$: $H(C^\bullet) \simeq H(D^\bullet)$ [Bo 80, §2, Déf. 3]. Most authors speak of a quasi-isomorphism or quism, but Bourbaki's term is more descriptive. Standard examples of homologisms are provided by resolutions. If C is a module, one can regard it as a complex $C^\bullet$ by putting $C^0 = C$ and $C^i = 0$ for all $i \neq 0$. If $0 \to C \xrightarrow{\epsilon} E^0 \to E^1 \to \ldots$ is for instance an injective resolution of C, and $E^\bullet$ is the complex $0 \to E^0 \to E^1 \to \ldots$, a homologism u: $C^\bullet \to E^\bullet$ is defined by $u = (u^i)$ with $u^0 = \epsilon$ and $u^i = 0$ for $i \neq 0$.

Given two complexes $C_\bullet$ and $D_\bullet$, we have the notion of the tensor product complex $C_\bullet \otimes_A D_\bullet$ [Bo 80, §4.1]. It is a descending complex with chain modules $(C_\bullet \otimes_A D_\bullet)_n = \oplus_{i+j=n} C_i \otimes_A D_j$. For the differential distinct conventions are in use, one of which is chosen by Bourbaki. We shall not make this explicit, since the reader is referred back for the proofs anyway. Similarly for $C_\bullet$, and $D^\bullet$ an ascending complex, $\mathrm{Homgr}_A(C_\bullet, D^\bullet)$ is defined [Bo 80, §5.1]. In what follows we shall only need this when $C_\bullet$ is concentrated in degree 0, in other words may be thought of as a module $C = C_0$, and it is easy to see that then $\mathrm{Homgr}_A(C_\bullet, D^\bullet) \simeq \mathrm{Hom}_A(C, D^\bullet)$ so that we may forget about the fancier notion.

1.2.1 PROPOSITION. Let u: $C^\bullet \to D^\bullet$ be a homologism of complexes and $F^\bullet$ a complex of flat modules with $F^i = 0$ for i sufficiently large. Then $1_{F^\bullet} \otimes u$: $F^\bullet \otimes_A C^\bullet \to F^\bullet \otimes_A D^\bullet$ is a homologism.

PROOF. This is [Bo 80, §4, Prop. 4].

1.2.2 PROPOSITION. Let A be a noetherian ring and C a finitely generated A-module. Suppose $F^\bullet$ is a complex of flat modules and $E^\bullet$ a complex of injective modules. Then there is an isomorphism of complexes

$F^\bullet \otimes_A \mathrm{Hom}_A(C,E^\bullet) \simeq \mathrm{Hom}_A(C,F^\bullet \otimes_A E^\bullet)$ and the complex $F^\bullet \otimes_A E^\bullet$ consists of injectives.

PROOF. For each pair of indices i and j, there is a natural homomorphism from $F^i \otimes_A \mathrm{Hom}_A(C,E^j)$ to $\mathrm{Hom}_A(C,F^i \otimes_A E^j)$ which is an isomorphism because C is finitely presented and F^i is flat [Bo 61a, Ch. 1, §2, Prop. 10]. It is straightforward to show that these maps commute with the differentials, hence yield the desired isomorphism of complexes. In fact, the compatibility is discussed in the greater generality of Homgr in [Bo 80, §5, 7, (16)].

For each i and j the isomorphism above extends to isomorphisms $F^i \otimes_A \mathrm{Ext}^n_A(C,E^j) \simeq \mathrm{Ext}^n_A(C,F^i \otimes_A E^j)$, $n \geq 0$, by [Bo 80, §6, Prop. 7c)] and, since $\mathrm{Ext}^n_A(C,E^j) = 0$ for $n \geq 1$, so is $\mathrm{Ext}^n_A(C,F^i \otimes_A E^j)$. Now it is enough to test injectivity for cyclic modules $C = A/\mathfrak{a}$, $\mathfrak{a}$ an ideal in A, so the $F^i \otimes_A E^j$ are all injective. Since over a noetherian ring every direct sum of injective modules is injective, all chain modules in the complex $F^\bullet \otimes_A E^\bullet$ are injective. For the two facts on injective modules which we have just used, the reader may for instance consult section 3.1 of this book.

1.2.3 PROPOSITION. Let $V_\bullet$ and $W_\bullet$ be complexes of vector spaces over a field k and suppose that either $V_i = 0$ for i sufficiently small or $W_j = 0$ for j sufficiently small. Then $H(V_\bullet) \otimes_k H(W_\bullet) \simeq H(V_\bullet \otimes_k W_\bullet)$. In other words, for each pair of indices i and j, there is an isomorphism of vector spaces $H_n(V_\bullet \otimes_k W_\bullet) \simeq \oplus_{i+j=n} H_i(V_\bullet) \otimes_k H_j(W_\bullet)$.

PROOF. Since k-vector spaces are certainly k-flat, this is a special case of the Künneth formula [Bo 80, §4, Th. 3, Cor. 4].

As in the rest of this book, (A,m,k) will be standard notation for a local ring A with maximal ideal m and residue class field $k = A/m$. The next theorem is taken from [Ba, Ch. IV, Prop. 2.7], see also [BS, Th. 4.1].

1.2.4 THEOREM. Let (A,m,k) be a noetherian local ring and $F^\bullet = 0 \to F^{-s} \to \ldots \to F^0 \to 0$ a finite complex of flat A-modules. Let C be an A-module, $E^\bullet$ an injective resolution of C and $I^\bullet$ the complex $F^\bullet \otimes_A E^\bullet$. Then $I^\bullet$ is a complex of injectives with $I^n = 0$ for $n < -s$, and $H^n(I^\bullet) \simeq H^n(F^\bullet \otimes_A C)$. Furthermore $H^n(\mathrm{Hom}_A(k,I^\bullet)) \simeq \oplus_{i+j=n} H^i(F^\bullet \otimes_A k) \otimes_k \mathrm{Ext}^j_A(k,C)$.

PROOF. We have seen in 1.2.2 that $I^\bullet$ is a complex of injectives, while $I^n = 0$ for $n < -s$ by definition of the tensor product of complexes. The homologism $C \to E^\bullet$ is preserved by $1_{F^\bullet} \otimes_A -$ according to 1.2.1, so $H^n(I^\bullet) \simeq H^n(F^\bullet \otimes_A C)$.

Next observe that there is an isomorphism of complexes $F^\bullet \otimes_A \mathrm{Hom}_A(k,E^\bullet) \simeq \mathrm{Hom}_A(k,I^\bullet)$ by 1.2.2. Now $\mathrm{Hom}_A(k,E^\bullet)$ is a complex of k-vector spaces, so $F^\bullet \otimes_A \mathrm{Hom}_A(k,E^\bullet) \simeq F^\bullet \otimes_A k \otimes_k \mathrm{Hom}_A(k,E^\bullet)$. We invoke 1.2.3 with $V^\bullet = F^\bullet \otimes_A k$ and $W^\bullet = \mathrm{Hom}_A(k,E^\bullet)$, since that isomorphism could equally well have been stated for ascending complexes - here $V^i = 0$ for $i > 0$. Observing that $H^j(W^\bullet) \simeq \mathrm{Ext}^j_A(k,C)$, we obtain the final isomorphism of the theorem.

2. ADIC TOPLOGIES AND COMPLETIONS

The homological conjectures which are discussed in this book are local and, what is more, need only be proved over a complete noetherian local ring. The structure of such rings, expressed in the Cohen Structure Theorems, is heavily used in their solution. Apart from these, we only need results from this chapter incidentally.

There are in the literature several excellent accounts of the topics in the title [AM], [Bo 61b, Ch. 3], [ZS], [Ma 86], but these do tend to concentrate all too soon on finitely generated modules over noetherian rings. Nevertheless, there exists a quite attractive theory at least for arbitrary modules over noetherian rings, and in some cases we may even waive the noetherian condition. Since this book features the "construction" of certain infinitely generated complete modules with good properties, cf. Theorems 5.2.3 and 9.1.1, and complete modules are drawing an increasing measure of attention [Ba], [Si], we have chosen to present an outline of this theory. In doing so we recall several standard results without proof, and for the less standard ones give either a proof, hints for a proof or a reference.

In the first section we show how the notion of purity can serve in the realm of adic topologies when the Artin-Rees Lemma is unavailable. On the other hand, a pure submodule is a poor man's direct summand, and this paves the way to our proof of Hochster's Direct Summand Theorem in equal characteristic, 10.3.5.

The second section discusses completion, briefly yet in somewhat more detail than we shall need.

In the third and final section then, Hensel's Lemma quickly leads to a proof that an equicharacteristic complete noetherian local ring contains a field of representatives.

2.1 INDUCED TOPOLOGIES AND PURITY

Let $\mathfrak{a}$ be an ideal in the ring A and M an A-module. The descending set of ideals $\mathfrak{a}^t$, $t \geq 0$, is taken as a basis of open neighbourhoods of the 0-element of A. This defines the $\mathfrak{a}$-adic topology of A which then is a topological ring. The module M becomes a topological module over A by taking the submodules $\mathfrak{a}^tM$ to form a basis of open neighbourhoods of 0 in M. To each $m \in M$ we can attach $v(m) = \sup\{t \mid m \in \mathfrak{a}^tM\}$. The values v(m) are nonnegative integers or ∞, and we can define $d(m,n) = 2^{-v(m-n)}$ for m and n in M. This function d is a pseudometric on the topological module M, which is a metric precisely when $\cap_{t=0}^{\infty} \mathfrak{a}^tM = 0$, or in other words, when the module M is Hausdorff in its $\mathfrak{a}$-adic topology. It is well known that $\mathfrak{a}$-adic completions, defined by an inverse limit construction in the next section, can also be described in terms of Cauchy sequences with respect to the pseudometric d, and sometimes we take this point of view.

A chain (M_t): $M = M_0 \supset M_1 \supset \ldots \supset M_t \supset \ldots$ of submodules is called a filtration of M. It is an $\mathfrak{a}$-filtration if $\mathfrak{a}M_t \subset M_{t+1}$ for all t, and a stable $\mathfrak{a}$-filtration if moreover $\mathfrak{a}M_t = M_{t+1}$ for all sufficiently large t. For instance, $(\mathfrak{a}^tM)$ is a stable $\mathfrak{a}$-filtration of M.

An $\mathfrak{a}$-filtration makes M into a topological A-module, and the topology defined by a stable $\mathfrak{a}$-filtration is just the $\mathfrak{a}$-adic topology of M. As is frequently the case, one of the fundamental results in the theory is humbly known as a lemma, the Artin-Rees Lemma:

2.1.1 THEOREM. Let $\mathfrak{a}$ be an ideal in a noetherian ring A and M a finitely generated A-module, with submodule N. There exists an integer $s \geq 0$ such that $\mathfrak{a}^t M \cap N = \mathfrak{a}^{t-s}(\mathfrak{a}^s M \cap N)$ for all $t \geq s$. In particular, the $\mathfrak{a}$-adic topology of N coincides with the topology induced by the $\mathfrak{a}$-adic topology of M.

To see that finite generation of M is essential, consider the example $A = \mathbf{Z}_{(p)}$, p a prime, $\mathfrak{a} = (p)$, $M = \mathbf{Q}$, $A = N \subset M$. Then the $\mathfrak{a}$-adic topology on M is the indiscrete topology hence far from Hausdorff. Thus the induced topology on N is also indiscrete and non Hausdorff. However, the $\mathfrak{a}$-adic toplogy on N is clearly Hausdorff, which also follows from the more general

2.1.2 PROPOSITION. Let $\mathfrak{a}$ be a proper ideal in a noetherian domain A. Then $\cap_{t=0}^{\infty} \mathfrak{a}^t = 0$, so A is Hausdorff in its $\mathfrak{a}$-adic topology.

This is a special case of the next result, Krull's celebrated Intersection Theorem. Though Krull's work is older, a particularly efficient proof using 2.1.1 is given in [AM, Th. 10.17].

2.1.3 THEOREM. Let $\mathfrak{a}$ be an ideal in a noetherian ring A and M a finitely generated A-module equipped with the $\mathfrak{a}$-adic topology. Then the closure of 0 in M is $\cap_{t=0}^{\infty} \mathfrak{a}^t M$ and consists of those $m \in M$ which are annihilated by some element of $1 + \mathfrak{a}$.

2.1.4 COROLLARY. If moreover $\mathfrak{a}$ is contained in the radical of A, then $\bigcap_{t=0}^{\infty} \mathfrak{a}^t M = 0$ and M is Hausdorff in its $\mathfrak{a}$-adic topology.

As we have seen, these results in general do not hold for infinitely generated modules, but we can sometimes replace the Artin-Rees Lemma by a strong condition imposed on a submodule: purity. Under a different name, this notion was used by Prüfer in the theory of infinite abelian groups, but it fell to P.M. Cohn to generalize it to modules [Cohn]. He also established some of its basic properties. Later on, Hochster and J.L. Roberts [HR 74], [HR 76] and Griffith [Gr 76], [Gr 78] realized its potential in commutative algebra.

2.1.5 DEFINITION. Let A be a ring, and $u: N \to M$ a homomorphism of A-modules. Then

(i) u is called pure if $u \otimes 1: N \otimes_A P \to M \otimes_A P$ is injective for every A-module P;

(ii) u is called ideally pure if $u \otimes 1: N \otimes_A A/\mathfrak{a} \to M \otimes_A A/\mathfrak{a}$ is injective for every ideal $\mathfrak{a}$;

(iii) u is called $\mathfrak{a}$-pure if $u \otimes 1: N \otimes_A A/\mathfrak{a} \to M \otimes_A A/\mathfrak{a}$ is injective for the ideal $\mathfrak{a}$.

Notice that these statements become increasingly weaker and that the first two imply that u is injective. Since every module is a direct limit of finitely presented submodules [Bo 80, §1, Prop. 7] and tensor products commute with direct limits, it is enough to test purity with finitely presented modules P. If N is an $\mathfrak{a}^t$-pure submodule of M for all sufficiently large t, then the $\mathfrak{a}$-adic topology on N coincides with the topology induced on N by the $\mathfrak{a}$-adic topology of M. The next result is easily verified by using properties of the tensor product.

2.1.6 PROPOSITION. Suppose $u: N \to F$ is a monomorphism into a flat A-module F.

(i) "u is pure"; "u is ideally pure"; "F/N is flat" are equivalent statements;

(ii) If conditions (i) are satisfied, N is also flat.

Suppose $\phi: L_0 \to L_1$ is a homomorphism between two free A-modules of finite rank. If we denote the linear dual $\mathrm{Hom}_A(-,A)$ by $-^*$, then $\phi^{**} \simeq \phi$, so that both $C = \mathrm{coker}(\phi)$ and $C' = \mathrm{coker}(\phi^*)$ can be any finitely presented A-module. The following neat observation is taken from [HR 76, Lemma 5.1]. As a corollary one obtained a useful characterization of purity.

2.1.7 PROPOSITION. Let $(M) = 0 \to M' \to M \to M'' \to 0$ be a short exact sequence of A-modules. Then, with the notation above, $\ker(C'\otimes_A M' \to C'\otimes_A M) \simeq \mathrm{coker}(\mathrm{Hom}_A(C,M) \to \mathrm{Hom}_A(C,M''))$.

PROOF. Tensor (M) with ϕ^* and notice that the Snake Lemma [Bo 80, §1, Prop. 2] applies to the resulting diagram. One obtains that $\ker(C'\otimes_A M' \to C'\otimes_A M) \simeq \mathrm{coker}(\ker(\phi^*\otimes_A M) \to \ker(\phi^*\otimes_A M''))$. Now use the isomorphisms $L_i^*\otimes_A N \simeq \mathrm{Hom}_A(L_i,N)$ which hold for any A-module N, $i = 0,1$, and the left exactness of $\mathrm{Hom}_A(-,N)$.

2.1.8 COROLLARY. The sequence (M) is pure (i.e. $M' \to M$ is pure) if and only if the mapping $\mathrm{Hom}_A(N,M) \to \mathrm{Hom}_A(N,M'')$ is surjective for every finitely presented module N. In case the module M" is finitely presented, purity of

(M) just means that (M) is split exact.

Thus a pure submodule is a kind of weak direct summand. Further properties of purity can be found in the papers mentioned. We now turn to an easy but basic example. Let A be any ring and let A_γ be the A-module A for every index γ in an arbitrary index set Γ. If A is noetherian, the free module $L = \oplus_{\gamma\in\Gamma} A_\gamma$ is an ideally pure submodule of the module $F = \Pi_{\gamma\in\Gamma} A_\gamma$. It even is a pure submodule by 2.1.6, since the product F is flat (even for a coherent ring, [Bo 61a, Ch. 1, §2, Exc. 12]). Keeping our notation, we find another pure submodule of F:

2.1.9 PROPOSITION. Let $\mathfrak{a}$ be an ideal in the noetherian ring A, which is assumed to be Hausdorff in its $\mathfrak{a}$-adic topology. Let $\bar{L}$ be the closure of L in F with respect to the $\mathfrak{a}$-adic topology. Then $\bar{L}$ is a pure submodule of F and hence flat.

PROOF. The second statement follows from the first by 2.1.6. Since L = F in case Γ is finite, we suppose Γ is infinite and seek to describe $\bar{L}$. A little reflection shows that an element (a_γ) of F is in $\bar{L}$ precisely when, for each $t \geq 1$, only finitely many a_γ's are not in $\mathfrak{a}^t$.

By 2.1.6 it is enough to test that for any given ideal $\mathfrak{b} \subset A$, there is an inclusion $\bar{L} \cap \mathfrak{b}F \subset \mathfrak{b}\bar{L}$, since the other inclusion is obvious. Choosing M = A and N = $\mathfrak{b}$ in 2.1.1, we have an integer $s \geq 0$ such that $\mathfrak{a}^t \cap \mathfrak{b} = \mathfrak{a}^{t-s}(\mathfrak{a}^s \cap \mathfrak{b}) \subset \mathfrak{a}^{t-s}\mathfrak{b}$ for all $t \geq s$. Let $b_1,\ldots,b_q$ be a set of generators for $\mathfrak{b}$. If $(a_\gamma) \in \bar{L} \cap \mathfrak{b}F$, there is a finite subset $\Gamma_0 \subset \Gamma$ such that $a_\gamma \in \mathfrak{a}^s$ for $\gamma \notin \Gamma_0$. For such an a_γ, if $a_\gamma \neq 0$, there is, by the Hausdorff property, a unique $t \geq s$ for which $a_\gamma \in \mathfrak{a}^t$, $a_\gamma \notin \mathfrak{a}^{t+1}$, and we can write $a_\gamma = \Sigma_{i=1}^q a_\gamma^{(i)} b_i$ with $a_\gamma^{(i)} \in \mathfrak{a}^{t-s}$. In case $a_\gamma = 0$, write $a_\gamma^{(i)} = 0$. Define an element $(b_\gamma) \in F$ by putting $b_\gamma = 0$ for $\gamma \in \Gamma_0$ and $b_\gamma = a_\gamma$ elsewhere. Then

$(a_\gamma)-(b_\gamma) \in \mathfrak{b}L$. Setting $a_\gamma^{(i)} = 0$ for $\gamma \in \Gamma_0$, we wish to show that the element $(a_\gamma^{(i)})$ is in $\overline{L}$. But for each $t \geq s$ we have $a_\gamma^{(i)} \notin \mathfrak{a}^t$ only if $a_\gamma \notin \mathfrak{a}^{t+s}$, which is the case for only finitely many γ's. Thus $(b_\gamma) \in \mathfrak{b}\overline{L}$ and we are through.

2.1.10 COROLLARY. The module L is a pure submodule of $\overline{L}$ and $L/\mathfrak{b}L \simeq \overline{L}/\mathfrak{b}\overline{L}$ for any ideal $\mathfrak{b}$ which is open in the $\mathfrak{a}$-adic topology.

PROOF. The first statement is clear since $L \subset F$ is pure. Thus for the second it is enough to show that $L + \mathfrak{b}\overline{L} = \overline{L}$. The ideal $\mathfrak{b}$ being open, it contains $\mathfrak{a}^t$ for some $t \geq 1$, and we need to prove that $L + \mathfrak{a}\overline{L} = \overline{L}$. This is obvious either from the description of $\overline{L}$ which we just used or from the definition of $\overline{L}$ as the topological closure and the fact that $\overline{L} \cap \mathfrak{a}F = \mathfrak{a}\overline{L}$ by purity.

These observations are due to Griffith [Gr 76, Lemma 1.7] and they led to

2.1.11 DEFINITION. Let M be a module over a ring A and $\mathfrak{a}$ an ideal. A free submodule $L \subset M$ is called $\mathfrak{a}$-basic if the extension is pure and L is dense in the $\mathfrak{a}$-adic topology of M.

Corollary 2.1.10 now states that L is an $\mathfrak{a}$-basic submodule of $\overline{L}$. Unlike Griffith, we do not require the Hausdorff property on M: (0) is a (p)-basic submodule of the $\mathbb{Z}_{(p)}$-module $\mathbb{Q}$. It is well known that over a local ring every finitely presented flat module is free. Two main results about the existence of basic submodules generalize this.

2.1.12 PROPOSITION.

(i) Every flat module over a local ring (A,m) possesses an $\mathfrak{m}$-basic submodule and is free when finitely generated;

(ii) Let $\mathfrak{a}$ be an ideal contained in the radical of a noetherian ring A. Suppose that M is an A-module with $\mathrm{Tor}_1^A(A/\mathfrak{a},M) = 0$ and such that $M/\mathfrak{a}M$ is $A/\mathfrak{a}$-free. Then M possesses an $\mathfrak{a}$-basic submodule.

Based on two different criteria for flatness, the existence was established in [BS, Cor. 3.6] and [Ba, Ch. I, Th. 4.10] respectively. We also refer to those papers for further results in this circle of ideas.

2.2 COMPLETIONS

For an ideal $\mathfrak{a}$ in the ring A and an A-module M there is a surjective inverse system of homomorphisms $M/\mathfrak{a}^{t+1}M \to M/\mathfrak{a}^tM$, $t \geq 1$, and one defines $\hat{M} = \varprojlim_t M/\mathfrak{a}^tM$. This $\hat{M}$ coincides with the topological completion of M in its $\mathfrak{a}$-adic topology. Here we speak of completion where Bourbaki writes "complété séparé". Thus our complete modules are always Hausdorff. The surjections $M \to M/\mathfrak{a}^tM$ give rise to a homomorphism τ_M: $M \to \hat{M}$ which is functorial. Also $\ker(\tau_M) = \cap_{t=1}^{\infty} \mathfrak{a}^tM$ which is the closure of (0) in the $\mathfrak{a}$-adic topology of M. In the ring $\hat{A}$ a basis of open neighbourhoods of (0) is defined by $(\mathfrak{a}^t)\hat{}$. In the resulting natural topology, $\hat{A}$ is a complete local ring and $\hat{M}$ is a topological $\hat{A}$-module in its natural topology. Tensoring τ_M over A with $\hat{A}$ and collapsing $\hat{A}\otimes_{\hat{A}}\hat{M}$ to $\hat{M}$ yields a functorial $\hat{A}$-homomorphism $\hat{A}\otimes_A M \to \hat{M}$ which is surjective if M is finitely generated.

In case A is noetherian and M finitely generated, the homomorphism just discussed is an isomorphism which means that if A is complete in its α-adic topology, so is M. Moreover $\hat{A}\otimes_A\alpha^t \simeq \hat{A}\alpha^t = (\hat{\alpha})^t \simeq (\alpha^t)^\wedge$ for $t \geq 1$. Also $A/\alpha^t \simeq \hat{A}/\hat{\alpha}^t$ hence $\hat{\hat{A}} = \hat{A}$ while $\hat{\alpha}$ is contained in the Jacobson radical of $\hat{A}$. All this is well known and clearly explained in [AM, Ch. 10] for instance.

2.2.1 PROPOSITION. The completion functor preserves surjections and, in case A is noetherian, is exact on the category of finitely generated A-modules. Thus over a noetherian ring its α-adic completion is flat.

PROOF. A short exact sequence of modules $0 \to M' \to M \to M'' \to 0$ gives rise to an inverse system of exact sequences

$$0 \to M'/(\alpha^t M \cap M') \to M/\alpha^t M \to M''/\alpha^t M'' \to 0.$$

Since the system is surjective, we find in the limit an exact sequence [AM, Prop. 10.12] $0 \to \tilde{M}' \to \hat{M} \to \hat{M}'' \to 0$ which proves the first statement. If A is noetherian and the module finitely generated, we know by 2.1.1 that the topology induced on M' is just its α-adic one, so that the completions $\tilde{M}'$ and $\hat{M}'$ coincide. The last statement follows from the isomorphism $\hat{A}\otimes_A M \simeq \hat{M}$ in case M is finitely generated and the fact that flatness is measured on such modules.

Suppose the ideal α in A is finitely generated and that M_γ, $\gamma \in \Gamma$, are an arbitrary set of A-modules. Put $M = \Pi_{\gamma\in\Gamma} M_\gamma$. Then $\alpha^t M = \Pi_{\gamma\in\Gamma} \alpha^t M_\gamma$ for all $t \geq 1$. In this setting the reader may easily prove [Ba, Ch. I, Lemma's 2.6 & 2.7]

2.2.2 LEMMA. Let A be noetherian and denote α-adic completion by $\hat{}$. Then

(i) If each M_γ is complete in its α-adic topology, so is M;

(ii) For an arbitrary A-module N, the $\hat{\mathfrak{a}}$-adic completion of the $\hat{A}$-module $\hat{A}\otimes_A N$ is isomorphic to $\hat{N}$.

For noetherian rings the second statement reduces the description of completions to completion of modules over complete rings. As we have seen, if N is finitely generated, then $\hat{A}\otimes_A N \simeq \hat{N}$ is already complete.

We now return to Proposition 2.1.9 in the case of a complete ring, and we retain its notation.

2.2.3 PROPOSITION. Let A be a noetherian ring which is complete in its $\mathfrak{a}$-adic topology. Then F is complete in this topology and $\hat{L} = \overline{L}$.

PROOF. The first statement is clear from 2.2.2 (i). The second one reflects the fact that the completion $\hat{L}$ of L can be identified with its topological closure in the complete F [Ke, Ch. 6, Th. 22], since the $\mathfrak{a}$-adic topology on L coincides with the topology which it inherits as a subspace of F.

2.2.4 COROLLARY. Let M be a module over a noetherian ring A satisfying either of the conditions in Proposition 2.1.12. Then its completion in the $\mathfrak{m}$- resp. $\mathfrak{a}$-adic topology is flat.

PROOF. By 2.1.12, M possesses a basic submodule L so that $\hat{M} = \hat{L}$. The module $\hat{A}\otimes_A L$ is $\hat{A}$-free and by 2.2.2 its completion is $\hat{L}$. The latter is $\overline{L}$ by 2.2.3 which is flat by 2.1.9.

2.2.5 THEOREM. Let $\mathfrak{a}$ be an ideal in the ring A and write ^ for completion in the $\mathfrak{a}$-adic topology. Suppose $\mathfrak{b}$ is an open ideal in this topology. Then the map $\tau_M\otimes_A A/\mathfrak{b}$ is split injective for every module M. It is even an isomorphism provided the ideal $\mathfrak{a}$ is finitely generated.

PROOF. Let $\sigma_t: \hat{M} \to M/\mathfrak{a}^tM$ be the canonical map stemming from the inverse limit, then $\sigma_t \circ \tau_M$ is just the projection $M \to M/\mathfrak{a}^tM$. Since $\mathfrak{b} \supset \mathfrak{a}^t$ for some $t \geq 1$, we can compose this with the projection $M/\mathfrak{a}^tM \to M/\mathfrak{b}M$. If we tensor with $A/\mathfrak{b}$, we obtain the identity on $M/\mathfrak{b}M$, which proves the first statement.

For the second statement it is now enough to prove surjectivity of $\tau_M \otimes_A A/\mathfrak{a}^t: M/\mathfrak{a}^tM \to \hat{M}/\mathfrak{a}^t\hat{M}$ for all $t \geq 1$. Suppose $\mathfrak{a} = (x_1,\ldots,x_n)$ and map $R = \mathbf{Z}[X_1,\ldots,X_n]$ to A by sending X_i to x_i, $i = 1,\ldots,n$. An A-module M is then also an R-module. Putting $I = (X_1,\ldots,X_n)$ we see that $I^tM = \mathfrak{a}^tM$ so that we may as well complete in the I-adic topology, thus reducing to the case of a noetherian ring which we again call A. Mapping a free module L onto M, we notice that it is enough to prove that $\tau_L \otimes_A A/\mathfrak{a}^t: L/\mathfrak{a}^tL \to \hat{L}/\mathfrak{a}^t\hat{L}$ is surjective. But for noetherian A we have seen that $A/\mathfrak{a}^t \simeq \hat{A}/\mathfrak{a}^t\hat{A}$, and direct sums are preserved by the tensor product, so we are through.

The result shows in particular, of course, that in case of a noetherian ring, the $\mathfrak{a}$-adic completion of every module is complete in its $\mathfrak{a}$-adic topology. This is not necessarily true for nonnoetherian rings, e.g. [Ba, Ch. I, §3]. For further results on completions and complete modules we refer to this thesis and to [BS] and [Si]. Here we only mention a useful consequence of 2.2.5 which was noticed by A.M.-Simon.

2.2.6 COROLLARY. Let $\mathfrak{b} \subset \mathfrak{a}$ be ideals in a ring A with $\mathfrak{b}$ finitely generated. Suppose the module M is complete in its $\mathfrak{a}$-adic topology. Then M is also $\mathfrak{b}$-adically complete.

PROOF. The factorizations $M \to M/\mathfrak{b}^t M \to M/\mathfrak{a}^t M$ extend to inverse systems so in the limit we obtain $\tau_M(\mathfrak{a}) = \psi \circ \tau_M(\mathfrak{b})$, where $\psi: \hat{M}(\mathfrak{b}) \to \hat{M}(\mathfrak{a})$, and further notation should be obvious. By assumption $\tau_M(\mathfrak{a})$ is an isomorphism, therefore M appears as a direct summand in $\hat{M}(\mathfrak{b})$. But $\mathfrak{b}$ is finitely generated, so $\hat{M}(\mathfrak{b})$ is complete in its $\mathfrak{b}$-adic topology. Therefore $\tau_M(\mathfrak{b})$ is an isomorphism, which is what we want.

2.3 LIFTING IN COMPLETE LOCAL RINGS

We shall need the simplest form of Hensel's Lemma and provide a quick proof.

2.3.1 PROPOSITION. Let (A,m) be a noetherian local ring, complete in its m-adic topology, and let $f \in A[X]$. Suppose $f(\alpha) \in m$ and $f'(\alpha) \notin m$ for a certain $\alpha \in A$. Then there exists an $a \in A$ with $a - \alpha \in m$, $f(a) = 0$ and $f'(a) \notin m$.

PROOF. We put $a_0 = \alpha$, and proceed by induction to construct $a_0, a_1, \ldots, a_t$ such that $f(a_t) \in m^{t+1}$, $f'(a_t) \notin m$ and $a_i - a_{i-1} \in m^i$ for $1 \le i \le t$. Clearly then the Cauchy sequence (a_t) has a unique limit $a \in A$ and, since the polynomial function $f: A \to A$ is continuous, $f(a) = 0$. Also $a - a_0 \in m$.

Suppose we have found $a_0, a_1, \ldots, a_{t-1}$ with the properties above. For any $b \in A$ we find $f(a_{t-1} + bf(a_{t-1})) - f(a_{t-1}) - f'(a_{t-1})bf(a_{t-1}) \in m^{t+1}$ by the Taylor formula $f(x + h) = f(x) + hf'(x) + h^2 g$. Since $f'(a_{t-1})$ is invertible by hypothesis, put $b = -f'(a_{t-1})^{-1}$ and $a_t = a_{t-1} - f'(a_{t-1})^{-1} f(a_{t-1})$.

Clearly a_t satisfies our requirements and we have proved Hensel's Lemma.

This basic result about the lifting of simple roots from k to A has many applications.

2.3.2 PROPOSITION. Suppose $(A,\mathfrak{m},k)$ is a noetherian local ring, complete in its $\mathfrak{m}$-adic topology. If the ring A is equicharacteristic, then for every maximal subfield $K \subset A$ the residue class field k is purely inseparable over the image of K in k.

PROOF. In case char A = 0, there is a copy of $\mathbf{Z}$ in A which is preserved by the residue class map. Thus $\mathbf{Z} \cap \mathfrak{m} = (0)$ and A contains a copy of $\mathbf{Q}$ which is again preserved by the residue class map. If char A = p > 0, then $\mathbf{F}_p \subset A$ and is preserved by going modulo $\mathfrak{m}$. Therefore A always contains a prime field. Order all subfields of A by inclusion and invoke Zorn's Lemma to establish a maximal one, say K.

Denoting going modulo $\mathfrak{m}$ by $\bar{\ }$, suppose k contains an $\bar{x}$ which is transcendental over $\bar{K}$, then $K[x] \cap \mathfrak{m} = (0)$ so that A contains the field K(x) which contradicts the maximality of K. Thus k is algebraic over $\bar{K}$. Suppose y is a simple root of an irreducible $\bar{f} \in \bar{K}[X]$. According to 2.3.1 there is a simple root x of $f \in K[X]$ with $\bar{x} = y$. This time K[x] is a field which intersects $\mathfrak{m}$ in (0), so K[x] extends K. This contradiction proves the result.

2.3.3 COROLLARY. If char A = 0, then $K + \mathfrak{m} = A$ and $K \cap \mathfrak{m} = (0)$ for every maximal subfield K of A.

We call such a field K a field of representatives of A. We

shall identify it with its isomorphic image $\bar{K} = k$.

Also in equal characteristic $p > 0$, there is always a field of representatives. Here however, not every maximal subfield of A fills the bill, as the example $A = \mathbb{F}_p(X)[[Y]]$, $K = \mathbb{F}_p(X^p+Y)$ shows. We need to proceed more circumspectly, since Hensel's Lemma no longer serves on account of inseparability.

2.3.4 LEMMA. Let (A,m,k) be a local ring of equal characteristic $p > 0$ and suppose $m^p = (0)$. Then A possesses a field of representatives.

PROOF. The image A^p of A under the Frobenius endomorphism $x \mapsto x^p$ is a field which is isomorphic to $A/m = k$. By Zorn's Lemma, we have a maximal subfield $K \subset A$ containing A^p and we claim that it is a field of representatives. For suppose $\bar{x} \in k\backslash\bar{K}$ for some $x \in A$. Since $x^p \in A^p \subset K$, the polynomial $X^p - x^p$ lives in $K[X]$, and it is irreducible because otherwise x would be in K and $\bar{x} \in \bar{K}$. But then $K[x] \supset K$ is a proper field extension and we have our contradiction.

2.3.5 THEOREM. A complete noetherian local ring (A,m,k) of equal characteristic $p > 0$ contains a field of representatives.

PROOF. Since $A = \varprojlim_t A/m^t$, its elements x are just coherent strings (x_t) such that $\phi_{t+1}(x_{t+1}) = x_t$, $t \geq 1$, where $\phi_{t+1}\colon A/m^{t+1} \to A/m^t$ are the canonical projections. Putting $K_1 = k$, our task therefore is to construct fields $K_t \subset A/m^t$ such that ϕ_{t+1} maps K_{t+1} isomorphically onto K_t.

We do this by induction. Since $p \geq 2$, Lemma 2.3.4 allows us to lift K_1 to a field of representatives K_2 in A/m^2. Suppose we have obtained $K_1,\ldots,K_t$, and consider the inverse image $R = \phi_{t+1}^{-1}(K_t) \subset A/m^{t+1}$ which is clearly a ring. Now $\ker(\phi_{t+1}) = m^t/m^{t+1}$ and it is in fact a prime ideal $\mathfrak{p}$ in

R with $\mathfrak{p}^2 = (0)$. An element $u \in R\backslash\mathfrak{p}$ maps onto a unit in A/m^t under ϕ_{t+1}, so $u \notin m/m^{t+1}$, hence u is a unit in A/m^{t+1}. The inverse of $\phi_{t+1}(u)$ lives in K_t, hence the inverse of u lives in R, so R is local with maximal ideal $\mathfrak{p}$. We now apply 2.3.4 to the situation $\phi_{t+1}|R\colon R \to K_t$ to obtain K_{t+1}. This finishes the proof.

The existence of fields of representatives was first proved by I.S. Cohen. The slick proof just given is due to A. Geddes [ZS, Vol. II, Ch. VIII, Th. 27]. Cohen also obtained results in mixed characteristic. These however are a bit harder to state and to prove, so we refer to the literature, [Bo 83, Ch. 9], [Ma 86, Ch. 10]. Besides these modern accounts, the reader is advised to consult Cohen's original treatment [Co], which contains a wealth of information.

3 INJECTIVE ENVELOPES AND MINIMAL INJECTIVE RESOLUTIONS

In this chapter we shall develop the theory of injective envelopes, describe minimal injective resolutions and treat the Matlis dual. The subject originated with the important paper [Matl 58]. An early categorical treatment can be found in [Ga]. The reader may also consult [SVa] and [Ma 86, §18].

We assume the reader knows some basic facts on injective modules, e.g. that in any diagram of A-module homomorphisms

$$\begin{array}{ccccc} 0 & \longrightarrow & M & \longrightarrow & N \\ & & \big\downarrow & & \\ & & I & & \end{array}$$

a homomorphism $N \to I$ exists making the diagram commutative if I is injective (for such an "injectivity test" it is, instead of $M \to N$, sufficient to look at inclusions $\mathfrak{a} \subset A$, $\mathfrak{a}$ an ideal of A), and also that each module M can be embedded in an injective module (i.e. the category of modules has enough injectives). As usual, A is a commutative ring.

3.1 THE INJECTIVE ENVELOPE OF A MODULE

3.1.1 DEFINITION. Let $0 \neq M \subset N$ be A-modules. The module N is called an essential extension of M, if $N' \cap M \neq 0$ for every nonnull submodule N' of N.

If $i: M \to N$ is an injective homomorphism, it is clear what is meant by saying that N is an essential extension of M. We need the following two lemmas:

3.1.2 LEMMA. Let $M \subset N \subset Q$ be A-modules. Then $M \to Q$ is an essential extension if and only if $M \to N$ and $N \to Q$ are both essential extensions.

PROOF. Left to the reader.

3.1.3 LEMMA. Let $M \subset N$ be A-modules. Then there exists a maximal essential extension of M in N.

PROOF. Let $M \subset N_1 \subset \ldots \subset N_i \subset \ldots \subset N$ be a chain of inclusions such that each N_i is an essential extension of M. A straightforward argument shows that $\bigcup_i N_i$ is an essential extension of M. The existence of a maximal essential extension of M in N then follows from Zorn's Lemma.

We now come to the main notion of this section, namely injective envelopes.

3.1.4 DEFINITION. Let $M \subset I$ be A-modules such that I is injective and an essential extension of M. Then I is called an injective envelope (or injective hull) of M.

Two important properties of injective envelopes are given in the following theorem.

3.1.5 THEOREM.

(i) Each A-module M has an injective envelope;

(ii) If I_1 and I_2 are injective envelopes of the same module M, then I_1 and I_2 are isomorphic.

PROOF. (i) Let J be an injective module such that $M \subset J$, let $I \subset J$ be a maximal essential extension of M in J, 3.1.3, and let I′ be a maximal submodule of J such that $I \cap I' = 0$. Then $I \to J/I'$ is an essential extension and, since J is injective, there exists a homomorphism $u: J/I' \to J$ making the following diagram commutative:

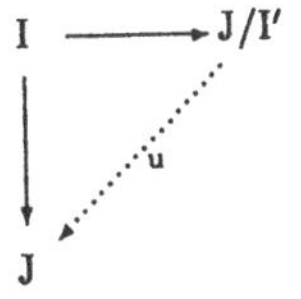

Now, if $\ker(u) \neq 0$, then $I \cap \ker(u) \neq 0$ (in J/I′), so the map $I \to J$ must have a proper kernel, contradiction. So $\ker(u) = 0$ and we have the chain of inclusions

$$0 \subset M \subset I \subset J/I' \subset J.$$

Since $M \subset I$ and $I \subset J/I'$ are essential extensions, Lemma 3.1.2 and the maximality of I finally show that $I = J/I'$, so $J = I \oplus I'$ and I is an *injective* essential extension of M.

(ii) Let $i_1: M \to I_1$ and $i_2: M \to I_2$ be injections of M into I_1 and I_2

respectively. Since I_2 is injective, there exists a homomorphism u such that $i_2 = u \circ i_1$. In the same way as above it is shown that ker(u) = 0, so $u(I_1)$ is an injective submodule of I_2, whence $I_2 = u(I_1) \oplus J$ with J injective. Since $M \to I_2$ is an essential extension and $i_2(M) \subset u(I_1)$, J must be 0. So $I_1 \simeq I_2$.

In the following we shall denote the injective envelope of M, determined up to isomorphism, by E(M).

3.1.6 EXERCISES.

1. Show that any homomorphism $f: M \to N$ can be extended to a homomorphism $\phi: E(M) \to E(N)$. Also show that ϕ is injective if and only if f is injective. Notice that ϕ is in general not uniquely determined by f, so that the injective hull is not functorial.

2. Let $i: M \to I$ be a homomorphism of A-modules and suppose that I is injective. Show that the following statements are equivalent:

(i) I is an injective envelope of M;

(ii) Whenever $j: M \to J$ is a monomorphism with J injective, then there exists a monomorphism $u: I \to J$ such that $j = u \circ i$;

(iii) If $u: M \to N$ is an essential extension, then there exists an essential extension $v: N \to I$ such that $i = v \circ u$.

3. Reformulate and prove Exercise 2 in case "I is injective" is replaced by "$M \to I$ is an essential extension".

3.1.7 PROPOSITION. Let $(M_k)_{k \in K}$ be a set of A-modules.

(i) If $N_k \supset M_k$ is an essential extension for each $k \in K$, then $\bigoplus_{k \in K} N_k$ is an essential extension of $\bigoplus_{k \in K} M_k$.

(ii) If K is finite or A is noetherian then $E(\bigoplus_{k \in K} M_k) = \bigoplus_{k \in K} E(M_k)$.

PROOF. (i) First assume $K = \{1,2\}$. Let $0 \neq Q \subset N_1 \oplus N_2$ and let p_i, $i = 1,2$, be the projection of $N_1 \oplus N_2$ onto N_i. If $p_1(Q) = 0$, then $Q \subset N_2$ and $Q \cap M_2 \neq 0$, so $Q \cap (M_1 \oplus M_2) \neq 0$. If $p_1(Q) \neq 0$, then $p_1(Q) \cap M_1 \neq 0$. Put $Q_1 = Q \cap p_1^{-1}(M_1) = Q \cap (M_1 \oplus N_2)$. If $p_2(Q_1) = 0$, then $Q_1 \subset N_1$, $Q_1 \cap M_1 \neq 0$ so $Q \cap (M_1 \oplus M_2) \neq 0$. If $p_2(Q_1) \neq 0$, then $p_2(Q_1) \cap M_2 \neq 0$. But then $p_2^{-1}(M_2) \cap Q_1 = \{(q_1,q_2) \in Q \subset N_1 \oplus N_2 \mid q_1 \in M_1 \text{ and } q_2 \in M_2\} = Q \cap (M_1 \oplus M_2) \neq 0$. This finishes the proof for $K = \{1,2\}$.

For K finite the proposition is easily proved by induction. If K is infinite and $Q \subset \bigoplus_{k\in K} N_k$, then each nonnull $q \in Q$ has the form $q = (q_k)_{k\in K'}$ with K′ finite. But now $Aq \cap (\bigoplus_{k\in K'} M_k) \neq 0$ so $Q \cap (\bigoplus_{k\in K} M_k) \neq 0$.

(ii) In view of (i) and Definition 3.1.4 it is sufficient to observe that the direct sum of the injective modules is again injective in the case that K is finite (obvious) or A is noetherian (obvious from an injectivity test on $\mathfrak{a} \subset A$).

3.2 DECOMPOSITION OF INJECTIVE MODULES

In this section A is a commutative noetherian ring. We begin by recalling that $\mathfrak{p} \in \operatorname{Spec} A$ is called an associated prime ideal of M ($\mathfrak{p} \in \operatorname{Ass} M$), if $\mathfrak{p} = \operatorname{Ann} m$ for some $m \in M$, $m \neq 0$. It is easy to prove that maximal elements in the set of ideals that annihilate some nonzero element of M are prime ideals, and since A is noetherian, that each nonnull module M has at least one associated prime ideal. Furthermore it is straightforward that the maximal elements among those ideals that annihilate some nonnull submodule of M, belong to Ass M.

3.2.1 DEFINITION. An A-module $M \neq 0$ is called an indecomposable module if it is not the direct sum of two proper submodules.

3.2.2 PROPOSITION. (Matlis) Suppose $I \neq 0$ is an injective A-module. Then the following statements are equivalent:

(i) I is indecomposable;

(ii) (0) is not the intersection of two nonnull submodules of I;

(iii) I is the injective envelope of any nonnull submodule of I;

(iv) The endomorphism ring of I is a local (not necessarily commutative) ring.

PROOF. The proofs (i) $\Rightarrow$ (iii) $\Rightarrow$ (ii) are left to the reader.
(iv) $\Rightarrow$ (i): If $I = I_1 \oplus I_2$ $(I_1, I_2 \neq 0)$, then for $f_1 = 1 \oplus 0$ and $f_2 = 0 \oplus 1$ we have $f_1 + f_2 = 1$ but neither f_1 nor f_2 is an isomorphism.
(ii) $\Rightarrow$ (iv): We need to show that for $f_1, f_2 \in \text{End } I$, if $f_1 + f_2 = 1$ then either f_1 or f_2 is an isomorphism. But if $f_1 + f_2 = 1$, then $\ker(f_1) \cap \ker(f_2) = 0$, so $\ker(f_1) = 0$ or $\ker(f_2) = 0$. Suppose $\ker(f_1) = 0$. Then $f_1(I) \simeq I$, so $f_1(I) \subset I$ is injective whence a direct summand of I. Now (ii) implies $f_1(I) = I$, so f_1 is bijective.

In the following two theorems it will be shown that each injective module is a direct sum of indecomposable submodules and that such a decomposition is unique up to isomorphism.

3.2.3 THEOREM. (Matlis) Every injective module I is a direct sum of indecomposable injective modules.

PROOF. Let I be an injective module and J a maximal submodule among those that are direct sums of indecomposable injective modules. The existence of such a J is standardly derived from Zorn's Lemma. Since the ring is noetherian, this J is injective, as observed in the proof of 3.1.7 (ii). Then $I = J \oplus J'$ for some injective module J'. We shall prove that, if $J' \neq 0$, then J' contains an indecomposable injective submodule, contradicting the maximality of J.

If $J' \neq 0$, then $\operatorname{Ass} J' \neq \emptyset$ and $A/\mathfrak{p} \subset J'$ for some prime ideal $\mathfrak{p}$. By Exercise 2. of 3.1.6 we can conceive of $E(A/\mathfrak{p})$ as contained in J'. By 3.2.2 we need to know that 0 is not the intersection of two nonnull submodules of $E(A/\mathfrak{p})$. But $E(A/\mathfrak{p})$ is an essential extension of $A/\mathfrak{p}$, and $\mathfrak{p} \neq \mathfrak{a} \cap \mathfrak{b}$ unless one of the ideals $\mathfrak{a}$ of $\mathfrak{b}$ equals $\mathfrak{p}$. Hence $E(A/\mathfrak{p})$ is indecomposable and we are through.

The second, important, theorem is in fact Azumaya's Decomposition Theorem, which states that if $M = \bigoplus_{i \in K} M_i$, where $\operatorname{End} M_i$ is a local ring for each $i \in K$, then M_i is indecomposable and this decomposition into indecomposables is unique up to permutation and isomorphism [AF, Th. 12.6]. For the reader's convenience we shall state here an injective module-version of this theorem and sketch a proof.

3.2.4 THEOREM. (Matlis) Let I be an injective module and let $I = \bigoplus_{i \in K} I_i$ and $I = \bigoplus_{j \in K'} J_j$ be two decompositions into indecomposables. Then there is a bijection $\sigma: K' \to K$ such that $I_{\sigma(j)} \simeq J_j$ for all $j \in K'$.

PROOF. We first show that for any J_j (say J_1) there exists a module I_{i_1} such that $I_{i_1} \simeq J_1$ and $I = J_1 \oplus (\bigoplus_{i \neq i_1} I_i)$. (Note $I = J_1 \oplus \dots$, not $I \simeq J_1 \oplus \dots$.) Let $p: I \to I$ be the endomorphism given by the projection onto J_1 along $\bigoplus_{j \neq 1} J_j$ and suppose $0 \neq m = \sum_{k=1}^{s} m_{i_k}$ ($m_{i_k} \in I_{i_k}$) satisfies $p(m) = m$

(i.e. $m \in J_1$). Then p and $p' = 1 - p$ are idempotents, $pp' = p'p = 0$ and $p + p' = 1$, so for each k $(1 \le k \le s)$ the composite maps

$$f_k: I_{i_k} \overset{\text{incl}}{\to} I \overset{p}{\to} I \overset{\text{proj}}{\to} I_{i_k} \text{ and}$$

$$f'_k: I_{i_k} \overset{\text{incl}}{\to} I \overset{p'}{\to} I \overset{\text{proj}}{\to} I_{i_k}$$

add to the identity on I_{i_k}, and by Proposition 3.2.2 (iv) at least one of them is an isomorphism.

If for each k the map f'_k is an isomorphism, then $\oplus_{k=1}^{s} f'_k : \oplus_{k=1}^{s} I_{i_k} \to \oplus_{k=1}^{s} I_{i_k}$ is an isomorphism and sends m to $p'(m) \ne 0$, contrary to $p(m) = m$ and $p'p = 0$. So on at least one module I_{i_k} (say I_{i_1}) the composition

$$I_{i_1} \overset{\text{incl}}{\to} I \overset{p}{\to} I \overset{\text{proj}}{\to} I_{i_1}$$

is an isomorphism with $\text{Im}(p) = J_1$ (in I), and J_1 is by $I \overset{\text{proj}}{\to} I_{i_1}$ sent isomorphically to I_{i_1}. Since $I \overset{\text{proj}}{\to} I_{i_1}$ sends $\underset{i \ne i_1}{\oplus} I_i$ to 0, it follows that $I = J_1 \oplus (\underset{i \ne i_1}{\oplus} I_i)$.

Now compare the decompositions $I = J_1 \oplus (\underset{i \ne i_1}{\oplus} I_i)$ and $I = J_1 \oplus (\underset{j \ne 1}{\oplus} J_j)$. For any $J_j \ne J_1$ (say J_2), let p_2 be the projection onto J_2 along $J_1 \oplus (\underset{j \ne 1,2}{\oplus} J_j)$. Then there is an $i \ne i_1$, say i_2, so that $p_2(I_{i_2}) = J_2$ by the same argument as above and $I = J_1 \oplus J_2 \oplus (\oplus I_i)$ is a next decomposition of I, $i \ne i_1, i_2$. Inductively, for each finite subset K" of K' we can find in this way an injection σ: K" $\to$ K such that $I_{\sigma(j)} \simeq J_j$, so if K is finite, then K' is finite and Card K $\ge$ Card K'.

In case Card K $= \infty$ it is easy to see that for each $i \in K$ there are only finitely many $j \in K'$ such that the projection $I \to J_j$ induces an isomorphism $I_i \simeq J_j$. Thus each $i \in K$ corresponds to a finite subset H_i of K' where the H_i cover K'. It follows that Card($\mathbb{N} \times K$) $\ge$ Card K', so again Card K $\ge$ Card K'. The opposite inequality Card K' $\ge$ Card K follows by interchanging K and K' in the above proof.

3.2.5 THEOREM.

(i) Each indecomposable injective module I satisfies $I = E(A/\mathfrak{p})$ for some $\mathfrak{p} \in \operatorname{Spec} A$ and each $E(A/\mathfrak{p})$ is indecomposable;

(ii) For each A-module M, $E(A/\mathfrak{p})$ is a direct summand of E(M) if and only if $\mathfrak{p} \in \operatorname{Ass} M$.

PROOF. (i) It has been shown in the proof of 3.2.3 that $E(A/\mathfrak{p})$ is indecomposable. Conversely suppose I is an indecomposable injective A-module. If $\mathfrak{p} \in \operatorname{Ass} I$, then $\mathfrak{p} = \operatorname{Ann} x$ for some $x \in I$ and $Ax \simeq A/\mathfrak{p}$. It follows that $I \supset E(Ax) \simeq E(A/\mathfrak{p})$, so $I \simeq E(A/\mathfrak{p})$.
(ii) If $E(A/\mathfrak{p})$ is a direct summand of E(M), then $A/\mathfrak{p} \cap M \neq 0$, so $\mathfrak{p} = \operatorname{Ann} m$ for $0 \neq m \in A/\mathfrak{p} \cap M$. Conversely, if $\mathfrak{p} \in \operatorname{Ass} M$, then $A/\mathfrak{p} \subset M$, so $E(A/\mathfrak{p}) \subset E(M)$ (Exercise 3.1.6 1), hence $E(A/\mathfrak{p})$ is a direct summand of E(M).

3.2.6 PROPOSITION. Each indecomposable injective module I has a single associated prime ideal, in fact $\operatorname{Ass} E(A/\mathfrak{p}) = \{\mathfrak{p}\}$.

PROOF. Let $\mathfrak{p} \in \operatorname{Ass} I$ and $\mathfrak{q} \in \operatorname{Ass} I$. Then $A/\mathfrak{p} \subset I$, $A/\mathfrak{q} \subset I$ and, since I is indecomposable, $A/\mathfrak{p} \cap A/\mathfrak{q} \neq 0$. Now for $0 \neq x \in A/\mathfrak{p} \cap A/\mathfrak{q}$ we have $\operatorname{Ann} x = \mathfrak{p}$ and $\operatorname{Ann} x = \mathfrak{q}$. So $\mathfrak{p} = \mathfrak{q}$.

3.2.7 PROPOSITION. Let $\mathfrak{p} \subset A$ be a prime ideal.

(i) For every $x \in A\backslash\mathfrak{p}$, multiplication by x is an automorphism on $E(A/\mathfrak{p})$;

(ii) For every $m \in E(A/\mathfrak{p})$ there exists a $t \geq 1$ such that $\mathfrak{p}^t m = 0$;

(iii) If $S \subset A$ is a multiplicative system, then $S^{-1}E(A/\mathfrak{p}) = E(A/\mathfrak{p})$ if $S \cap \mathfrak{p} = \emptyset$, and $S^{-1}E(A/\mathfrak{p}) = 0$ otherwise.

PROOF. (i) Since Ass $E(A/\mathfrak{p}) = \{\mathfrak{p}\}$ and $x \notin \mathfrak{p}$, multiplication by x is injective on $E(A/\mathfrak{p})$, so $\mathrm{im}(.x) \simeq E(A/\mathfrak{p})$, and im(.x) is injective module whence a direct summand of $E(A/\mathfrak{p})$. But $E(A/\mathfrak{p})$ is indecomposable, so $\mathrm{im}(.x) = E(A/\mathfrak{p})$.

(ii) Each prime ideal minimal above Ann m belongs to Ass $E(A/\mathfrak{p})$. But Ass $E(A/\mathfrak{p}) = \{\mathfrak{p}\}$, so Ass $Am = \{\mathfrak{p}\}$ and the result follows.

(iii) Is clear from (i) and (ii).

In the following $E(A/\mathfrak{p})$ is given an $A_{\mathfrak{p}}$-module structure by defining $x/s.m = \alpha_s^{-1}(xm)$, $m \in E(A/\mathfrak{p})$, $x \in A$, $s \in A\backslash\mathfrak{p}$. Here the map α_s is simply multiplication by s on $E(A/\mathfrak{p})$ as in 3.2.7 (i).

3.2.8 PROPOSITION. Let $\mathfrak{p} \in \mathrm{Spec}\, A$. Then

(i) $\mathrm{Hom}_A(M,E(A/\mathfrak{p})) \simeq \mathrm{Hom}_{A_{\mathfrak{p}}}(M,E(A/\mathfrak{p}))$ for each $A_{\mathfrak{p}}$-module M;

(ii) $E_{A_{\mathfrak{p}}}(A_{\mathfrak{p}}/\mathfrak{p}A_{\mathfrak{p}}) \simeq E_A(A/\mathfrak{p})$.

PROOF. (i) Let $f \in \mathrm{Hom}_A(M,E(A/\mathfrak{p}))$. Then $sf(a/s.m) = f(am) = af(m)$, and $f(a/s.m) = a/s.f(m)$, so f is $A_{\mathfrak{p}}$-linear.

(ii) $E(A/\mathfrak{p})$ is injective as an $A_{\mathfrak{p}}$-module since $\mathrm{Hom}_A(-,E(A/\mathfrak{p})) \simeq \mathrm{Hom}_{A_{\mathfrak{p}}}(-,E(A/\mathfrak{p}))$ is an exact functor. Since $A/\mathfrak{p} \to A_{\mathfrak{p}}/\mathfrak{p}A_{\mathfrak{p}}$ is an essential extension, one has $E(A/\mathfrak{p}) = E(A_{\mathfrak{p}}/\mathfrak{p}A_{\mathfrak{p}})$.

3.2.9 LEMMA. Let $\mathfrak{a} \subset A$ be an ideal and I an injective A-module. Then $\mathrm{Hom}_A(A/\mathfrak{a},I)$ is an injective $A/\mathfrak{a}$-module. Moreover, if $\mathfrak{p} \in \mathrm{Spec}\, A$ and $\mathfrak{a} \subset \mathfrak{p}$, then $\mathrm{Hom}_A(A/\mathfrak{a},E(A/\mathfrak{p})) \simeq E_{\bar{A}}(\bar{A}/\bar{\mathfrak{p}})$, where $\bar{A} = A/\mathfrak{a}$. If $\mathfrak{a} \not\subset \mathfrak{p}$, then $\mathrm{Hom}_A(A/\mathfrak{a},E(A/\mathfrak{p})) = 0$.

PROOF. $\mathrm{Hom}_A(A/\mathfrak{a},I) \simeq$ the submodule of I annihilated by $\mathfrak{a}$, and in the diagram of $A/\mathfrak{a}$-modules

$$\begin{array}{ccccc} 0 & \longrightarrow & M & \longrightarrow & N \\ & & \downarrow & & \\ & & \mathrm{Hom}_A(A/\mathfrak{a}, I) & & \end{array}$$

there exists a map $u: N \to \mathrm{Hom}_A(A/\mathfrak{a}, I)$ making the diagram commutative, namely the same map u as in

$$\begin{array}{ccccc} 0 & \longrightarrow & M & \longrightarrow & N \\ & & \downarrow & \swarrow_{u} & \\ & & I & & \end{array}$$

where M and N are considered as A-modules (with $\mathfrak{a} \subset \mathrm{Ann}\, M \cap \mathrm{Ann}\, N$). For the second statement note that $\mathrm{Hom}_A(A/\mathfrak{a}, -)$ preserves essential extensions. $\mathrm{Hom}_A(A/\mathfrak{a}, E(A/\mathfrak{p})) \neq 0 \Leftrightarrow \mathfrak{a} \subset \mathfrak{p}$ is left to the reader.

3.2.10 LEMMA. Suppose A is a domain and Q(A) its fraction field. Then Q(A) is the injective envelope of A.

PROOF. Take $\mathfrak{p} = (0)$ in 3.2.8 (ii) and observe that a field is injective over itself.

3.2.11 PROPOSITION. Let $\mathfrak{p} \in \mathrm{Spec}\, A$. Then $\mathrm{Hom}_A(A/\mathfrak{p}, E(A/\mathfrak{p})) \simeq E_{A/\mathfrak{p}}(A/\mathfrak{p}) \simeq k(\mathfrak{p})$, where $k(\mathfrak{p})$ is the residue class field of $A_{\mathfrak{p}}$.

PROOF. Apply Lemmas 3.2.9 and 3.2.10.

3.2.12 COROLLARY. For a local ring (A, m, k) we have $\mathrm{Hom}_A(k, E(k)) \simeq k$.

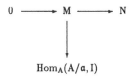

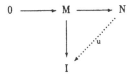

PROOF. Take $\mathfrak{p} = \mathfrak{m}$ in the proposition. We leave to the reader to verify that the isomorphism $k \simeq \mathrm{Hom}_A(k, E(k))$ is given by $a \mapsto .a =$ multiplication by a on k.

Though the $E(A/\mathfrak{p})$ behave well from a formal point of view, we should like to know their structure more closely. To this end, see the recent note [Kuc].

3.3 MINIMAL INJECTIVE RESOLUTIONS

We still assume the ring A is commutative and noetherian. In this section we shall consider "minimal" injective resolutions of A-modules and derive some important properties, mainly following Bass' seminal paper [Bas].

3.3.1 DEFINITION. A minimal resolution of an A-module M is an exact sequence

$$(*) \qquad 0 \to M \xrightarrow{\epsilon} E^0 \xrightarrow{d^0} E^1 \xrightarrow{d^1} E^2 \to \cdots$$

where $E^0 = E(M)$, $E^1 = E(\mathrm{coker}(\epsilon))$ and $E^i = E(\mathrm{coker}(d^{i-2}))$ for $i \geq 2$.

Each E^i is uniquely determined up to isomorphism and in fact two such minimal resolutions are isomorphic. Write $E^i(M)$ for the i-th module of the nonaugmented complex $E^{\bullet}$ attached to (*). So far, everything is perfectly general. In the noetherian case, as a consequence of Theorems

3.2.4 and 3.2.5, each $E^i(M)$ can be written as $E^i(M) = \bigoplus_{\mathfrak{p}\in \mathrm{Spec} A} \mu^i(\mathfrak{p},M)E(A/\mathfrak{p})$, where μE denotes μ copies of E (μ may be infinite, and we do not distinguish between different infinite cardinals). The importance of the integers $\mu^i(\mathfrak{p},M)$ shows in Theorem 3.3.3. First

3.3.2 PROPOSITION. Let $S \subset A$ be a multiplicatively closed set.

(i) If I is an injective A-module, then $S^{-1}I$ is an injective $S^{-1}A$-module.

(ii) If $E^\bullet$ is a minimal injective resolution of an A-module M, then $S^{-1}E^\bullet$ is a minimal injective resolution of $S^{-1}M$ over $S^{-1}A$. As a consequence, $\mu^i_A(\mathfrak{p},M) = \mu^i_{S^{-1}A}(S^{-1}\mathfrak{p},S^{-1}M)$ provided $S \cap \mathfrak{p} = \emptyset$.

PROOF. (i) We show that $\mathrm{Ext}^1_{S^{-1}A}(N,S^{-1}I) = 0$ for any finitely generated $S^{-1}A$-module N. If M is a finitely generated A-module such that $N = S^{-1}M$, then $\mathrm{Ext}^1_A(M,I) = 0$ and $\mathrm{Ext}^1_{S^{-1}A}(S^{-1}M,S^{-1}I) = S^{-1}\mathrm{Ext}^1_A(M,I)$ (since M is finitely generated and A is noetherian [Bo 80, §6, Prop 10]), so the result follows. For nonnoetherian rings the conclusion may fail [Dad].
(ii) Localization being an exact functor, we only need to show that localization preserves essential extensions. This is true for noetherian rings, as was proved by Bass in a rather general setting. A quick proof using the Artin-Rees Lemma 2.1.1 runs as follows. If $M \subset N$ is an essential extension of A-modules, we need to show $N' \cap S^{-1}M \neq 0$ for every nonnull submodule N′ of $S^{-1}N$. We may clearly assume N′ is finitely generated and therefore $N' = S^{-1}D$ for some finitely generated A-submodule $D \subset N$. Then $D \cap M$ is finitely generated, so $S^{-1}(D \cap M) = N' \cap S^{-1}M = 0$ would imply $s(D \cap M) = 0$ for some $s \in S$. Applying the Artin-Rees Lemma to the principal ideal (s) and the submodule $D \cap M$ of D, we choose k large enough so that $(s)^nD \cap M = (s)^nD \cap (D \cap M) = (s)^{n-k}((s)^kD \cap (D \cap M))$ for $n > k$. The right hand term is 0 by assumption, while the left one is not, because $(s)^nD \neq 0$

and M is essential in N. This proves the proposition.

3.3.3 THEOREM. (Bass) Let $\mathfrak{p} \in \mathrm{Spec}(A)$, $k(\mathfrak{p}) = A_\mathfrak{p}/\mathfrak{p}A_\mathfrak{p}$ and let M be an A-module. Then

$$\mu^i(\mathfrak{p},M) = \dim_{k(\mathfrak{p})}\mathrm{Ext}^i_{A_\mathfrak{p}}(k(\mathfrak{p}),M_\mathfrak{p}) = \dim_{k(\mathfrak{p})}\mathrm{Ext}^i_A(A/\mathfrak{p},M)_\mathfrak{p}.$$

In particular, if M is finitely generated, then all integers $\mu^i(\mathfrak{p},M)$ are finite.

PROOF. By Proposition 3.3.2 it is sufficient to prove that $\mu^i(m,M) = \dim_k \mathrm{Ext}^i_A(k,M)$, where (A,m,k) is a noetherian local ring. Let $0 \to M \to E^0 \overset{d^0}{\to} E^1 \to \ldots$ be a minimal injective resolution of M. Then the homologies of $0 \to \mathrm{Hom}(k,E^0) \to \mathrm{Hom}(k,E^1) \to \ldots$ are $\mathrm{Ext}^i_A(k,M)$.

Write $E^i = \mu^i(m,M)E(A/m)\oplus(\bigoplus_{\mathfrak{p}\neq m} \mu^i(\mathfrak{p},M)E(A/\mathfrak{p}))$. Then $\mathrm{Hom}_A(k,E^i) \simeq \mathrm{Hom}_A(k,\mu^i(m,M)E(A/m)) \simeq \mu^i(m,M)\mathrm{Hom}_A(k,E(A/m))$ since $\mathrm{Hom}_A(k,E(A/\mathfrak{p})) = 0$ for $\mathfrak{p} \neq m$. From $\mathrm{Hom}_A(k,E(k)) \simeq k$ it follows that $\mu^i(m,M) = \dim_k \mathrm{Hom}_A(k,E^i)$.

In order to show finally that $\mu^i(m,M) = \dim_k \mathrm{Ext}^i_A(k,M)$, we prove that in the complex $\mathrm{Hom}_A(k,E^\bullet)$ all boundary morphisms are null. Let $0 \neq f \in \mathrm{Hom}_A(k,E^i)$. Then $mf = 0$ and $\mathrm{im}(f) \simeq k$. Now $\ker(d^i) \to E^i$ is an essential extension, so $\mathrm{im}(f) \cap \ker(d^i) \neq 0$ and, since $\mathrm{im}(f)$ is a simple module, $\mathrm{im}(f) \subset \ker(d^i)$, which implies that $d^i(\mathrm{im}(f)) = 0$. So the boundary homomorphism $\mathrm{Hom}(k,E^i) \to \mathrm{Hom}(k,E^{i+1})$ is the null map.

The integers $\mu^i(\mathfrak{p},M)$ are often called the Bass numbers of M w.r.t. $\mathfrak{p}$ and they will figure extensively in other chapters, notably in Chapter 10 (Gorenstein rings). Several beautiful results are known for Gorenstein rings (i.e. rings with finite injective dimension over themselves) such as, when $\mathfrak{p}_0 \subsetneq \mathfrak{p}_1 \subsetneq \cdots \subsetneq \mathfrak{p}_d$ is a chain of prime ideals of maximal length in A, then $\mu^i(\mathfrak{p}_i,A) = 1$ and $\mu^j(\mathfrak{p}_i,A) = 0$ for $i \neq j$, one of the highlights of Bass' paper [Bas], see 10.1.6. The following proposition, where A is not necessarily Gorenstein, but commutative and noetherian as usual, is preliminary to this result.

3.3.4 PROPOSITION. Let M be a finitely generated A-module and $\mathfrak{p} \subset \mathfrak{q} \in \operatorname{Spec} A$ two distinct prime ideals with no other prime ideal "between" them. Then $\mu^i(\mathfrak{p},M) \neq 0$ implies $\mu^{i+1}(\mathfrak{q},M) \neq 0$.

PROOF. By localizing at $\mathfrak{q}$, we reduce to a local ring $(A,\mathfrak{m},k)$ in which $\mathfrak{p}$ is a submaximal prime. Choose $x \in \mathfrak{m}\setminus\mathfrak{p}$, put $B = A/\mathfrak{p}$, $C = B/xB = A/(\mathfrak{p},x)$ and suppose $\mu^i(\mathfrak{p},M) = \dim_{k(\mathfrak{p})}\operatorname{Ext}_A^i(B,M)_\mathfrak{p} \neq 0$. Now the exact sequence $0 \to B \xrightarrow{\cdot x} B \to C \to 0$ yields that $\operatorname{Ext}_A^i(B,M) \xrightarrow{\cdot x} \operatorname{Ext}_A^i(B,M) \to \operatorname{Ext}_A^{i+1}(C,M)$ is exact, so by Nakayama $\operatorname{Ext}_A^{i+1}(C,M) \neq 0$.

Consider the module C. From $\ell(C) < \infty$ it follows that there is a filtration $k = C_0 \subset C_1 \subset \ldots \subset C_n = C$ such that $C_r/C_{r-1} \simeq k$ $(1 \leq r \leq n)$. Now for each r we have exact sequences $0 \to C_{r-1} \to C_r \to k \to 0$ and $\operatorname{Ext}_A^{i+1}(k,M) \to \operatorname{Ext}_A^{i+1}(C_r,M) \to \operatorname{Ext}_A^{i+1}(C_{r-1},M)$. If $\operatorname{Ext}_A^{i+1}(k,M) = 0$ then we find a chain of inclusions $\operatorname{Ext}_A^{i+1}(C_n,M) \subset \operatorname{Ext}_A^{i+1}(C_{n-1},M) \subset \ldots \subset \operatorname{Ext}_A^{i+1}(C_0,M) = 0$ (since $C_0 = k$), contradicting $\operatorname{Ext}_A^{i+1}(C,M) \neq 0$. So $\operatorname{Ext}_A^{i+1}(k,M) \neq 0$, whence $\mu^{i+1}(\mathfrak{m},M) \neq 0$.

Another useful result on minimal injective resolutions is the following. Introduce the convention that for any ascending complex $C^\bullet$ the shifted complex $C^\bullet[n]$ shall have C^{r+n} in degree r. For the notion of regular element see 5.1.1.

3.3.5 THEOREM. Let $E^\bullet$ be the associated complex of a minimal injective resolution of an A-module M and suppose $x \in A$ is both M-regular and A-regular. Then $\operatorname{Hom}_A(A/(x),E^\bullet[1])$ yields a minimal injective resolution of the $A/(x)$-module M/xM.

PROOF. It follows from Lemma 3.2.9 that the $\mathrm{Hom}_A(A/(x),E^i)$ are injective $A/(x)$-modules. Now Ass E^0 = Ass M and x is M-regular implies $\mathrm{Hom}_A(A/(x),E^0) = 0$ and then, putting $K = \mathrm{coker}(\epsilon: M \to E^0)$, we find $\mathrm{Hom}_A(A/(x),K) \simeq \mathrm{Ext}_A^1(A/(x),M)$. Furthermore x is A-regular entails that $\mathrm{pd}_A A/(x) = 1$, so $H^i(\mathrm{Hom}_A(A/(x),E^\bullet)) = \mathrm{Ext}_A^i(A/(x),M) = 0$ for $i \geq 2$. It follows that $0 \to \mathrm{Ext}_A^1(A/(x),M) \to \mathrm{Hom}_A(A/(x),E^1) \to \mathrm{Hom}_A(A/(x),E^2) \to \ldots$ is an injective $A/(x)$-resolution of $\mathrm{Ext}_A^1(A/(x),M)$ which is minimal since $\mathrm{Hom}_A(A/(x),-)$ preserves essential extensions.

The only thing left to do is to show that $\mathrm{Ext}_A^1(A/(x),M) \simeq M/xM$. For this, apply $\mathrm{Hom}_A(-,M)$ to the exact sequence $0 \to A \xrightarrow{\cdot x} A \to A/(x) \to 0$, which yields the long exact sequence

$$0 \to 0 \to \mathrm{Hom}_A(A,M) \xrightarrow{\cdot x} \mathrm{Hom}_A(A,M) \to \mathrm{Ext}_A^1(A/(x),M) \to 0$$

which is clearly isomorphic to

$$0 \to 0 \to M \xrightarrow{\cdot x} M \to \mathrm{Ext}_A^1(A/(x),M) \to 0;$$ so the theorem follows.

The next easy corollary is in fact an important "change of rings" result on injective dimension (which we write as id).

3.3.6 COROLLARY. In the situation of Theorem 3.3.5,

(i) $\mathrm{Ext}_A^{i+1}(N,M) \simeq \mathrm{Ext}_{A/(x)}^i(N,M/xM)$ for each $A/(x)$-module N and all $i \geq 0$; hence $\mathrm{id}_A M \geq \mathrm{id}_{A/(x)} M/xM + 1$;

(ii) If (A,m,k) is local and M is finitely generated, $\mathrm{id}_A M = \sup\{i \mid \mathrm{Ext}_A^i(k,M) \neq 0\}$ and $\mathrm{id}_A M = \mathrm{id}_{A/(x)} M/xM + 1$.

PROOF. (i) The proof uses the isomorphism $\mathrm{Hom}_B(M \otimes_A N,T) \simeq \mathrm{Hom}_A(M,\mathrm{Hom}_B(N,T))$ where the modules M, N and T are taken over the rings A, both A and B, and B respectively. For the modules M and N of the corollary, let $E^\bullet$ be as in 3.3.5. Then $\mathrm{Ext}_A^{i+1}(N,M) \simeq H^{i+1}\mathrm{Hom}_A(N\otimes_{A/(x)}A/(x),E^\bullet) \simeq$

$\simeq H^{i+1}\mathrm{Hom}_{A/(x)}(N,\mathrm{Hom}_A(A/(x),E^\bullet)) = \mathrm{Ext}^i_{A/(x)}(N,M/xM)$ which proves the first contention. The second follows.

(ii) The first statement is clear from 3.3.3 and 3.3.4. Since $x \in \mathfrak{m}$, the field k is an A/(x)-module and the second statement is then clear by (i).

We end this section with two result which have nothing to do with minimal injective resolutions, but we fit them in here because they also use the Ext-functors.

3.3.7 LEMMA. Let N be a finitely generated module over a noetherian ring A and let $M = \varinjlim_j M_j$ for some directed system J. Then $\mathrm{Ext}^i_A(N,M) \simeq$

$\varinjlim_j \mathrm{Ext}^i_A(N,M_j)$ for all i, naturally in N.

PROOF. Resolve N in terms of finitely generated free modules, and let $L_\bullet$ be the associated complex with $H_0(L_\bullet) = N$. Observe that $\mathrm{Hom}_A(A,M_j) \simeq M_j$ so that for each of the chain modules $\mathrm{Hom}_A(L_i,M) \simeq$ $\varinjlim_j \mathrm{Hom}_A(L_i,M_j)$. Furthermore the exact functor $\varinjlim_j$ commutes with homology, so $\mathrm{Ext}^i_A(N,M) = H_i\mathrm{Hom}_A(L_\bullet,M) \simeq H_i(\varinjlim_j \mathrm{Hom}_A(L_\bullet,M_j)) \simeq \varinjlim_j H_i\mathrm{Hom}_A(L_\bullet,M_j) =$

$\varinjlim_j \mathrm{Ext}^i_A(N,M_j)$.

We shall use this result in 4.1.8, but here only mention

3.3.8 PROPOSITION. Over a noetherian ring, a direct limit of injective modules is injective.

PROOF. In the introduction to this chapter we mentioned that to test injectivity of a module I, it suffices to prove that every ideal $\mathfrak{a}$

in A induces a surjection $\mathrm{Hom}_A(A,I) \to \mathrm{Hom}_A(\mathfrak{a},I)$. Now if I is a direct limit of injectives, then $\mathrm{Ext}^1_A(A/\mathfrak{a},I) = 0$ by the lemma, so this is certainly true.

3.4 MATLIS DUALITY

In this section the ring $(A,\mathfrak{m},k)$ is assumed to be local noetherian. The $\mathfrak{m}$-adic completion $\varprojlim_t A/\mathfrak{m}^t$ of A will be denoted by $\hat{A}$, and $\hat{M}$ stands for $\varprojlim_t M/\mathfrak{m}^t M$ for each $\hat{A}$-module M. We write E is the injective envelope E(k) of the residue class field k.

3.4.1 DEFINITION. The functor $\mathrm{Hom}_A(-,E)$ is called the Matlis duality functor, to be denoted as $-^\vee$, $M^\vee = \mathrm{Hom}_A(M,E)$ is called the Matlis dual of M for each A-module M and $M^{\vee\vee}$ stands for $(M^\vee)^\vee$.

3.4.2 PROPOSITION. Matlis duality affords a faithfully exact contravariant functor and $\mathrm{Ann}\, M^\vee = \mathrm{Ann}\, M$ for every module M.

PROOF. Contravariance and exactness are clear from the definition. For faithfulness, consider $0 \neq m \in M$ in an A-module M. Then the canonical map $Am \simeq A/\mathrm{Ann}\, m \to A/\mathfrak{m} \to E$ extends to a nonnull homomorphism $M \to E$. So $M^\vee \neq 0$ if $M \neq 0$. For the second assertion we need to prove that $\mathrm{Ann}\, M^\vee \subset \mathrm{Ann}\, M$ since the other inclusion is obvious. So take $a \notin \mathrm{Ann}\, M$ and an $m \in M$ with $am \neq 0$. The above construction with am instead of m yields an $f \in M^\vee$ such that $f(am) \neq 0$; hence $af \neq 0$, and $a \notin \mathrm{Ann}\, M^\vee$.

For each module M there exists a functorial homomorphism $M \to M^{\vee\vee}$ defined by $m \mapsto$ the map which assigns to each $\phi \in \mathrm{Hom}_A(M,E)$ the element $\phi(m) \in E$. It is straightforward that $A^{\vee\vee} \simeq \mathrm{Hom}_A(E,E)$ and that the homomorphism $A \to A^{\vee\vee} \simeq \mathrm{Hom}_A(E,E)$ corresponds with $a \mapsto$ multiplication on E by a, written .a as before.

If M is finitely generated, then $M^{\vee\vee} \simeq M\otimes_A\mathrm{Hom}_A(E,E)$, because the isomorphism is obviously true for $M = A^n$ and one knows that $\mathrm{Hom}_A(-,E)$ is contravariant exact. Moreover the homomorphism $M \to M^{\vee\vee}$ can be described by $m \mapsto m\otimes \mathrm{id}_E$ and $am \mapsto m\otimes(.a)$. In general, the map $M \to M^{\vee\vee}$ is always injective, as follows from the proof of 3.4.2. In order to see that it is not always surjective, take a k-vectorspace V and write it as a direct sum $\oplus$ k taken over some index set whose cardinality is the dimension of V. Then $\mathrm{Hom}_A(V,E) \simeq \Pi\, \mathrm{Hom}_A(k,E) \simeq \Pi\, k$ by 3.2.11 with $\mathfrak{p} = \mathfrak{m}$. But $\Pi\, k \simeq \mathrm{Hom}_k(V,k)$ so that $V^{\vee}$ is isomorphic to the linear dual of the vector space V. It follows that the canonical map $V \to V^{\vee\vee}$ is only an isomorphism when $\dim V < \infty$.

3.4.3 DEFINITION. An A-module M is called reflexive if the canonical map $M \to M^{\vee\vee}$ is an isomorphism.

In order to study this concept, we recall that a full subcategory $\mathcal{S}$ of A-modules is called a Serre subcategory of the category of A-modules if for each exact sequence of A-modules

$$0 \to M' \to M \to M'' \to 0,$$

M belongs to $\mathcal{S}$ iff both M' and M" belong to $\mathcal{S}$. Examples of a Serre subcategory are the category of artinian A-modules and the category of noetherian A-modules. Also the reflexive A-modules form a Serre subcategory, as stated in

3.4.4 LEMMA.

(i) Let M be an A-module. Then each submodule $M' \subset M$ and each factor module $M'' = M/M'$ is reflexive iff M is reflexive.

(ii) Put $M'^{\perp} = \{\phi \in M^{\vee} \mid \phi(M') = 0\}$. If M is reflexive, then the correspondence $M' \mapsto M'^{\perp}$ is 1-1 between all submodules of M and all submodules of $M^{\vee}$, and reverses inclusions.

PROOF. (i) Apply the Snake Lemma to the injection of $0 \to M' \to M \to M'' \to 0$ into its Matlis bi-dual and observe that $M' \to (M')^{\vee\vee}$ and $M'' \to (M'')^{\vee\vee}$ are isomorphisms iff $M \to M^{\vee\vee}$ is an isomorphism. (ii) Notice that $M'^{\perp} = \ker(M^{\vee} \to (M')^{\vee})$ and the statement follows from (i) by dualizing once more.

Useful reflexive modules are the finite length ones. We already know that $k \simeq k^{\vee}$ by 3.2.11, and this implies even more:

3.4.5 PROPOSITION. An A-module M has finite length iff $M^{\vee}$ has, and in this case $\ell(M) = \ell(M^{\vee})$. Such a module is reflexive.

PROOF. Apply 3.4.4 (i) to exact sequences arising from a composition series of M resp. $M^{\vee}$ to prove that M is reflexive. The equality of the lengths then follows for instance from 3.4.4 (ii).

This result allows one to tie up the Matlis bi-dual with completion.

3.4.6 THEOREM. For any noetherian local ring A, there is a canonical isomorphism $A^{\vee\vee} \simeq \hat{A}$.

PROOF. Write $E_t = \{e \in E \mid m^t e = 0\}$ for $t \geq 0$. Then E can be filtered as $E \supset \ldots \supset E_t \supset \ldots \supset E_1 \supset E_0 = 0$ and 3.2.7 (ii) tells us that $E = \bigcup_t E_t = \varinjlim_t E_t$. Since $A^\vee = \mathrm{Hom}_A(A,E) \simeq E$, we obtain $A^{\vee\vee} \simeq \mathrm{Hom}_A(E,E) = \mathrm{Hom}_A(\varinjlim_t E_t, E) \simeq \varprojlim_t \mathrm{Hom}_A(E_t,E)$. But $E_t \simeq \mathrm{Hom}_A(A/m^t,E) = (A/m^t)^\vee$, so $A^{\vee\vee} \simeq \varprojlim_t \mathrm{Hom}_A((A/m^t)^\vee,E) = \varprojlim_t (A/m^t)^{\vee\vee} \simeq \varprojlim_t A/m^t \simeq \hat{A}$. Notice that this yields an isomorphism of rings.

3.4.7 COROLLARY. $M^{\vee\vee} \simeq \hat{M}$ if M is a finitely generated A-module.

PROOF. $M^{\vee\vee} \simeq M \otimes_A \mathrm{Hom}_A(E,E) \simeq M \otimes_A \hat{A}$ and $M \otimes_A \hat{A} \simeq \hat{M}$ if M is finitely generated. Note that $M^{\vee\vee} \simeq M$ if A is a complete noetherian local ring.

In the following an $\hat{A}$-module structure on the module E is defined by letting $\hat{A}$ act on E through the isomorphism $\hat{A} \simeq A^{\vee\vee} \simeq \mathrm{Hom}_A(E,E)$.

3.4.8 THEOREM. $E = E(\hat{A}/\hat{m})$ as an $\hat{A}$-module.

PROOF. We leave to the reader to verify that E (injective and indecomposable as an A-module) is also $\hat{A}$-injective and $\hat{A}$-indecomposable, while $\mathrm{Ass}_{\hat{A}} E = \{\hat{m}\}$ is obvious.

Extending the A-module structure on E to an $\hat{A}$-module structure can easily be achieved for a larger class of A-modules as follows:

3.4.9 PROPOSITION. Let M be an A-module and suppose for each $m \in M$ there is an integer $t \geq 1$ such that $m^t m = 0$. Then M admits a unique $\hat{A}$-module structure extending the A-module structure and satisfying $\hat{m}^t m = 0$ for t sufficiently large (depending on m).

PROOF. If $\mathfrak{m}^t m = 0$ and $\hat{a} \in \hat{A}$, then put $\hat{a}m = am$ where a is chosen in A such that $\hat{a} - a \in \mathfrak{m}^t\hat{A}$. We leave to the reader to verify that this $\hat{A}$-module structure is well-defined and is unique.

Proposition 3.4.9 has two consequences for E. The first one is that the $\hat{A}$-module structure on E described in the proof of 3.4.9 by its uniqueness coincides with the $\hat{A}$-module structure described above 3.4.8. The second consequence is that the set of the A-submodules and the set of $\hat{A}$-submodules of E are the same.

3.4.10 THEOREM. Suppose A is complete. Then E is reflexive and there is a 1-1, inclusion reversing, correspondence between all submodules $M \subset E$ and all ideals $\mathfrak{a} \subset A$ given by $M \mapsto \operatorname{Ann} M$.

PROOF. Since A is complete, $E^\vee \simeq A$, so that $E^{\vee\vee} \simeq A^\vee \simeq E$. As a submodule of E, the module M is reflexive and $M \mapsto M^\perp$ describes a 1-1 correspondence by 3.4.4. But $M^\perp \subset E^\vee \simeq A$ is just the ideal Ann M.

3.4.11 COROLLARY. For an arbitrary noetherian local ring $(A,\mathfrak{m},k)$

(i) The module E is artinian;

(ii) A module M is artinian iff $M \subset E^n$ for some $n \geq 1$;

(iii) The dual $M^\vee$ of a noetherian M is artinian and vice versa.

PROOF. (i) By the observation preceding the theorem, we may assume A to be complete in its $\mathfrak{m}$-adic toplogy. But then all submodules of E correspond uniquely to all ideals of the noetherian ring A. Since this correspondence reverts inclusions, item (i) is clear.

(ii) The "if" part being a consequence of (i), assume that M is artinian.

Consider all maps $\phi: M \to E^n$ where n ranges over all positive integers. Take a ϕ whose kernel is minimal; this exists because of the artinian condition. If $0 \neq m \in \ker(\phi)$, then we obtain a nonnull homomorphism $M \to E$ as in the proof of 3.4.2. If we call this map ψ, the kernel of $(\phi,\psi): M \to E^n \oplus E$ is properly contained in $\ker(\phi)$, contradicting the minimality of the latter. So ϕ is injective.

(iii) A surjection $A^n \to M$ onto a noetherian M turns into an injection $M^\vee \to A^{n\vee} \simeq E^n$, so $M^\vee$ is artinian by (ii). In turn, an injection $M^\vee \to E^n$ gives rise to a surjection $(A^{\vee\vee})^n \to M^{\vee\vee}$ which means that $A^n \to M$ must be a surjection since the bi-dual is exact and faithful.

For complete local rings the situation is quite satisfactory:

3.4.12 COROLLARY. Let A again be complete. Then a module M is artinian iff $M^\vee$ is noetherian and M is noetherian iff $M^\vee$ is artinian. Both types of module are reflexive.

PROOF. Since A is complete, $A^\vee \simeq E$ and $E^\vee \simeq A$; both these modules are reflexive. A noetherian module is a homomorphic image of some A^n, while an artinian one is a submodule of some E^n. The result follows.

The following theorem describes reflexive modules more precisely.

3.4.13 THEOREM. (Enochs [En, Prop. 1.3]) Let A be complete. Then an A-module M is reflexive if and only if it contains a finitely generated submodule N such that M/N is artinian.

PROOF. "If" follows from 3.4.4 and 3.4.12. For the opposite

implication, if $\mathfrak{m} \in \mathrm{Ass}\, M$, take $N = 0$, else take $N = \mathfrak{m}m_a \subset M$ for some nonnull $m_1 \in M$. Now consider M/N and let V be the k-vectorspace $\{m \in M/N \mid \mathfrak{m}m = 0\}$. Then V is relexive by 3.4.4 and finitely generated by the remark just above 3.4.3. If M/N is essential over V, then M/N is contained in E^n for some n; if not, we enlarge N and V by adding $\mathfrak{m}m_2 \neq 0$ to N for a suitably chosen $m_2 \in M$ and again define V as the set $\{m \in M/N \mid \mathfrak{m}m = 0\}$, and so on. If this process does not stop, then M contains a submodule N' such that M/N' contains an infinite dimensional k-vectorspace V', contrary to 3.4.4, so the process must stop. At the last stage, M/N contains a finite dimensional k-vector space V such that M/N is essential over V and hence artinian.

To finish this chapter we present two Ext-Tor dualities involving the Matlis dual functor and describe a relation between the Bass numbers μ^i and the Betti numbers β_i. As is well known, for a local noetherian ring A and an A-module M, the Betti numbers $\beta_i(M)$ are defined as $\beta_i(M) = \dim_k \mathrm{Tor}_i^A(k,M)$. If

$$\cdots \xrightarrow{d_3} L_2 \xrightarrow{d_2} L_1 \xrightarrow{d_1} L_0 \to M \to 0$$

is a minimal free resolution of a finitely generated M (i.e. $\mathrm{im}(d_i) \subset \mathfrak{m}L_{i-1}$ or equivalently rk L_i is minimal for each i and $d_i \otimes 1_k = 0$ [Se, Ch. IV, App. I]), then each L_i consists of $\beta_i(M)$ copies of A, as is easily seen by tensoring with k. The numbers μ^i were defined in 3.3.1. In the following (and further on in the book) $\mu^i(\mathfrak{m},M)$ is abbreviated to $\mu^i(M)$. We emphasize that Bass and Betti numbers are nonnegative integers or ∞; we do not distinguish infinite cardinalities. The Bass numbers are related to the Betti numbers in several ways. E.g. in Chapter 10 we shall prove that $\mathrm{pd}_A M < \infty$ (projective dimension) iff $\mathrm{id}_A M < \infty$ for a finitely generated A-module M over a d-dimensional local Gorenstein ring A, and that in this case $\mu^i(M) = \beta_{d-i}(M)$. A more general result involving the Matlis dual is 3.4.15 below.

3.4.14 PROPOSITION. Let N and M be A-modules. Then

(i) $\mathrm{Tor}_i^A(N,M)^\vee \simeq \mathrm{Ext}_A^i(N,M^\vee)$;

(ii) $\mathrm{Tor}_i^A(N,M^\vee) \simeq \mathrm{Ext}_A^i(N,M)^\vee$ provided N is finitely generated.

PROOF. (i) Let F^i and G^i be the functors $\mathrm{Hom}_A(\mathrm{Tor}_i^A(-,M),E)$ and $\mathrm{Ext}_A^i(-,\mathrm{Hom}_A(M,E))$ respectively. Since $\mathrm{Hom}_A(-,E)$ is exact and contravariant, $F = (F^i)$ and $G = (G^i)$ both form a right connected exact sequence of contravariant functors. For N arbitrary, $\mathrm{Hom}_A(N\otimes_A M,E) \simeq \mathrm{Hom}_A(N,\mathrm{Hom}_A(M,E))$ describes a well known adjunction [Bo 62, §4, Cor. de Prop. 1], [St 78, 1.7.12 (b)], so we have an isomorphism $\sigma^0\colon F^0 \to G^0$. In order to extend this to an isomorphism $\sigma\colon F \to G$ by the appropriate dual of 1.1.3, we need to know that $F^iP = G^iP = 0$ for $i \geq 1$ and every projective module P, and this is obvious.

(ii) This proceeds by another dual of 1.1.3 affecting left connected exact sequences of covariant functors. This time at the 0-level we have an isomorphism $N\otimes_A\mathrm{Hom}_A(M,E) \simeq \mathrm{Hom}_A(\mathrm{Hom}_A(N,M),E))$ for finitely generated N, since this isomorphism is obviously true for $N = A^n$ and both functors are right exact in N. Again, the higher functors vanish on projectives.

3.4.15 COROLLARY. $\beta_i(M) = \mu^i(M^\vee)$ and $\mu^i(M) = \beta_i(M^\vee)$ for any A-module M.

PROOF. Take $N = k$ in the proposition. Then in (i)

$$\mathrm{Tor}_i^A(k,M)^\vee \simeq (\beta_i(M).k)^\vee \text{ and } \mathrm{Ext}_A^i(k,M^\vee) \simeq \mu^i(M^\vee).k.$$

So (i) implies $\beta_i(M) = \mu^i(M^\vee)$ since $k \simeq k^\vee$. The second identity follows easily from (ii).

4. LOCAL COHOMOLOGY AND KOSZUL COMPLEXES

Both subjects are well known and standard tools in commutative algebra. Nevertheless, it is not easy to find all that we shall need covered in one single text, see however the recent book [HIO, Ch. VII]. Since both topics will play an important role in later chapters, we have decided to present a bare-bones account. Thus proofs will be merely outlined or sketched, and in some cases we shall just provide the reader with a reference.

Local cohomology was introduced by Grothendieck, using the language of sheaves, in [Gro 67]. In his hands and those of others it became an important technique in commutative algebra [HK], [Sc 82a], algebraic geometry [Ha], [Li], the theory of invariants [HR 74], [HR 76], analytic geometry [ST] and the theory of singularities [Gre]. In section 4.1 we content ourselves with a quick introduction by way of commutative algebra which is due to Sharp [Sh 70].

Koszul complexes or exterior algebra complexes have been with us even longer. They and related complexes play a key part not only in commutative algebra but for instance in complex algebraic geometry and in differential geometry (de Rham complex). Here we give a frankly utilitarian account, not emphasizing the exterior algebra structure like in [Bo 80], see section 4.2.

In section 4.3 finally we consider limits of Koszul complexes and explain how these serve to read off local cohomology.

4.1 LOCAL COHOMOLOGY

To define local cohomology and show that it can also be written as a direct limit of Ext-functors, one only need assume one's ring to be commutative - though noncommutative versions have been given [Go, Ch. 60]. For further results though, we shall assume the ring to be noetherian so as to be able to use the structure of the indecomposable injectives.

4.1.1 DEFINITION. Let $\mathfrak{a}$ be an ideal in a ring A and M an A-module. Put $L_{\mathfrak{a}}(M)$ to be the set of all elements $m \in M$ which are annihilated by some positive power of the ideal $\mathfrak{a}$.

Observing that $\{m \in M \mid \mathfrak{a}^t m = 0\} \simeq \mathrm{Hom}_A(A/\mathfrak{a}^t, M)$ we see by taking the limit over t, that $L_{\mathfrak{a}}(M)$ is functorially isomorphic to $\varinjlim_t \mathrm{Hom}_A(A/\mathfrak{a}^t, M)$. It is immediate that $L_{\mathfrak{a}}$ is a left exact covariant additive functor from A-modules to A-modules, so that if we take its right derived functors $R^i L_{\mathfrak{a}}$, then $R^0 L_{\mathfrak{a}} = L_{\mathfrak{a}}$. These $R^i L_{\mathfrak{a}}$ are now defined to be the local cohomology functors $H^i_{\mathfrak{a}}$, for which we have by [CE, Ch. V, Prop. 4.1]

4.1.2 PROPOSITION. The local cohomology functors $H^i_{\mathfrak{a}}$ form an exact connected right sequence of functors as defined in 1.1.1.

We have already seen that $L_{\mathfrak{a}} \simeq H^0_{\mathfrak{a}} \simeq \varinjlim_t \mathrm{Ext}^0_A(A/\mathfrak{a}^t, -)$. In order to extend this to the higher derived functors, first notice that the $\varinjlim_t \operatorname{Ext}^i_A(A/\mathfrak{a}^t,-)$ form an exact connected right sequence of functors because the direct limit is an exact functor. By 1.1.3 it is enough to show that both $H^i_{\mathfrak{a}}$ and $\varinjlim_t \operatorname{Ext}^i_A(A/\mathfrak{a}^t,-)$ vanish on injective modules for $i \geq 1$. As derived functors, this is clear for the $H^i_{\mathfrak{a}}$. The other case is equally obvious, since each of the $\operatorname{Ext}^i_A(A/\mathfrak{a}^t,-)$, $i \geq 1$, vanishes on injectives. Hence

4.1.3. PROPOSITION. The sequence of functors $H^i_{\mathfrak{a}}$ and $\varinjlim_t \operatorname{Ext}^i_A(A/\mathfrak{a}^t,-)$ are naturally isomorphic, exact connected right sequences of functors.

We prove a couple of change of rings results for local cohomology, quite often useful. For the first, consider a homomorphism of rings $f\colon A \to B$ and if $\mathfrak{a}$ is an ideal in A write $\mathfrak{a}^e$ for the corresponding ideal $f(\mathfrak{a})B$ in B. Now any B-module N can be viewed as an A-module by restriction of scalars, so we may consider both $H^i_{\mathfrak{a}}(N)$ and $H^i_{\mathfrak{a}^e}(N)$, which are easily seen to be isomorphic for $i = 0$. One may ask whether $H^i_{\mathfrak{a}}$ and $H^i_{\mathfrak{a}^e}$ form naturally isomorphic exact connected right sequences of functors. This is true in the noetherian case.

4.1.4 PROPOSITION. Let $f\colon A \to B$ be a homomorphism of noetherian rings, and $\mathfrak{a}$ an ideal in A. Then $H^i_{\mathfrak{a}}$ and $H^i_{\mathfrak{a}^e}$ form naturally isomorphic exact connected right sequences of functors from B-modules to A-modules.

PROOF. As before, we need only prove that both functors vanish on injective B-modules for $i \geq 1$. This is clear for $H^i_{\mathfrak{a}^e}$. Since $L_{\mathfrak{a}}$ commutes with direct sums, so does each $H^i_{\mathfrak{a}}$ [CE, Ch. V, Th. 9.4]. If I is an injective B-module, it is a direct sum of indecomposable injective B-modules by 3.2.3, so we need only prove that $H^i_{\mathfrak{a}}(I) = 0$ for an indecomposable injective B-module, which is of the form $E_B(B/\mathfrak{q})$ for some prime $\mathfrak{q}$ in B, 3.2.5. To inspect the behaviour of $E_B(B/\mathfrak{q})$ as an A-module, put $\mathfrak{q}^c = f^{-1}(\mathfrak{q}) \subset A$. If $\mathfrak{a} \not\subset \mathfrak{q}^c$, then there is an $a \in \mathfrak{a}$ with $f(a) \notin \mathfrak{q}$, so by 3.2.7, multiplication by a is an automorphism on I, hence on $H^i_{\mathfrak{a}}(I)$ for all $i \geq 0$. For any $m \in H^i_{\mathfrak{a}}(I)$, there is a $t \geq 1$ such that $\mathfrak{a}^t m = 0$. In particular $a^t m = 0$, which shows that $m = 0$.

We may therefore assume that $\mathfrak{a} \subset \mathfrak{q}^c$. Now take a minimal injective A-resolution $0 \to I \to E^0_A(I) \to \ldots \to E^i_A(I) \to \ldots$ of I. An indecomposable injective direct summand of $E^i_A(I)$ is of type $E_A(A/\mathfrak{p})$, where $\mathfrak{p}$ is a prime in A, and it can only occur if $\mathfrak{p} \in \mathrm{Supp}_A I$, see 3.3.3. In this case $\mathfrak{q}^c \subset \mathfrak{p}$ (since otherwise there exists a $q \in \mathfrak{q}^c \backslash \mathfrak{p}$, such that each element of I is killed by some power of q). Thus $\mathfrak{a} \subset \mathfrak{p}$, and since this is true for every such $\mathfrak{p}$, each element of $E^i_A(I)$ is killed by a power of $\mathfrak{a}$, for $E^i_A(I)$ is a direct sum of indecomposables by 3.2.3. Hence $L_{\mathfrak{a}}(E^i_A(I)) = E^i_A(I)$ and it follows that $H^i_{\mathfrak{a}}(I) = 0$ for $i \geq 1$.

A companion result which goes from A-modules to B-modules requires a flatness assumption.

4.1.5 PROPOSITION. Let $f\colon A \to B$ be a homomorphism of rings with A noetherian. Let $\mathfrak{a}$ be an ideal in A and $\mathfrak{a}^e$ its extension to B. If B is a flat A-module, then $B \otimes_A H^i_{\mathfrak{a}}(-)$ and $H^i_{\mathfrak{a}^e}(B \otimes_A -)$ form naturally isomorphic exact connected right sequences of functors from A-modules to B-modules.

PROOF. Let M be an A-module, then there is a chain of natural isomorphisms

$$B \otimes_A H^i_{\mathfrak{a}}(M) \simeq B \otimes_A (\varinjlim_t \mathrm{Ext}^i_A(A/\mathfrak{a}^t, M)) \simeq \varinjlim_t (B \otimes_A \mathrm{Ext}^i_A(A/\mathfrak{a}^t, M)) \simeq$$

$$\varinjlim_t \mathrm{Ext}^i_B(B \otimes_A A/\mathfrak{a}^t, B \otimes_A M) \simeq \varinjlim_t \mathrm{Ext}^i_B(B/(\mathfrak{a}^e)^t, B \otimes_A M) \simeq H^n_{\mathfrak{a}^e}(B \otimes_A M).$$

Here we have used 4.1.3 for the first and the last isomorphism. The second one is true because tensor products commute with direct limits, [Bo 62, §6,

Prop. 12]. For the third we need that B is A-flat, A noetherian and $A/\mathfrak{a}^t$ finitely generated, [Bo 80, §6, Prop. 10 b)]. Naturality follows from the naturality in 4.1.3, once applied to the ideal $\mathfrak{a}$, and the other time to $\mathfrak{a}^e$.

4.1.6 COROLLARY. If $\mathfrak{a}$ is an ideal and S a multiplicative system in a noetherian ring A, then $S^{-1}H^i_{\mathfrak{a}}(-)$ and $H^i_{S^{-1}\mathfrak{a}}(S^{-1}-)$ form two naturally isomorphic exact connected right sequences of functors from A-modules to $S^{-1}\mathfrak{a}$-modules.

4.1.7 COROLLARY. If A is noetherian and $\hat{A}$ is its completion with respect to an ideal $\mathfrak{a} \subset A$, then $\hat{A}\otimes_A H^i_{\mathfrak{a}}(-)$ and $H^i_{\hat{\mathfrak{a}}}(\hat{A}\otimes_A -)$ form two naturally isomorphic exact connected right sequences of functors from A-modules to $\hat{A}$-modules.

PROOF. By 2.2.1 the completion $\hat{A}$ is A-flat.

Next we show that local cohomology commutes with direct limits in the noetherian case.

4.1.8 PROPOSITION. Let $\mathfrak{a}$ be an ideal in a noetherian ring A and M_j a direct system of A-modules with respect to some directed set. If $M = \varinjlim_j M_j$, then $H^i_{\mathfrak{a}}(M) \simeq \varinjlim_j H^i_{\mathfrak{a}}(M_j)$ for all i.

PROOF. By Lemma 3.3.7, $\mathrm{Ext}^i_A(A/\mathfrak{a}^t, M) \simeq \varinjlim_j \mathrm{Ext}^i_A(A/\mathfrak{a}^t, M_j)$ for $t \geq 1$ and these isomorphisms commute with the maps induced by the canonical projections $A/\mathfrak{a}^{t+s} \to A/\mathfrak{a}^t$. Therefore $H^i_{\mathfrak{a}}(M) \simeq \varinjlim_t \mathrm{Ext}^i_A(A/\mathfrak{a}^t, M) \simeq \varinjlim_t \varinjlim_j \mathrm{Ext}^i_A(A/\mathfrak{a}^t, M_j)$. However, direct limits commute with each other, [Bo 62, §6, Prop. 7], so the latter is isomorphic to $\varinjlim_j \varinjlim_t \mathrm{Ext}^i_A(A/\mathfrak{a}^t, M_j) \simeq \varinjlim_j H^i_{\mathfrak{a}}(M_j)$.

For the local cohomology we wish to consider $h^-_\mathfrak{a}$ in the notation of 1.1.1. $h^-_\mathfrak{a}(M) = \inf\{i \mid H^i_\mathfrak{a}(M) \neq 0\}$. Thus to each ideal $\mathfrak{a} \subset A$ and module M is attached a nonnegative integer or ∞. Notice that in the next lemma we need no noetherian condition.

4.1.9 LEMMA. Let $\mathfrak{a}$ be an ideal in the ring A and M an A-module. Let E be an injective envelope of M and $M' = E/M$. Then $H^i_\mathfrak{a}(M) \simeq H^{i-1}_\mathfrak{a}(M')$ for $i \geq 2$ and also for $i = 1$ if $L_\mathfrak{a}(M) = 0$.

PROOF. The first statement follows from the long exact sequence of local cohomology, since $H^i(E) = 0$ for the injective module E, $i \geq 1$. For the second, it is enough to show that $H^0_\mathfrak{a}(M) = L_\mathfrak{a}(M) = 0$ implies $H^0_\mathfrak{a}(E) = L_\mathfrak{a}(E) = 0$. This in turn is an easy consequence of the envelope property of E with respect to M.

4.1.10 COROLLARY. In the situation above, if $h^-_\mathfrak{a}(M) \neq 0$, then $h^-_\mathfrak{a}(E) = \infty$ and $h^-_\mathfrak{a}(M') = h^-_\mathfrak{a}(M) - 1$.

PROOF. Immediate from the lemma.

In further chapters of this book, we shall mostly consider local cohomology with respect to the maximal ideal in a noetherian local ring. For this purpose we record

4.1.11 PROPOSITION. Let $(A,\mathfrak{m},k)$ be a noetherian local ring and M a finitely generated A-module. Then $H^i_\mathfrak{m}(M)$ is an artinian A-module for every i.

PROOF. Take a minimal injective resolution of M and consider the concomitant complex $E^\bullet$ with chain modules $E^i(M)$ as in 3.3.1. Thus $E^i(M) = \oplus\ \mu^i(\mathfrak{p},M)E(A/\mathfrak{p})$, the sum being taken over all primes $\mathfrak{p}$ in A. We have already seen that $L_{\mathfrak{m}} = H^0_{\mathfrak{m}}$ commutes with direct sums, so that $L_{\mathfrak{m}}(E^i(M)) \simeq \oplus\ \mu^i(\mathfrak{p},M)L_{\mathfrak{m}}(E(A/\mathfrak{p}))$. Furthermore $L_{\mathfrak{m}}(E(A/\mathfrak{p})) = 0$ for $\mathfrak{p} \neq \mathfrak{m}$ and $L_{\mathfrak{m}}(E(k)) = E(k)$ by virtue of 3.2.7. Thus the i-th chain module of $L_{\mathfrak{m}}(E^\bullet)$ consists of $\mu^i(\mathfrak{m},M)$ copies of E(k); since the $\mu^i(\mathfrak{m},M)$ are finite by 3.3.3 and E(k) is artinian by 3.4.11, this is artinian. Now the artinian modules form a Serre subcategory of all the A-modules, so the homology $H^iL_{\mathfrak{m}}(E^\bullet) = H^i_{\mathfrak{m}}(M)$ is artinian.

4.2 KOSZUL COMPLEXES

As mentioned at the beginning of this chapter, the subject is both venerable and important. We only treat the simplest case and the basic facts which we shall need.

Let $x = x_1,\ldots,x_n$ be an arbitrary sequence of elements in a ring A. Let $\alpha = (i_1,\ldots,i_p)$, $1 \le i_1 < \ldots < i_p \le n$, be an ascending sequence of integers, where $0 \le p \le n$. We define $K_p(x)$ to be a free A-module on a basis (e_α), one for each such α, its rank is therefore $\binom{n}{p}$. To record that we are really dealing with an exterior algebra, we also write $e_\alpha = e_{i_1}\wedge\ldots\wedge e_{i_p}$, and we define a module homomorphism $d_p\colon K_p(x) \to K_{p-1}(x)$ by

$d_p(e_\alpha) = \Sigma_{j=1}^p(-1)^{j+1}x_{i_j}e_{i_1}\wedge\ldots\wedge\hat{e}_{i_j}\wedge\ldots\wedge e_{i_p}$, where ^ means deleting the item.

It is easy to verify that $d_{p-1}\circ d_p = 0$, so that we obtain a finite complex

$$K_\bullet(x) = 0 \to K_n(x) \xrightarrow{d_n} K_{n-1}(x) \to \ldots \to K_1(x) \xrightarrow{d_1} K_0(x) \to 0$$

of finitely generated free modules.

4.2.1 DEFINITIONS. For any A-module M, we define $K_\bullet(x,M)$ to be the complex $K_\bullet(x)\otimes_A M$ and $K^\bullet(x,M)$ to be $\mathrm{Hom}_A(K_\bullet(x),M)$. We write their p-th homologies as respectively $H_p(x,M)$ and $H^p(x,M)$.

We speak of the covariant or homological Koszul complex $K_\bullet(x,M)$ of x on M and similarly the contravariant of cohomological $K^\bullet(x,M)$. In both cases their chain modules are finite direct sums of copies of M and are nonnull at most between degrees 0 and n. If we call $\mathfrak{a}$ the ideal generated by $x_1,\dots,x_n$ and write ${}_{\mathfrak{a}}M$ for the submodule annihilated by $\mathfrak{a}$ then it is not difficult to see that $H_0(x,M) \simeq M/\mathfrak{a}M \simeq H^n(x,M)$ and $H_n(x,M) \simeq {}_{\mathfrak{a}}M \simeq H^0(x,M)$, which suggests a kind of duality we shall describe in 4.2.8. Notice also that $K_\bullet(x,A) \simeq K_\bullet(x)$ and we allow ourselves the notations $H_p(x,A)$ or $H_p(x)$ for its homology in degree p. Similarly for the contravariant complex.

For a single element $z \in A$, the complex $K_\bullet(z)$ is just the complex $0 \to A \xrightarrow{.z} A \to 0$ concentrated in degrees 0 and 1 with $H_0(z) = A/(z)$ and $H_1(z) = {}_{(z)}A$. These complexes are the building blocks of a general Koszul complex as follows.

4.2.2 THEOREM. Let $x = x_1,\dots,x_n$ be a sequence of elements in a ring A. Then $K_\bullet(x) \simeq K_\bullet(x_1)\otimes_A\dots\otimes_A K_\bullet(x_n)$.

PROOF. The statement being tautological for n = 1 , put $y = x_1,\dots,x_{n-1}$ and assume that $K_\bullet(y)$, with boundary maps d_p, is isomorphic to $K_\bullet(x_1)\otimes_A\dots\otimes_A K_\bullet(x_{n-1})$. In the tensor product complex $C_\bullet = K_\bullet(y)\otimes_A K_\bullet(x_n)$, the chain modules C_p can be identified with $K_p(y)\oplus K_{p-1}(y)$ and the boundary map ∂_p is described by the matrix

$$\begin{pmatrix} d_p & (-1)^{p-1}.x_n \\ 0 & d_{p-1} \end{pmatrix}$$

where we follow the conventions of [Bo 80, §4, 1]. There is an isomorphism ϕ_p

between C_p and $K_p(x)$ defined by setting $\phi_p(e_\alpha) = e_\alpha$ and $\phi_p(e_\beta) = e_\beta \wedge e_n$ for basis elements of $K_p(y)$ resp. $K_{p-1}(y)$. One easily verifies that these ϕ_p's commute with the respective boundary operators, so that we obtain an isomorphism $\phi_\bullet$: $K_\bullet(y) \otimes_A K_\bullet(x_n) \to K_\bullet(x)$, which finishes the proof.

Keeping the notation of this demonstration, we state

4.2.3 PROPOSITION. For each A-module M, there are exact sequences

$$\ldots \to H_p(y,M) \xrightarrow{.x_n} H_p(y,M) \to H_p(x,M) \to H_{p-1}(y,M) \xrightarrow{.x_n} \ldots \text{ and}$$

$$\ldots \xrightarrow{.x_n} H^{p-1}(y,M) \to H^p(x,M) \to H^p(y,M) \xrightarrow{.x_n} H^p(y,M) \to \ldots .$$

PROOF. The complex $C_\bullet$ may also be described as the mapping cone of the morphism of complexes $.x_n\bullet$: $K_\bullet(y) \to K_\bullet(y)$. True, the sign conventions differ from the ones chosen in [Bo 80, §2, 6] but serve equally well. It is easy to check that the split short exact sequences

$$0 \to K_p(y) \xrightarrow{(1\ 0)} C_p \xrightarrow{\binom{0}{1}} K_{p-1}(y) \to 0$$

commute with the boundary maps so that we get a split short exact sequence of complexes $0 \to K_\bullet(y) \to C_\bullet \to K_\bullet(y)[-1] \to 0$. Here we have used the standard notation for a shifted complex as in 3.3.5. We now apply the additive functors $- \otimes_A M$ resp. $\mathrm{Hom}_A(-,M)$ to again obtain split short exact sequences of complexes. These give rise to long exact homology sequences as above, except that the multiplications are alternately by plus or minus x_n. But this does not affect exactness. The reader may like to make explicit the other maps in the sequences.

By splitting into short exact sequences one obtains

4.2.4 COROLLARY. There are short exact sequences

(i) $0 \to H_0(x_n, H_p(y,M)) \to H_p(x,M) \to H_1(x_n, H_{p-1}(y,M)) \to 0;$

(ii) $0 \to H^1(x_n, H^{p-1}(y,M)) \to H^p(x,M) \to H^0(x_n, H^p(y,M)) \to 0.$

4.2.5 PROPOSITION. The functors $H_p(x,-)$ form an exact connected left sequence, and the $H^p(x,-)$ an exact connected right sequence of functors from A-modules to A-modules.

PROOF. Since the chain modules $K_p(x)$ are free, the functors $K_p(x)\otimes_A-$ and $\mathrm{Hom}_A(K_p(x),-)$ are exact. Thus a short exact sequence of modules gives rise to two short exact sequences of complexes. The proposition follows from their associated long exact homology sequences.

4.2.6 PROPOSITION. If $x = x_1,\dots,x_n \in A$, then the map $.x_{i\bullet}: K_\bullet(x) \to K_\bullet(x)$ is homotopic to 0 for $i = 1,\dots.,n$.

PROOF. By symmetry, it is enough to show that the above is true for $i = 1$. In fact, we shall construct a chain homotopy s of $K_\bullet(x)$ of degree 1, such that $d_{p+1}\circ s^p + s^{p-1}\circ d_p$ is multiplication on $K_p(x)$ by x_1 for all p. We define s^p on a basis element e_α of $K_p(x)$ by putting $s^p(e_\alpha) = 0$ if $\alpha = (1,i_2,\dots,i_p)$, $s^p(e_\alpha) = e_1\wedge e_\alpha$ otherwise. We leave it to the reader to check that s has the desired property.

4.2.7 COROLLARY. If $\mathfrak{a} = (x_1,\dots,x_n)$ and the A-module M is annihilated by $\mathfrak{b}$, then $(\mathfrak{a}+\mathfrak{b})H_p(x,M) = (\mathfrak{a}+\mathfrak{b})H^p(x,M) = 0$ for all p.

PROOF. Since each $.x_{i\bullet}: K_\bullet(x) \to K_\bullet(x)$ is homotopic to 0, the same holds for the map $.x_{i\bullet}$ on the complexes $K_\bullet(x,M) = K_\bullet(x)\otimes_A M$ and $K^\bullet(x,M) = \mathrm{Hom}_A(K_\bullet(x),M)$, so that $\mathfrak{a}$ annihilates their homology. The ideal $\mathfrak{b}$ even annihilates their chain modules, so the corollary follows.

We now discuss a duality between the covariant and the contravariant Koszul complexes. Let $\alpha = (i_1,\ldots,i_p)$ be a subsequence of $(1,\ldots.,n)$; we define $\bar{\alpha}$ to be its "complement". Thus if e_α is a basis element of $K_p(x)$, $e_{\bar{\alpha}}$ is an element of $K_{n-p}(x)$ and this assignment is seen to define an isomorphism between these two free modules. Let $\sigma(\alpha)$ be the sign of the permutation which takes $(1,\ldots,n)$ to $(\alpha,\bar{\alpha})$. If M is an A-module, each element of $K_p(x,M)$ is uniquely a sum of elements $e_\alpha \otimes m_\alpha$, where $m_\alpha \in M$ and the sum is over the basis elements e_α of $K_p(x)$. As to $K^p(x,M)$, define f_α^m to be the map from $K_p(x)$ to M which takes e_α to m and all the other basis elements to 0. Each element of $K^p(x,M)$ is uniquely a sum of such f_α^m. The boundary maps d_p^M and d_M^p in $K_\bullet(x,M)$ resp. $K^\bullet(x,M)$ are easily made explicit from those in $K_\bullet(x)$.

4.2.8 THEOREM. For every module M, there is an isomorphism of complexes

$$\begin{array}{ccccccccccc}
0 & \longrightarrow & K_n(x,M) & \longrightarrow \cdots \longrightarrow & K_p(x,M) & \xrightarrow{d_p^M} & K_{p-1}(x,M) & \longrightarrow \cdots \longrightarrow & K_0(x,M) & \longrightarrow & 0 \\
 & & \downarrow \phi_n & & \downarrow \phi_p & & \downarrow \phi_{p-1} & & \downarrow \phi_0 & & \\
0 & \longrightarrow & K^0(x,M) & \longrightarrow \cdots \longrightarrow & K^{n-p}(x,M) & \xrightarrow[d_M^{n-p+1}]{} & K^{n-p+1}(x,M) & \longrightarrow \cdots \longrightarrow & K^n(x,M) & \longrightarrow & 0
\end{array}$$

In particular, $H_p(x,M) \simeq H^{n-p}(x,M)$ for all p.

PROOF. For a basis element $e_\alpha \otimes m$ in $K_p(x,M)$, we define $\phi_p(e_\alpha \otimes m) = (-1)^{\binom{p}{2}} \sigma(\alpha) f_{\bar{\alpha}}^m$. In order to check that $\phi_{p-1} \circ d_p^M = d_M^{n-p+1} \circ \phi_p$, we need only compare their effects on such an $e_\alpha \otimes m$ and show that they are the same. This verification, where the main worry is to keep track of the signs, we leave to the reader. Here is an instance where introduction of the full machinery of exterior algebra would have rendered the sign convention more natural [Bo 80, §9, 1].

One sometimes needs to know the behaviour of Koszul complexes under ring extensions, as we investigated for local cohomology in 4.1.4 - 4.1.7.

4.2.9 PROPOSITION. Let $x = x_1,\ldots,x_n$ be a sequence of elements in a ring A and M an A-module. Let $f: A \to B$ be a ring homomorphism and $fx = fx_1,\ldots,fx_n$, then $B\otimes_A K_\bullet(x,M) \simeq K_\bullet(fx,B\otimes_A M)$ and $B\otimes_A K^\bullet(x,M) \simeq K^\bullet(fx,B\otimes_A M)$. If moreover B is a flat A-module, then also $B\otimes_A H_p(x,M) \simeq H_p(fx,B\otimes_A M)$ and $B\otimes_A H^p(x,M) \simeq H^p(fx,B\otimes_A M)$.

PROOF. First $B\otimes_A K_\bullet(x,A) \simeq K_\bullet(fx,B)$ because their chain modules and boundary maps can be identified. Then $K^\bullet(fx,B\otimes_A M) = \mathrm{Hom}_B(K_\bullet(fx,B),B\otimes_A M) \simeq \mathrm{Hom}_B(B\otimes_A K_\bullet(x,A),B\otimes_A M) \simeq B\otimes_A \mathrm{Hom}_A(K_\bullet(x,A),M) = B\otimes_A K^\bullet(x,M)$, where the last isomorphism depends on the fact that the chain modules of $K_\bullet(x,A)$ are free and finitely generated. The proof is similar for $K_\bullet$. If B is flat, $B\otimes_A$- is an exact functor, so commutes with taking homology.

Even easier is the next proposition.

4.2.10 PROPOSITION. Let $x = x_1,\ldots,x_n \in A$ and let F and M be A-modules. Then $F\otimes_A K_\bullet(x,M) \simeq K_\bullet(x,F\otimes_A M)$ and $F\otimes_A K^\bullet(x,M) \simeq K^\bullet(x,F\otimes_A M)$. If F is flat, then $F\otimes_A H_p(x,M) \simeq H_p(x,F\otimes_A M)$ and $F\otimes_A H^p(x,M) \simeq H^p(x,F\otimes_A M)$ for all p.

The following pretty result is used in multiplicity theory. Let $x = x_1,\ldots,x_n$ be elements in a ring A which generate an ideal $\mathfrak{a}$. It is easy to see that for each $s \geq 0$

$$0 \to \mathfrak{a}^s K_n(x) \to \mathfrak{a}^{s+1}K_{n-1}(x) \to \dots \to \mathfrak{a}^{s+n}K_0(x) \to 0$$

is a subcomplex of $K_\bullet(x)$, which we call ${}^sK_\bullet(x)$.

4.2.11 PROPOSITION. If A is noetherian, there exists an $s_0 \geq 0$, such that ${}^sK_\bullet(x)$ is exact for $s \geq s_0$.

Since we do not need the fact in this book, we refer to [EF] for the proof. This heavily depends on the Artin-Rees Lemma 2.1.1, and it is easy to see that the result fails for nonnoetherian rings. There are also such statements for finitely generated modules, as well as contravariant versions. Compare with Lemma 4.3.3.

4.3 LIMITS OF KOSZUL COMPLEXES AND LOCAL COHOMOLOGY

In this section, we introduce a hierarchy of Koszul complexes attached to a set of elements $x = x_1,\dots,x_n$ in a ring A and construct maps between them. In the limit we obtain complexes whose homology is shown to yield the local cohomology modules with respect to the ideal generated by the x's, at least in the noetherian case.

Let $x = x_1,\dots,x_n \in A$. For $t \geq 1$ we write $x(t) = x_1^t,\dots,x_n^t$, $\mathfrak{a}_t = (x_1^t,\dots,x_n^t)$, and consider the Koszul complex $K_\bullet(x(t))$ which we abbreviate to $K_\bullet^{(t)}$; for boundary maps we write $d_p^{(t)}$. With obvious notation, the group $H_0^{(t)}$ is just $A/\mathfrak{a}_t$ so we have a homomorphism $H_0^{(t+s)} \to H_0^{(t)}$ given by the canonical projection $A/\mathfrak{a}_{t+s} \to A/\mathfrak{a}_t$ for all $s \geq 0$, but also a map $H_0^{(t)} \to H_0^{(t+s)}$ given by $x_1^s\dots x_n^s\colon A/\mathfrak{a}_t \to A/\mathfrak{a}_{t+s}$. Both maps actually stem from morphisms of complexes as follows. Define $\theta_p^{t,s}\colon K_p^{(t+s)} \to K_p^{(t)}$ where

$\theta_p^{t,s}(e_\alpha) = x_\alpha^s e_\alpha$ when $\alpha = (i_1,\ldots,i_p)$, $1 \le i_1 < \ldots < i_p \le n$; here we have written x_α for the product $x_{i_1}\ldots x_{i_p}$. Similarly we define $\eta_p^{t,s}\colon K_p^{(t)} \to K_p^{(t+s)}$ by putting $\eta_p^{t,s}(e_\alpha) = x_{\bar\alpha}^s e_\alpha$.

4.3.1 PROPOSITION. The maps $\theta_\bullet^{t,s}\colon K_\bullet^{(t+s)} \to K_\bullet^{(t)}$ and $\eta_\bullet^{t,s}\colon K_\bullet^{(t)} \to K_\bullet^{(t+s)}$ are morphisms of complexes, such that $H_0(\theta_\bullet^{t,s})$ and $H_0(\eta_\bullet^{t,s})$ are the maps just mentioned.

PROOF. We verify that $\eta_{p-1}^{t,s} \circ d_p^{(t)} = d_p^{(t+s)} \circ \eta_p^{t,s}$, leaving the map $\theta_\bullet^{t,s}$ to the reader. Well,

$$\eta_{p-1}^{t,s}(d_p^{(t)}(e_\alpha)) = \Sigma_{j=1}^p(-1)^{j+1}x_{i_j}^t\eta_{p-1}^{t,s}(e_{i_1}\wedge\ldots\wedge\hat e_{i_j}\wedge\ldots\wedge e_{i_p}) =$$
$$\Sigma_{j=1}^p(-1)^{j+1}x_{i_j}^t.x_1^s\ldots\hat x_{i_1}^s\ldots x_{i_j}^s\ldots\hat x_{i_p}^s\ldots x_n^s.e_{i_1}\wedge\ldots\wedge\hat e_{i_j}\wedge\ldots\wedge e_{i_p} =$$
$$\Sigma_{j=1}^p(-1)^{j+1}x_1^s\ldots\hat x_{i_1}^s\ldots\hat x_{i_j}^s\ldots\hat x_{i_p}^s\ldots x_n^s.x_{i_j}^{t+s}.e_{i_1}\wedge\ldots\wedge\hat e_{i_j}\wedge\ldots\wedge e_{i_p} =$$
$$d_p^{(t+s)}(x_1^s\ldots\hat x_{i_1}^s\ldots\hat x_{i_p}^s\ldots x_n^s.e_{i_1}\wedge\ldots\wedge e_{i_p}) = d_p^{(t+s)}(\eta_p^{t,s}(e_\alpha)).$$

Observe that the $\theta_\bullet^{t,s}$ and the $\eta_\bullet^{t,s}$ form an inverse resp. direct system of complexes $K_\bullet^{(t)}$. For any A-module M we get $\mathrm{Hom}_A(\theta_\bullet^{t,s},1_M)$ which makes $K^\bullet(x(t),M) = \mathrm{Hom}_A(K_\bullet^{(t)},M)$ and $\eta_\bullet^{t,s}\otimes 1_M$ which makes $K_\bullet(x^{(t)},M) = K_\bullet^{(t)}\otimes_A M$ into a direct system of complexes. We denote their direct limits by $K_\infty^\bullet(x,M)$ and $K_\bullet^\infty(x,M)$ respectively and also write $K_\infty^\bullet(x)$ for $K_\infty^\bullet(x,A)$ and $K_\bullet^\infty(x)$ for $K_\bullet^\infty(x,A)$. We use corresponding notations for their homology.

There is an autoduality which we express in

4.3.2 THEOREM. For every $x = x_1,\ldots,x_n \in A$ and A-module M there are isomorphisms $H_\infty^{n-p}(x,M) \simeq H_p^\infty(x,M) \simeq H_p(K_\bullet^\infty(x)\otimes_A M) \simeq H^{n-p}(K_\infty^\bullet(x)\otimes_A M)$.

PROOF. Notice that the map $\phi_\bullet$ in Theorem 4.2.8 does not depend on t and provides an isomorphism between direct systems of complexes.

Therefore, taking the direct limit we obtain the duality 4.2.8, now between $K_\bullet^\infty(x,M)$ and $K_\infty^\bullet(x,M)$. Taking homology, we get the first isomorphism above. In particular $K_\bullet^\infty(x)$ and $K_\infty^\bullet(x)$ are in fact, except for reversed indices, isomorphic complexes, whence the last isomorphism. To show finally that $H_p^\infty(x,M) \simeq H_p(K_\bullet^\infty(x)\otimes_A M)$, it is sufficient to know that tensor products commute with direct limits.

We now wish to establish an isomorphism between $H_\infty^p(x,M)$ and the local cohomology group $H_\alpha^p(M)$ where $\alpha = (x_1,\dots,x_n)$, provided A is noetherian. We need a preliminary result about the $\theta_\bullet^{t,s}$.

4.3.3 LEMMA. Let A be a noetherian ring and $t \geq 1$ a given integer. There exists an $s_0 \geq 0$ such that, for $s \geq s_0$, the map $H_p(\theta_\bullet^{t,s})\colon H_p^{(t+s)} \to H_p^{(t)}$ is the null morphism for $p \geq 1$.

PROOF. Put $\alpha = (x_1,\dots,x_n)$ and choose r large enough so that $\alpha^r \subset \alpha_t = (x_1^t,\dots,x_n^t)$. For each of the complexes $K_\bullet^{(t)}$ write $Z_p^{(t)} = \ker(d_p^{(t)})$ for the cycles and $B_p^{(t)} = \operatorname{im}(d_{p+1}^{(t)})$ for the boundaries in $K_p^{(t)}$. By the Artin-Rees Lemma 2.1.1, there exists an $s_1 \geq 0$, such that for $s \geq s_1$ we have $Z_p^{(t)} \cap \alpha^s K_p^{(t)} \subset \alpha^{s-s_1}(Z_p^{(t)} \cap \alpha^{s_1} K_p^{(t)})$. Put $r+s_1 = s_0$, then for $s \geq s_0$ there are inclusions

$$Z_p^{(t)} \cap \alpha^s K_p^{(t)} \subset \alpha^r(Z_p^{(t)} \cap \alpha^{s_1} K_p^{(t)}) \subset \alpha^r Z_p^{(t)} \subset \alpha_t Z_p^{(t)} \subset B_p^{(t)};$$

the last inclusion is a consequences of 4.2.7, applied to the Koszul complex $K_\bullet^{(t)}$.

We now shall prove that $H_p(\theta_\bullet^{t,s}) = 0$ for $s \geq s_0$, $p \geq 1$. Since $\theta_\bullet^{t,s}$ is a morphism of complexes, we have $\theta_p^{t,s}(Z_p^{(t+s)}) \subset Z_p^{(t)}$, and we need to show that actually $\theta_p^{t,s}(Z_p^{(t+s)}) \subset B_p^{(t)}$. Well, $\theta_p^{t,s}(K_p^{(t+s)}) \subset \alpha^s K_p^{(t)}$ - this is the point where we use $p \geq 1$ - so $\theta_p^{t,s}(Z_p^{(t+s)}) \subset Z_p^{(t)} \cap \alpha^s K_p^{(t)} \subset B_p^{(t)}$, and we are through.

4.3.4 THEOREM. Let $x = x_1,\dots,x_n$ be elements in a noetherian ring A which generate the ideal $\mathfrak{a}$. Then $H^p_\infty(x,M) \simeq H^p_\mathfrak{a}(M)$ for every module M.

PROOF. First observe that each chain module $K^{(t)}_p$ is free, so that a short exact sequence of modules $0 \to M' \to M \to M'' \to 0$ functorially gives rise to a short exact sequence of corresponding complexes $\mathrm{Hom}_A(K^{(t)}_\bullet, M)$ for each $t \geq 1$. Since the direct limit is exact, we get a short exact sequence $0 \to K^\bullet_\infty(x,M') \to K^\bullet_\infty(x,M) \to K^\bullet_\infty(x,M'') \to 0$. The associated long exact homology sequence shows that the $H^p_\infty(x,-)$ form an exact connected right sequence of functors on A-modules.

We need to prove that $H^0_\infty(x,M) \simeq H^0_\mathfrak{a}(M)$ and that $H^p_\infty(x,I) = 0$ for every injective module I and $p \geq 1$. Since $\mathrm{Hom}_A(-,M)$ is left exact, $H^0_\infty(x,M) \simeq \varinjlim_t \mathrm{Hom}_A(A/\mathfrak{a}_t, M)$, where the limit is taken over the canonical projection $A/\mathfrak{a}_{t+s} \to A/\mathfrak{a}_t$. Since the systems $\mathfrak{a}_t$ and $\mathfrak{a}^t$ are cofinal, we find $H^0_\infty(x,M) \simeq L_\mathfrak{a}(M) = H^0_\mathfrak{a}(M)$. Now for an injective module I the functor $\mathrm{Hom}_A(-,I)$ is exact, and since $\varinjlim$ is exact, both functors commute with taking homology. Therefore $H^p_\infty(x,I) \simeq \varinjlim_t \mathrm{Hom}_A(H^{(t)}_p(x),I)$ where the limit is induced by the maps $H_p(\theta^{t,s}_\bullet)$. By the previous lemma, this limit is 0 for $p \geq 1$.

Next we give an explicit construction for the complex $K^\bullet_\infty(x)$. For $\alpha = (i_1,\dots,i_p)$, $1 \leq i_1 < \dots < i_p \leq n$, write A_{x_α} for the module A localized at the element $x_\alpha = x_{i_1}\cdots x_{i_p}$ and write $C^p = \oplus A_{x_\alpha}$, where the sum is taken over all such α's. Notice that if $\beta = (k_1,\dots,k_{p+1})$ and $\alpha = (k_1,\dots,\hat{k}_j,\dots,k_{p+1})$, then A_{x_β} is a further localization of A_{x_α}.

Now define $\delta^p\colon C^p \to C^{p+1}$ on the component $A_{x_\alpha} \to A_{x_\beta}$ as $\delta^p(y) = (-1)^{j+1}y/1$ if α and β are as above, and 0 otherwise, for $y \in A_{x_\alpha}$. Clearly $\delta^{p+1}\circ\delta^p = 0$ and we have an ascending complex $C^\bullet$.

4.3.5 PROPOSITION. There is a canonical isomorphism $K_\infty^\bullet(x) \simeq C^\bullet$.

PROOF. We use the fact that the direct limit of the system $A \xrightarrow{.x_\alpha} A \xrightarrow{.x_\alpha} A \xrightarrow{.x_\alpha} \ldots$ is, as a ring, isomorphic to A_{x_α}, where the element a in the n-th copy of A $(n \geq 0)$ is identified with a/x_α^n.

Consider now in $K_{(t)}^p(x)$ and $K_{(t)}^{p+1}(x)$ the one-dimensional free components generated by f_α^1 resp. f_β^1, notation as above 4.2.8, where α and β are as defined. It is easy to see that the limit of the direct system

$$\begin{array}{ccccccc}
\cdots & \longrightarrow & A & \xrightarrow{.x_\alpha^s} & A & \longrightarrow & \cdots \\
 & & \Big\downarrow{\scriptstyle (-1)^{j+1}.x_{k_j}^t} & & \Big\downarrow{\scriptstyle (-1)^{j+1}.x_{k_j}^{s+t}} & & \\
\cdots & \longrightarrow & A & \xrightarrow[.x_\beta^s]{} & A & \longrightarrow & \cdots
\end{array}$$

is just the map δ^p on the components $A_{x_\alpha} \to A_{x_\beta}$.

4.3.6 COROLLARY. If $x = x_1,\ldots,x_n$ are elements in a noetherian ring A which generate the ideal $\mathfrak{a}$, then $H_{\mathfrak{a}}^p(M) \simeq H^p(C^\bullet \otimes_A M)$ for every module M and every p.

4.3.7 EXERCISE. Give a similar description of the complex $K_\bullet^\infty(x)$.

This description of the local cohomology modules in terms of the complex $K_\infty^\bullet(x)$ or $C^\bullet$ is often convenient. For instance, $H_{\mathfrak{a}}^n(M)$ is isomorphic to the direct limit of the system

$$M/\mathfrak{a}M \xrightarrow{.x_1\ldots x_n} M/\mathfrak{a}_2M \to \ldots \to M/\mathfrak{a}_tM \xrightarrow{.x_1\ldots x_n} M/\mathfrak{a}_{t+1}M \to \ldots$$

which we shall examine in the next chapter.

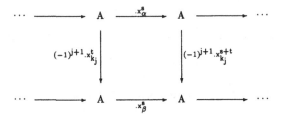

We conclude by remarking that the above treatment of local cohomology allows a more geometric interpretation in terms of Čech cohomology. This is the point of view taken in [Gr 67] and [Ha], but we do not need this. A lot more is known about both local cohomology and Koszul complexes, for which we refer e.g. to [Se, Ch. IV], [Sc 82] and [HIO, Ch. VII]

5 (PRE-) REGULAR SEQUENCES AND DEPTH

The theory of regular sequences became prominent in the middle fifties in the hands of such authors as Auslander-Buchsbaum, Rees, Serre and Kaplansky. It provides a link between ideal-theoretic properties of a ring and its modules on the one hand and homological concepts on the other. The reason is that, while reflecting certain ideal-theoretic conditions, regular sequences behave well with respect to homological invariants. These sequences are usually treated for finitely generated modules over noetherian rings. In section 5.1 we take a more general and formal point of view and consider the notion of a pre-regular sequence, a weak form of regular sequence. In section 5.2 we prove that after completion the two notions coincide, which permits an immediate proof of a theorem in Bourbaki. In section 5.3 then, we introduce the standard notion of depth and tie it up with an important variant, the Ext-depth. The well-known results in the noetherian case will be easy consequences of our more general considerations.

5.1 (PRE-) REGULAR SEQUENCES

Suppose A is a ring and M an A-module. We shall often work with a fixed sequence $x_1,\ldots,x_n$ of elements in A and standardly write $\mathfrak{a}$ for the ideal $(x_1,\ldots,x_n)$.

5.1.1 DEFINITION. Such a sequence is called an M-sequence (or regular on M) if $\mathfrak{a}M \neq M$ and x_i is a non zerodivisor on $M/(x_1,\ldots,x_{i-1})M$ for $i = 1,\ldots,n$.

The first non degeneracy condition is added to avoid trivial cases; the second is what really matters; it expresses that the endomorphism $.x_i$ is injective on the module $M/(x_1,\ldots,x_{i-1})M$. We sometimes shall write x for the sequence $x_1,\ldots,x_n$ and say that M is x-regular. When M = A we just speak of a regular sequence in A.

5.1.2 EXAMPLES.

1) The sequence $X = X_1,\ldots,X_n$ is regular in both the rings $A[X_1,\ldots,X_n]$ and $A[[X_1,\ldots,X_n]]$. In fact, a regular sequence in certain respects behaves like a set of indeterminates, as we shall see.

2) Let k be a field and put $A = k[X,Y,Z]$. Then X, Y(X-1), Z(X-1) is an A-sequence but Y(X-1), Z(X-1), X is not.

3) For k as above, put $A = k[[X,Y]]$. Let E be the injective hull of the A-module A/(Y); then X acts automorphically on E and Y nilpotently on every element of E (3.2.7). Hence X, Y is a regular sequence on $M = A \oplus E$ but Y, X is not. Here the module E is far from being separated in the (X,Y)-adic topology of the local ring A.

4) An example of this phenomenon with separated module M is described in [BS, Ex. 3.11 a)] . There Z, X, Y is an M-sequence but X, Y, Z is not.

What these examples do not contradict is that if M is a

finitely generated module over a noetherian local ring and x is a regular sequence on M, then it is an M-sequence in any order. This fact is well known [Se, Ch. IV, Cor. de Prop. 3], [Ka 74, Th. 119] and forms one of the cornerstones of the usefulness of regular sequences. We shall obtain a proof as spin-off from our study of the infinitely generated case. Though we cannot in general permute regular sequences, certain operations are allowed. The following proposition is often helpful. A proof can be found in [No 76, 5.1] and [La, Prop. 1] following an exercise in [Ka 74, Ch. 3, Exc. 12]. We invite the reader to provide his own proof of the successive statements.

5.1.3 PROPOSITION. Let A be a ring, M an A-module, $x = x_1,\dots,x_n \in A$.

a) If M is x-regular, then x_i is regular on $M/(x_1,\dots,x_{i-1},x_{i+1},\dots,x_n)M$ for $i = 1,\dots,n$.

b) If x, y and z are elements of A with $(x,y,z)M \neq M$, then yz,x is an M-sequence iff both y,x and z,x are.

c) Let $x_1,\dots,x_{i-1},x_{i+1},\dots,x_n$, y and z be elements of A, generating an ideal $\mathfrak{c}$ and assume that $\mathfrak{c}M \neq M$. Then the sequence $x_1,\dots,x_{i-1},yz,x_{i+1},\dots,x_n$ is M-regular iff both $x_1,\dots,x_{i-1},y,x_{i+1},\dots,x_n$ and $x_1,\dots,x_{i-1},z,x_{i+1},\dots,x_n$ are.

d) A sequence $x_1^{\upsilon_1},\dots,x_n^{\upsilon_n}$ is M-regular iff $x_1,\dots,x_n$ is M-regular. Here $\upsilon = \upsilon_1,\dots,\upsilon_n$ is a sequence of positive integers.

e) All permutations of a sequence $x_1,\dots,x_n$ are M-regular iff all subsequences of $x_1,\dots,x_n$ in the given order are M-regular.

As mentioned, regular sequences behave much as a set of indeterminates in a polynomial ring. We shall pursue this theme in order to derive certain results on such sequences.

5.1.4 PROPOSITION. Let $R = \mathbf{Z}[X_1,\dots,X_n]$ and $J \subset R$ be an ideal generated by monomials $m_1,\dots,m_q$ in $X_1,\dots,X_n$. If $m_1,\dots,m_p$, $1 \le p \le q$, contain the indeterminate X_1, and $m_{p+1},\dots,m_q$ do not, then $(J : X_1)$ is generated by the monomials $X_1^{-1}m_1,\dots,X_1^{-1}m_p,m_{p+1},\dots,m_q$. In case $(J : X_1) = J$, then $J = (m_{p+1},\dots,m_q)$.

PROOF. Let $f \in (J : X_1)$, then $X_1 f = \Sigma_{i=1}^q f_i m_i$, say, with $f_i \in R$. For each i write $f_i(X_1,\dots,X_n) = f_i(0,X_2,\dots,X_n) + X_1 g_i(X_1,\dots,X_n)$ and substitute in the previous equation. Putting $X_1 = 0$ we find $0 = \Sigma_{i=p+1}^q f_i(0,X_2,\dots,X_n)m_i$. Thus $X_1 f = \Sigma_{i=1}^p f_i m_i + \Sigma_{i=p+1}^q X_1 g_i m_i$, hence $f = \Sigma_{i=1}^p X_1^{-1} f_i m_i + \Sigma_{i=p+1}^q g_i m_i$, which proves our first contention. In case $(J : X_1) = J$, we can continue with the new generators and eventually obtain $J = (X_1^{-\upsilon_1}m_1,\dots,X_1^{-\upsilon_p}m_p,m_{p+1},\dots,m_q)$, $\upsilon_i \ge 1$, $i = 1,\dots,p$, such that X_1 no longer occurs in these generators. Then $X_1^{-\upsilon_1}m_1 \in J$, so $X_1^{-\upsilon_1}m_1 = \Sigma_{i=1}^p h_i m_i + \Sigma_{i=p+1}^q h_i m_i$ for certain $h_i \in R$, $i = 1,\dots,q$. Now $X_1^{-\upsilon_1}m_1$ does not contain X_1, but $m_1,\dots,m_p$ all do. Hence $\Sigma_{i=1}^p h_i m_i = 0$ and $X_1^{-\upsilon_1}m_1$ is expressed in terms of $m_{p+1},\dots,m_q$. Similarly for $X_1^{-\upsilon_2}m_2,\dots,X_1^{-\upsilon_p}m_p$, so $J = (m_{p+1},\dots,m_q)$.

5.1.5 COROLLARY. Let $J \subset R$ be an ideal generated by monomials. Then J is generated by a subset of $X_1,\dots,X_n$ if and only if, for $i = 1,\dots,n$, either $X_i \in J$ or $(J : X_i) = J$.

PROOF. Only if being clear, let us assume that $X_1,\dots,X_r$ are not in J but $X_{r+1},\dots,X_n$ are, $0 \le r \le n$. Since $(J : X_i) = J$ for $i = 1,\dots,r$, successive application of the proposition shows that J is generated by monomials only involving $X_{r+1},\dots,X_n$. Then patently $J = (X_{r+1},\dots,X_n)$.

5.1.6 COROLLARY. Let $J \subset R$ be an ideal generated by monomials and M an R-module. Let $p \ge 0$ be an integer. Suppose $\mathrm{Tor}_p^R(R/I,M) = 0$ for every ideal

$I \supset J$ of the shape $I = (X_{i_1}, \ldots, X_{i_k})$, $1 \leq i_1 < \ldots < i_k \leq n$. Then $\mathrm{Tor}_p^R(R/J,M) = 0$. Similar statements hold for $\mathrm{Ext}_R^p(R/J,M)$ and $\mathrm{Ext}_R^p(M,R/J)$.

PROOF. Suppose $\mathrm{Tor}_p^R(R/J,M) \neq 0$; the collection of ideals $J' \supset J$ with J' generated by monomials in $X_1, \ldots, X_n$ and $\mathrm{Tor}_p^R(R/J',M) \neq 0$ contains a maximal specimen I by noetherian induction. The endomorphism $.X_1$ on R/I gives rise to a short exact sequence

$$0 \to R/(I : X_1) \to R/I \to R/(X_1,I) \to 0.$$

The associated long exact Tor-sequence and the maximality of I in combination with 5.1.4 assert that either $(I : X_1) = I$ or $(X_1,I) = I$. In view of 5.1.5, I is of the shape $I = (X_{i_1}, \ldots, X_{i_k})$, hence $\mathrm{Tor}_p^R(R/I,M) = 0$. This contradiction proves that $\mathrm{Tor}_p^R(R/J,M) = 0$. The proof for the Ext's is similar.

This corollary is useful in the proof of the next theorem, although one can manage with less. It is essential however for proving 5.1.8. Before stating the next theorem, let us introduce some notation. For a vector $\upsilon = \upsilon_1, \ldots, \upsilon_n$, we write x^υ for the monomial $x_1^{\upsilon_1} \ldots x_n^{\upsilon_n}$. In case all the υ_i are equal, say t, we write $x^t = x_1^t \ldots x_n^t$. Thus $x^1 = x_1^1 \ldots x_n^1$ stands for the product of the x's, not for their sequence, hence $x^1 \neq x$. Moreover we write $\mathfrak{a}_t$ for the ideal $(x_1^t, \ldots, x_n^t)$, so that here $\mathfrak{a}_1 = \mathfrak{a}$. This notation is a little bit misleading, since of course $\mathfrak{a}_t$ depends on the choice of generators $x_1, \ldots, x_n$. Notice that $x^t\mathfrak{a} \subset \mathfrak{a}_{t+1}\mathfrak{a}^{nt-t}$.

We now fix a sequence x in a ring A and a module M and relate the notion of regularity to the Koszul homology.

5.1.7 THEOREM. Consider the statements

(i) The module M is x-regular;

(ii) $H_p(x,M) = 0$ for all $p \geq 1$;

(iii) $H_1(x,M) = 0$;

(iv) The maps $.x^t\colon M/\alpha M \to M/\alpha_{t+1}\alpha^{nt-t-1}M$ are injective for all $t \geq 1$.

Then (i) ⇒ (ii) ⇒ (iii) ⇒ (iv).

In case $n = 1$, we interpret α^{-1} to mean A. Notice that conditions (ii), (iii) and (iv) are independent of the order of the x_i's, but (i) is not.

PROOF. (i) ⇒ (ii): The statement is obvious for $n = 1$; suppose it has been proved for 1, 2,..., n-1. Since x_n is a non zerodivisor on $M/(x_1,\ldots,x_{n-1})M \simeq H_0(x_1,\ldots,x_{n-1},M)$, the short exact sequences of 4.2.4 (i) show that $H_p(x_1,\ldots,x_n,M) = 0$ for $p \geq 1$.

(ii) ⇒ (iii): Of course.

(iii) ⇒ (iv): Let $R = \mathbb{Z}[X_1,\ldots,X_n]$ and apply the same notational conventions to the sequence of indeterminates X_i as we have done to the x's. In particular, write $I = (X_1,\ldots,X_n)$ and $I_t = (X_1^t,\ldots,X_n^t)$. The sequences

$$0 \to R/I \overset{.X^t}{\to} R/I_{t+1}.I^{nt-t-1} \to R/I^{nt} \to 0$$

are exact for all $t \geq 1$. We wish to show that they remain exact under $-\otimes_R M$, so it is enough to know that $\mathrm{Tor}_1^R(R/I^r,M) = 0$ for $r \geq 1$. Here M is turned into an R-module through the ring homomorphism $R \to A$ determined by $X_i \mapsto x_i$. Since I^r is generated by all monomials in $X_1,\ldots,X_n$ of degree r, by 5.1.6 it is enough to show that $\mathrm{Tor}_1^R(R/I,M) = 0$ or rather that $\mathrm{Tor}_1^R(\mathbb{Z},M) = 0$. Now $K_\bullet(X,R)$ is a free resolution of the R-module $\mathbb{Z}$ in view of (i) ⇒ (ii) (Hilbert syzygy theorem), while $K_\bullet(X,R)\otimes_R M \simeq K_\bullet(X,R)\otimes_R A\otimes_A M \simeq K_\bullet(x,A)\otimes_A M = K_\bullet(x,M)$. Hence $\mathrm{Tor}_1^R(\mathbb{Z},M) = H_1(x,M)$ which is 0 by assumption. This finishes the proof.

It should be pointed out that the implications in the theorem are all strict [Ba, Ch. V, §4] although they may be reversed in certain good cases, 5.2.4, 5.2.5 and 5.2.6. We next prove an old result of Kaplansky's [Ka 62, Th. 1] which provides us with a host of ideals of finite projective dimension in any ring, once we have regular sequences.

5.1.8 THEOREM. Let $x_1,\ldots,x_n \in A$ and all its permutations form a regular sequence and suppose the ideal $\mathfrak{c} \subset A$ is generated by monomials in the x's. Then $\mathrm{pd}_A A/\mathfrak{c} \le n$.

PROOF. Let $R = \mathbf{Z}[X_1,\ldots,X_n]$ and make A into an R-algebra by mapping $1 \mapsto 1$, $X_i \mapsto x_i$, $i = 1,\ldots,n$. Consider for each monomial generator of $\mathfrak{c}$ the corresponding monomial in $X_1,\ldots,X_n$; they generate an ideal $J \subset R$.

We first wish to show that $\mathrm{pd}_R R/J \le n$ or that $\mathrm{Ext}^p_R(R/J,M) = 0$ for any R-module M and $p > n$. By 5.1.6 it is enough to prove that $\mathrm{Ext}^p_R(R/I,M) = 0$ for all ideals $I \supset J$ of the shape $I = (X_{i_1},\ldots,X_{i_k})$, $1 \le i_1 < \ldots < i_k \le n$. If we write $X = X_{i_1},\ldots,X_{i_k}$, then the Koszul complex $K_\bullet(X,R)$ is a finite free resolution of R/I of length $k \le n$, so that the statement for the Ext's is true.

Now let $P_\bullet$ be a projective R-resolution of R/J of length $\le n$. If $\mathrm{Tor}^R_p(R/J,A) = 0$ for $p \ge 1$, then $P_\bullet \otimes_R A$ provides a projective A-resolution of $A/\mathfrak{c}$ of the same length. Again by 5.1.6 it is enough to show that $\mathrm{Tor}^R_p(R/I,A) = 0$ for the ideals I described above, or that the Koszul complex $K_\bullet(x',A)$ is exact except at the 0-th place, where we have written $x' = x_{i_1},\ldots,x_{i_k}$. But $x_1,\ldots,x_n$ form an A-sequence in any order, so by 5.1.3 e) the sequences $x_{i_1},\ldots,x_{i_k}$ are all regular in A. Therefore $H_p(x',A) = 0$ for $p \ge 1$ by 5.1.7, and we are through.

From now on we shall write $\mathfrak{R}(\mathfrak{c})$ for the radical of the ideal $\mathfrak{c}$.

5.1.9 COROLLARY. [Ka 62, Th. 2] Suppose A is a $\mathbb{Z}$-torsion free module. If in addition $x_i \in \mathfrak{R}(\mathfrak{c})$ for $i = 1,\ldots,n$, then $pd_A A/\mathfrak{c} = n$.

The proof [Va, Th. 2.6], which goes by localizing and using minimal resolutions, we leave to the reader.

We now return to Theorem 5.1.7; in order to explore (iv) further, we write $B_x(M)$ for the union of the kernels of the composite maps $M \to M/\mathfrak{a}M \xrightarrow{\cdot x^t} M/\mathfrak{a}_{t+1}\mathfrak{a}^{nt-t-1}M$; these kernels form an ascending sequence of submodules of M as t goes to infinity. In other words, $B_x(M)$ (and we shall write B(M) when x is not in doubt) consists of those $m \in M$ with $x^t m \in \mathfrak{a}_{t+1}\mathfrak{a}^{nt-t-1}M$ for some $t \geq 1$. We add one more item of notation. If $\upsilon = \upsilon_1,\ldots,\upsilon_n$ is a vector of nonnegative exponents, we write $|\upsilon| = \upsilon_1 + \ldots + \upsilon_n$ for the total degree. In case $n = 1$, conditions (i) and (iii) in the following proposition are easily seen to be equivalent, while (ii) should be deleted. So let us suppose $n \geq 2$.

5.1.10 PROPOSITION. The following are equivalent:

(i) $m \in B(M)$;

(ii) m is one of the m_υ's in a relation $\Sigma_{|\upsilon|=s}\, x^\upsilon m_\upsilon = 0$ for some $s \geq 1$, all $m_\upsilon \in M$;

(iii) m is one of the m_υ's in a relation $\Sigma_{|\upsilon|=s}\, x^\upsilon m_\upsilon \in \mathfrak{a}^{s+1}M$ for some $s \geq 1$, all $m_\upsilon \in M$.

PROOF. The implications (i) ⇒ (ii) ⇒ (iii) are clear. For (ii) ⇒ (i), let m be one of the m_υ's in a relation $\Sigma_{|\upsilon|=s}\, x^\upsilon m_\upsilon = 0$, say m_υ. Put $\tau = \tau_1,\ldots,\tau_n$ where $\tau_i = (\max_{j=1}^n \upsilon_j) - \upsilon_i$, $i = 1,\ldots,n$. Multiplying our identity by x^τ, we find that $m_\upsilon \in B(M)$.

For (iii) ⇒ (i), suppose that m is one of the m_υ in a relation $\Sigma_{|\upsilon|=s} x^\upsilon m_\upsilon \in \mathfrak{a}^{s+1}M$. An arbitrary element of $\mathfrak{a}^{s+1}M$ can be written as $\Sigma_{|\upsilon|=s} x^\upsilon n_\upsilon$ with all $n_\upsilon \in \mathfrak{a}M$. We do so with our element: $\Sigma_{|\upsilon|=s} x^\upsilon m_\upsilon = \Sigma_{|\upsilon|=s} x^\upsilon n_\upsilon$. Thus $\Sigma_{|\upsilon|=s} x^\upsilon(m_\upsilon - n_\upsilon) = 0$ and the implication (ii) ⇒ (i) yields that all $m_\upsilon - n_\upsilon$ are in B(M). Since $\mathfrak{a}M \subset B(M)$ and $n_\upsilon \in \mathfrak{a}M$, we find that all m_υ are in B(M).

In order to further study B(M), we consider the graded rings $(A/\mathfrak{a})[X_1,\ldots,X_n]$ and $G_\mathfrak{a}(A) = \oplus_{t=0}^\infty \mathfrak{a}^t/\mathfrak{a}^{t+1}$, the associated graded ring of the ideal $\mathfrak{a}$. There is a surjective homomorphism of the former to the latter given by $X_i \mapsto x_i$. One can extend this to a surjective homomorphism of graded modules β_M: $(A/\mathfrak{a})[X_1,\ldots,X_n]\otimes_A M \to G_\mathfrak{a}(M)$. It is easily seen that every homogeneous element of the former module can be written as $\xi = \Sigma_{|\upsilon|=s} X^\upsilon \otimes m_\upsilon$ and that it is 0 if and only if all $m_\upsilon \in \mathfrak{a}M$. On the other hand $\beta_M(\xi) = \Sigma_{|\upsilon|=s} x^\upsilon m_\upsilon \bmod \mathfrak{a}^{s+1}M$.

We next introduce another submodule of M, to wit Q(M) (or $Q_x(M)$ when necessary), as the collective kernel of the maps $M \to M/\mathfrak{a}M \xrightarrow{x^t} M/\mathfrak{a}_{t+1}M$, $t \geq 1$. In other words $Q(M) = \{m \in M|\ x^t m \in \mathfrak{a}_{t+1}M$ for some $t \geq 1\}$. Then obviously $\mathfrak{a}M \subset B(M) \subset Q(M)$. We write $\mathfrak{b}$ for the ideal B(A) and $\mathfrak{q}$ for Q(A). It is easy to see that always $\mathfrak{b}M \subset B(M)$ and $\mathfrak{q}M \subset Q(M)$. Also B and Q are functorial, i.e. if f: M → N is A-linear then $f(B(M)) \subset B(N)$ and $f(Q(M)) \subset Q(N)$.

As a remark aside, let us examine these submodules as we take powers of the x's. When $x = x_1,\ldots,x_n$, write $x(t) = x_1^t,\ldots,x_n^t$. We leave to the reader a proof of

5.1.11 PROPOSITION. For every A-module M, there are descending chains of submodules of M

$B_x(M) = B_{x(1)}(M) \supset \ldots \supset B_{x(t)}(M) \supset B_{x(t+1)}(M) \supset \ldots$ and

$Q_x(M) = Q_{x(1)}(M) \supset \ldots \supset Q_{x(t)}(M) \supset Q_{x(t+1)}(M) \supset \ldots$.

In the situation of 5.1.7 (iv), which is condition (ii) below, matters simplify considerably.

5.1.12 THEOREM. Let all notations be taken with respect to $x = x_1,\ldots,x_n \in A$ and let M be an A-module, then the next three conditions are equivalent:

(i) β_M is an isomorphism;

(ii) $\mathfrak{a}M = B(M)$;

(iii) $\mathfrak{a}M = Q(M)$.

PROOF. Since we need only test injectivity of the graded homomorphism β_M on homogeneous elements of $(A/\mathfrak{a})[X_1,\ldots,X_n] \otimes_A M$, the equivalence of (i) and (ii) is a consequence of the equivalence of items (i) and (iii) in 5.1.10. Furthermore (iii) obviously implies (ii). So suppose $\mathfrak{a}M = B(M)$ and take $m \in Q(M)$, say with $x^t m \in \mathfrak{a}_{t+1}M$. Since $B(M) = Q(M)$ when $n = 1$, assume $n \geq 2$. We shall prove by induction on s that $x^t m \in \mathfrak{a}_{t+1}\mathfrak{a}^s M$ for all $0 \leq s \leq nt-t-1$. The case $s = 0$ is our assumption and the case $s = nt-t-1$ our goal, because then $m \in B(M)$ by definition of the latter. So suppose we know that $x^t m \in \mathfrak{a}_{t+1}\mathfrak{a}^s M$ for some given s, $0 \leq s < nt-t-1$. If we write $\Gamma = \{\upsilon \in \mathbb{N}^n \mid \upsilon_i \geq t+1$ for at least one i, $1 \leq i \leq n$, and $|\upsilon| = s+t+1\}$, then the elements of $\mathfrak{a}_{t+1}\mathfrak{a}^s M$ are just of the form $\Sigma_{\upsilon\in\Gamma}\, x^\upsilon m_\upsilon$. Hence we can write $x^t m = \Sigma_{\upsilon\in\Gamma}\, x^\upsilon m_\upsilon$. We wish to prove that all $m_\upsilon \in \mathfrak{a}M$, for then $x^t m \in \mathfrak{a}_{t+1}\mathfrak{a}^{s+1}M$ and we have taken the induction step. Now $\Sigma_{\upsilon\in\Gamma}\, x^\upsilon m_\upsilon = x^t m \in \mathfrak{a}^{nt}M$ while $s < nt-t-1$ implies that $s+t+1 < nt$. Hence $\Sigma_{\upsilon\in\Gamma}\, x^\upsilon m_\upsilon \in \mathfrak{a}^{s+t+2}M$. In view of 5.1.10 all $m_\upsilon \in B(M)$. By assumption however $\mathfrak{a}M = B(M)$, so that all $m_\upsilon \in \mathfrak{a}M$.

Condition (iii) of 5.1.7 also has an interpretation in terms of graded modules. Write $R_\mathfrak{a}(M) = \oplus_{t=0}^{\infty} \mathfrak{a}^t M$ and let I be the ideal in $A[X_1,\ldots,X_n]$ generated by all elements $x_iX_j - x_jX_i$, $1 \le i < j \le n$. There is a surjective graded A-homomorphism α_M: $A[X_1,\ldots,X_n]/I \otimes_A M \to R_\mathfrak{a}(M)$ determined by $\alpha_M(\bar{f}\otimes m) = f(x_1,\ldots,x_n)$ where f is a homogeneous polynomial in $A[X_1,\ldots,X_n]$. Then condition (iii) of 5.1.7 is equivalent with α_M being an isomorphism. Since we do not use this fact, we leave its proof to the reader who may also consult [Bo 80, §9, Th. 1]. In fact that theorem of Bourbaki is just our Theorem 5.1.7 with condition (iv) replaced by "β_M is an isomorphism". There is also a connection with Rees' note [Re 85].

5.1.13 DEFINITION. If the sequence x satisfies (iv) in 5.1.7 and $M \neq \mathfrak{a}M$ for a module M we say that x is a pre-regular sequence on M or that M is x-pre-regular.

In the equivalent formulation "β_M is an isomorphism" this condition was called quasiregular by Grothendieck in [Gro 64, Ch. 0, Déf. 15.1.7]. We shall justify our choice of name in the next section.

We have seen that x-pre-regularity of a module M can be described either by saying that all maps $.x^t$: $M/\mathfrak{a}_x M \to M/\mathfrak{a}_{x(t+1)}\mathfrak{a}_x^{nt-t-1}M$ are injective (definition) or that all maps $.x^t$: $M/\mathfrak{a}_x M \to M/\mathfrak{a}_{x(t+1)}M$ are injective (5.1.12 (iii)). The next results show that the latter condition implies injectivity and unicity for more general determinantal maps. We therefore study this circle of ideas in a slightly more general form.

If $\mathfrak{a}_y = (y_1,\ldots,y_n) \subset (x_1,\ldots,x_n) = \mathfrak{a}_x$ are ideals in the ring A, we can write the equations expressing the y's in the x's in vector notation: $\underline{y} = T\underline{x}$. Here $\underline{x}$ resp. $\underline{y}$ stand for the column vector of the x's resp. y's while T is an n×n-matrix with entries in A which is usually not uniquely determined. If $\tilde{T}$ is the adjoint matrix of T, then $\tilde{T}T = T\tilde{T} = \Delta I_n$ where $\Delta = \det T$ and I_n is the n-square identity matrix. Thus $\Delta \underline{x} = \tilde{T}T\underline{x} = \tilde{T}\underline{y}$ hence multiplication by Δ yields a map which we again call $.\Delta\colon A/\mathfrak{a}_x \to A/\mathfrak{a}_y$ and hence for any given A-module M an A-homomorphism $.\Delta\colon M/\mathfrak{a}_x M \to M/\mathfrak{a}_y M$. The example which we have already considered is where $y = x(t+1)$ and $T = \mathrm{diag}(x_1^t,\dots,x_n^t)$, then $\Delta = x^t$. The purport of the next proposition and its corollaries is that these maps $.x^t$ to a large extent control the behaviour of the general determinantal maps. To understand the proofs of the last four results in this section, the reader may find it helpful to draw diagrams of the maps involved, which we omit for reasons of space.

5.1.14 PROPOSITION. Suppose $\underline{y} = T\underline{x} = T'\underline{x}$ and write $\Delta' = \det T'$. Then $\Delta - \Delta' \in \mathfrak{b}_y$.

PROOF. We make the transition from T to T′ by a single row at a time, and are therefore reduced to showing: if $T = (\tau_{ij})$ and $T' = (\tau'_{ij})$ only differ in their last row, then $\Delta - \Delta' \in \mathfrak{b}_y$. Now $\Delta - \Delta' = \det S$ where the n-square matrix $S = (\sigma_{ij})$ equals T (and T′) in the first n-1 rows, while $\sigma_{nj} = \tau_{nj} - \tau'_{nj}$, $j = 1,\dots,n$. Since $S\underline{x} = \underline{z}$ with $z = (y_1,\dots,y_{n-1},0)$, we find that $(\Delta - \Delta')\mathfrak{a}_x \subset \mathfrak{a}_z$. But $y_n \in \mathfrak{a}_x$, so $(\Delta - \Delta')y_n = a_1y_1 + \dots + a_{n-1}y_{n-1}$ with $a_1,\dots,a_{n-1} \in A$, and by 5.1.10 this implies the result. With a little more care one may prove that actually $y^1(\Delta - \Delta') \in \mathfrak{a}_{y(2)}\mathfrak{a}_y^{n-2}$.

The next corollary expresses functorial behaviour of the determinantal maps with respect to the submodules Q(M).

5.1.15 COROLLARY. $\Delta(Q_x(M)) \subset Q_y(M)$.

PROOF. Take $m \in Q_x(M)$, then for some $t \geq 1$ we know that

$x^t m \in \mathfrak{a}_{x(t+1)}M$. Since $\mathfrak{a}_y \subset \mathfrak{a}_x \subset \mathfrak{R}(\mathfrak{a}_x) = \mathfrak{R}(\mathfrak{a}_{x(t+1)})$, there exists an $s \geq 1$ such that $\mathfrak{a}_{y(s+1)} \subset \mathfrak{a}_{x(t+1)}$. Let S be an n-square matrix such that $y(s+1) = S\underline{x}(t+1)$, and call its determinant Γ. Then $\underline{y}(s+1) = (\mathrm{diag}(y_1^s,\dots,y_n^s).T)\underline{x} = (S.\mathrm{diag}(x_1^t,\dots,x_n^t))\underline{x}$. By the proposition, $(y^s\Delta - \Gamma x^t)m \in \mathfrak{b}_{y(s+1)}M$. On the other hand

$$\Gamma x^t m \in \mathfrak{a}_{y(s+1)}M \subset \mathfrak{b}_{y(s+1)}M \subset \mathfrak{q}_{y(s+1)}M \subset Q_{y(s+1)}(M).$$

Thus $y^s\Delta m \in Q_{y(s+1)}(M)$. Therefore $(y^{s+1})^u.y^s\Delta m \in \mathfrak{a}_{y((s+1)(u+1))}M$ for some $u \geq 1$. But $(y^{s+1})^u y^s = y^{su+s+u}$ and we see that $y^{su+s+u}.\Delta m \in \mathfrak{a}_{y(su+s+u+1)}M$ which means that $\Delta m \in Q_y(M)$.

5.1.16 COROLLARY. Suppose in addition that $\mathfrak{a}_x \subset \mathfrak{R}(\mathfrak{a}_y)$. Then $(\mathfrak{a}_y M : \Delta) \subset Q_x(M)$.

PROOF. There exists a $t \geq 1$ such that $x_i^{t+1} \in \mathfrak{a}_y$ for $i = 1,\dots,n$. Suppose $S\underline{y} = \underline{x}(t+1)$, and let the n×n-matrix S have determinant Γ. Then $ST\underline{x} = \underline{x}(t+1) = \mathrm{diag}(x_1^t,\dots,x_n^t)\underline{x}$, and $\Gamma\Delta - x^t \in \mathfrak{b}_{x(t+1)}$ by 5.1.14. Now let $m \in M$ be such that $\Delta m \in \mathfrak{a}_y M$, then $\Gamma\Delta m \in \mathfrak{a}_{x(t+1)}M \subset \mathfrak{b}_{x(t+1)}M$, so that $x^t m \in \mathfrak{b}_{x(t+1)}M \subset \mathfrak{q}_{x(t+1)}M \subset Q_{x(t+1)}M$. Therefore $(x^{t+1})^s.x^t m \in \mathfrak{a}_{x((t+1)(s+1))}(M)$, i.e. $m \in Q_x(M)$.

5.1.17 PROPOSITION. Let $\mathfrak{a}_y = (y_1,\dots,y_n) \subset (x_1,\dots,x_n) = \mathfrak{a}_x$ be ideals in A, and let $T\underline{x} = \underline{y}$ with $\det T = \Delta$. If the module M is y-pre-regular, then the map $.\Delta\colon M/\mathfrak{a}_x M \to M/\mathfrak{a}_y M$ does not depend on the choice of the matrix T. If M is x-pre-regular and if $\mathfrak{a}_x$ and $\mathfrak{a}_y$ have the same radical, then Δ is injective.

PROOF. If also $T'\underline{x} = \underline{y}$ with $\Delta' = \det T'$, then $\Delta - \Delta' \in \mathfrak{b}_y$ by 5.1.14. Now $\mathfrak{b}_y M \subset B_y(M)$ and the latter submodule equals $\mathfrak{a}_y M$ in the y-pre-regular M. This proves the first statement. In the second case, we have

$(\mathfrak{a}_y M : \Delta) \subset Q_x(M)$ in view of 5.1.16. But M is x-pre-regular, so $Q_x(M) = \mathfrak{a}_x M$ which means injectivity of the map Δ.

We should like to point out that there is some similarity with considerations in [SS, (1.2)] and [Ku 86, E, 18-21].

5.2 PRE-REGULARITY UNDER COMPLETION; CONNECTIONS WITH LOCAL COHOMOLOGY.

We begin by further investigating pre-regularity of a module. If $x = x_1,\ldots,x_n$ is a sequence of elements in a ring A and if μ and υ are elements of $\mathbb{N}^n$, recall that multiplication by x^υ on an A-module M induces a map $.x^\upsilon\colon M/\mathfrak{a}_\mu M \to M/\mathfrak{a}_{\mu+\upsilon}M$, where $\mathfrak{a}_\tau$, $\tau \in \mathbb{N}^n$, is the ideal generated by $x_1^{\tau_1},\ldots,x_n^{\tau_n}$, and $x^\upsilon = x_1^{\upsilon_1}\ldots x_n^{\upsilon_n}$. As observed at the end of the previous chapter, we have natural limit mappings $h_\mu\colon M/\mathfrak{a}_\mu M \to H^n_{\mathfrak{a}}(M)$, making the diagrams

$$\begin{array}{ccc} M/\mathfrak{a}_\mu M & \xrightarrow{\;.x^\upsilon\;} & M/\mathfrak{a}_{\mu+\upsilon} M \\ & {\scriptstyle h_\mu}\searrow \quad \swarrow{\scriptstyle h_{\mu+\upsilon}} & \\ & H^n_{\mathfrak{a}}(M) & \end{array}$$

commutative. Bearing in mind that $\mu > 0$ means that all μ_i are greater than zero, we have the following

5.2.1 PROPOSITION. In the situation above, the following statements are equivalent:

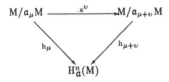

(i) M is x-pre-regular;

(ii) h_1 is injective, where h_1: $M/\mathfrak{a}M \to H^n_{\mathfrak{a}}(M)$ is the canonical limit map, and $\mathfrak{a}M \neq M$;

(iii) The maps x^{υ}: $M/\mathfrak{a}_{\mu}M \to M/\mathfrak{a}_{\mu+\upsilon}M$ and h_{μ} are injective for all μ, $\upsilon \in \mathbf{N}^n$, $\mu > 0$, and $\mathfrak{a}M \neq M$.

PROOF. Since the composite map $M/\mathfrak{a}M \stackrel{.x^t}{\to} M/\mathfrak{a}_{t+1}M \stackrel{h_{t+1}}{\to} H^n_{\mathfrak{a}}(M)$ is just h_1: $M/\mathfrak{a}M \to H^{\mathfrak{a}}_n(N)$, we see that (ii) ⇒ (i). Furthermore, the implication (iii) ⇒ (ii) is obviously true and the only thing left to prove is (i) ⇒ (iii). In addition to the notation we agreed upon after 5.1.6 and above, we fix an n-tuple of nonnegative integers $\mu = \mu_1,\dots,\mu_n$ and define $\Gamma(r) = \{\gamma \in \mathbf{N}^n |\ \gamma_i < \mu_i,\ i = 1,\dots,n \text{ and } |\gamma| = r\}$, where r is a positive integer. Define $\mathfrak{a}(r)$ as the ideal generated by all x^{γ} with $\gamma \in \Gamma(r)$ together with the ideal $\mathfrak{a}_{\mu}$. Evidently $\mathfrak{a}(r) = \mathfrak{a}^r + \mathfrak{a}_{\mu}$, so the $\mathfrak{a}(r)$ form a descending chain of ideals with $\mathfrak{a}(1) = \mathfrak{a} = \mathfrak{a}_1$ and $\mathfrak{a}(r) = \mathfrak{a}_{\mu}$ for $r > |\mu|-n$, while $\mathfrak{a}(r)\mathfrak{a}(q) \subset \mathfrak{a}(r+q)$.

In order to prove that (i) ⇒ (iii), let $x^{\upsilon}m \in \mathfrak{a}_{\mu+\upsilon}M$. If we put $\rho = (|\upsilon|,\dots,|\upsilon|)$, we obtain $x^{\rho}m = x^{\rho-\upsilon}.x^{\upsilon}m \in x^{\rho-\upsilon}.\mathfrak{a}_{\mu+\nu}M \subset \mathfrak{a}_{|\upsilon|+1}M$ (the last inclusion follows from the fact that $\mu > 0$). Using (i), we can conclude that $m \in \mathfrak{a}M = \mathfrak{a}(1)M$. By induction, we shall show that $m \in \mathfrak{a}(r)M$ for all r. In particular, once $r > |\mu|-n$, we obtain that $m \in \mathfrak{a}_{\mu}M$.

Assume therefore that $m \in \mathfrak{a}(r)M$, say $m = \Sigma_{\gamma\in\Gamma(r)}\ x^{\gamma}m_{\gamma} + \Sigma_{j=1}^n\ x_j^{\mu_j}m_j$. It will be enough to show that each $m_{\gamma} \in \mathfrak{a}(1)M$. Since $x^{\upsilon}m \in \mathfrak{a}_{\mu+\upsilon}M$, we see that $x^{\upsilon}.\Sigma_{\gamma\in\Gamma(r)}\ x^{\gamma}m_{\gamma} \in \mathfrak{a}_{\mu+\upsilon}M$. Thus for a fixed $\gamma \in \Gamma(r)$, we find that $x^{\gamma+\upsilon}m_{\gamma} \in (x^{\upsilon}\mathfrak{c} + \mathfrak{a}_{\mu+\upsilon})M$ where $\mathfrak{c}$ is the ideal generated by all x^{σ}, $\sigma \in \Gamma(r)\backslash\{\gamma\}$. However, each $x^{\sigma} \in (x_1^{\gamma_1+1},\dots,x_n^{\gamma_n+1})$, since for some i we must have $\sigma_i > \gamma_i$. Therefore, $x^{\gamma+\upsilon}m_{\gamma} \in x^{\upsilon}.(x_1^{\gamma_1+1},\dots,x_n^{\gamma_n+1})M + \mathfrak{a}_{\mu+\upsilon}M \subset (x_1^{\upsilon_1+\gamma_1+1},\dots,x_n^{\upsilon_n+\gamma_n+1})M$ because $\gamma_i < \mu_i$, $i = 1,\dots,n$. Now (i) again yields $m_{\gamma} \in \mathfrak{a}M = \mathfrak{a}(1)M$. Hence $m \in \mathfrak{a}(r+1)M$ and we are through.

As a first direct consequence of the above proposition we state the following

5.2.2 EXERCISE. Let A be a ring, $x = x_1,\dots,x_n \in A$, M an A-module. Let $\tau \in \mathbf{N}^n$, $\tau > 0$, and consider $y = x_1^{\tau_1},\dots,x_n^{\tau_n}$. Then if x is M-pre-regular, so is y.

The next result plays a key role in our treatment of pre-regularity and justifies the term: pre-regular becomes regular under completion.

5.2.3 THEOREM. Let $x = x_1,\dots,x_n$ be elements in a ring A which generate an ideal $\mathfrak{a}$ and let M be an A-module with $\mathfrak{a}M \neq M$. Let $y = y_1,\dots,y_n$ be an arbitrary permutation of the x's and write $\hat{M}$ for the $\mathfrak{a}$-adic completion of M. If M is x-pre-regular, the module $\hat{M}$ is y-regular. Moreover for each k, $0 \le k \le n$, the submodule $(y_1,\dots,y_k)\hat{M}$ is closed in the $\mathfrak{a}$-adic topology of $\hat{M}$ and the factor module $\hat{M}/(y_1,\dots,y_k)\hat{M}$ is the completion in its $\mathfrak{a}$-adic topology of $M/(y_1,\dots,y_k)M$.

PROOF. For integers $1 \le k \le n$ and $t \ge 1$ write $N_{k,t} = (y_1,\dots,y_{k-1},y_k^t,\dots,y_n^t)M$, $N'_{k,t} = (y_1,\dots,y_{k-1},y_k^{t+1},y_{k+1}^t,\dots,y_n^t)M$ and $M_{k,t} = (y_1,\dots,y_{k-1})M + \mathfrak{a}^tM$. Clearly $N'_{k,t} \subset N_{k,t} \subset M_{k,t} \subset M$, and all these systems of submodules define descending filtrations on M for $t \to \infty$. The following diagrams commute

$$0 \longrightarrow M/N_{k,t} \xrightarrow{\;\cdot y_k\;} M/N'_{k,t} \longrightarrow M/N_{k+1,t} \longrightarrow 0$$

$$\downarrow \pi_{k,t} \qquad\qquad \downarrow \pi'_{k,t} \qquad\qquad \downarrow \pi_{k+1,t}$$

$$M/M_{k,t} \xrightarrow{\;\cdot y_k\;} M/M_{k,t} \longrightarrow M/M_{k+1,t} \longrightarrow 0$$

in which the vertical surjections are induced by the identity on M. Exactness
on the right of both rows is immediate; since pre-regularity is impervious
to permutation of the x's, injectivity of the second map in the top row is
guaranteed by 5.2.1.

For fixed k these diagrams form an inverse system indexed by t
and we take the inverse limit. All the systems being inversely surjective,
the limit of the top sequence remains exact. However, for all t we have
$M_{k,nt} \subset N_{k,t} \subset M_{k,t}$ so that $\pi_k = \varprojlim_t \pi_{k,t}$ is an isomorphism, and also
$M_{k,n(t+1)} \subset N'_{k,t} \subset M_{k,t}$ so that $\pi'_k = \varprojlim_t \pi'_{k,t}$ is an isomorphism. Writing
$M_k = \varprojlim_t M/M_{k,t}$ we find the exact sequence $0 \to M_k \xrightarrow{\;\cdot y_k\;} M_k \to M_{k+1} \to 0$ as the
limit of the bottom row. In particular, $M_1 = \hat{M}$, so by induction
$M_{k+1} = \hat{M}/(y_1,\ldots,y_k)\hat{M}$. Moreover $\hat{M}/\mathfrak{a}\hat{M} \simeq M/\mathfrak{a}M \neq 0$ by 2.2.5, so we have shown
that $\hat{M}$ is y-regular.

As for the topological assertions, we have $M_{k+1} = \varprojlim_t M/M_{k+1,t} \simeq$
$\varprojlim_t \overline{M}_{k+1}/\mathfrak{a}^t\overline{M}_{k+1}$ where we have put $\overline{M}_{k+1} = M/(y_1,\ldots,y_k)M$, so M_{k+1} is the $\mathfrak{a}$-adic
completion of $\overline{M}_{k+1}$ and is then complete by 2.2.5. The module
$M_{k+1} = \hat{M}/(y_1,\ldots,y_k)\hat{M}$ is therefore certainly Hausdorff, so that its 0-element
is a closed point. The inverse image $(y_1,\ldots,y_k)\hat{M}$ under the continuous
residue class map $\hat{M} \to M_{k+1}$ is hence closed, and the theorem has been proved.

As a corollary we obtain a result which was proved by
Grothendieck and by Bourbaki in a different manner, [Gro 64, Ch. 0, Prop.

15.1.9] and [Bo 80, §9, Th. 1]. Setting and notation will be as in the theorem.

5.2.4 COROLLARY. Suppose that the modules $M/(x_1,\ldots,x_{k-1})M$ are all Hausdorff in the $\mathfrak{a}$-adic topology, $k = 1,\ldots,n$. Then for the sequence $x = x_1,\ldots,x_n$ and the module M all four statements of Theorem 5.1.7 are equivalent.

PROOF. We need to show that x_k acts injectively on the module $M/(x_1,\ldots,x_{k-1})M$. Now Hausdorffness means that this module maps injectively into its completion, so it is enough to show that x_k acts injectively on this completion. But this completion is none other than the module $\hat{M}_k$ in the proof of 5.2.3 ($x = y$) and we have seen that x_k acts injectively.

5.2.5 COROLLARY. Let A be a noetherian ring, $x = x_1,\ldots,x_n$ be elements in A and M a finitely generated A-module. Suppose that the ideal $\mathfrak{a}$ generated by the x's is in the Jacobson radical of A. Then all four statements of 5.1.7 are equivalent. In particular x-pre-regularity is equivalent to x-regularity on M.

Indeed, the topological conditions of 5.2.4 are satisfied according to 2.1.4. As another corollary to 5.2.3, we find

5.2.6 COROLLARY. In the situation of 5.2.3, let $z = z_1,\ldots,z_m$ be elements in $\mathfrak{a}$. Then the following statements are equivalent:

(i) M is z-pre-regular;

(ii) $\hat{M}$ is z-pre-regular;

(iii) $\hat{M}$ is z-regular;

(iv) Statements (i), (ii) and (iii) are true for all permutations of the z's.

PROOF. Writing $(z_1,\ldots,z_m) = \mathfrak{b}$, observe that $\hat{M}$ is also complete in its $\mathfrak{b}$-adic topology by 2.2.6. Then (i) ⇒ (iii) is contained in 5.2.3, (iii) ⇒ (ii) is in 5.1.7. For (ii) ⇒ (i), note that we have commutative diagrams

$$\begin{array}{ccc} M/\mathfrak{b}M & \xrightarrow{.x^t} & M/\mathfrak{b}_{t+1}M \\ {\scriptstyle \tau_M\otimes_A A/\mathfrak{b}}\downarrow & & \downarrow{\scriptstyle \tau_M\otimes_A A/\mathfrak{b}_{t+1}} \\ \hat{M}/\mathfrak{b}\hat{M} & \xrightarrow[.x^t]{} & \hat{M}/\mathfrak{b}_{t+1}\hat{M} \end{array}$$

Since by 2.2.5 the vertical maps are bijective, the implication follows immediately. Pre-regularity does not suffer from permuting z's, so 5.2.3 takes care of the equivalence with (iv).

We shall finish this section with results on the structure of $H^n_{\mathfrak{a}}(M)$ for an x-pre-regular module M. In passing, we mention a notion dual to regular sequences: coregular sequences, see e.g. [Matl 74], [Matl 78] or [Ba]. As one can find in the above references (although the first one is not explicit about pre-regularity and only treats the ring case), if M is an x-pre-regular module, then $H^n_{\mathfrak{a}}(M)$ is x-coregular. We shall not pursue this subject further. Instead we prove the following proposition with its corollaries which shows a relation between $\mathfrak{a}$-adic completion and $H^n_{\mathfrak{a}}(M)$.

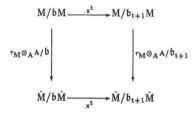

5.2.7 PROPOSITION. Let $x = x_1,\ldots,x_n$ be elements in a ring A which generate an ideal $\mathfrak{a}$, and let M be an x-pre-regular A-module. Then $\mathrm{Hom}_A(A/\mathfrak{a}_\mu, H^n_\mathfrak{a}(M)) \simeq M/\mathfrak{a}_\mu M$ for all $\mu \in \mathbb{N}^n$, $\mu > 0$.

PROOF. Since $h_\mu\colon M/\mathfrak{a}_\mu M \to H^n_\mathfrak{a}(M)$ is monic by 5.2.1, it is enough to prove that $\mathrm{im}(h_\mu) = \{v \in H^n_\mathfrak{a}(M) \mid \mathfrak{a}_\mu v = 0\}$. One of the inclusions being obvious, we shall take an element $v \in H^n_\mathfrak{a}(M)$ such that $\mathfrak{a}_\mu v = 0$ and we shall prove that $v \in \mathrm{im}(h_\mu)$. By definition of direct limit, $v = h_{\mu+\upsilon}(m \bmod \mathfrak{a}_{\mu+\upsilon}M)$ for some $\upsilon \in \mathbb{N}^n$ and $m \in M$. Obviously, $h_{\mu+\upsilon}$ being monic by 5.2.1, $\mathfrak{a}_\mu v = 0$ implies that $\mathfrak{a}_\mu m \subset \mathfrak{a}_{\mu+\upsilon}M$. If we can prove that this condition forces that m is congruent to $x^\upsilon\tilde{m} \bmod \mathfrak{a}_{\mu+\upsilon}M$ for some $\tilde{m} \in M$, we are done, since we then have $v = h_{\mu+\upsilon}(x^\upsilon\tilde{m} \bmod \mathfrak{a}_{\mu+\upsilon}M) = h_\mu(\tilde{m} \bmod \mathfrak{a}_\mu M) \in \mathrm{im}(h_\mu)$. Note that $x_j^{\mu_j}m \in \mathfrak{a}_{\mu+\upsilon}M$, $1 \le j \le n$, and hence $m \in \mathfrak{a}_{\mu+\upsilon}M + x_j^{\upsilon_j}M$ (by pre-regularity, 5.2.1), which implies that $m \equiv x_j^{\upsilon_j}m_j \bmod \mathfrak{a}_{\mu+\upsilon}M$ for some $m_j \in M$.

We shall now show by induction on k that $m \equiv x_1^{\upsilon_1}\ldots x_k^{\upsilon_k}\tilde{m}_k \bmod \mathfrak{a}_{\mu+\upsilon}M$ for some $\tilde{m}_k \in M$. For $k = 1$ we just have proved the statement so let $1 < k \le n$. By hypothesis, we may assume that $m \equiv x_1^{\upsilon_1}\ldots x_{k-1}^{\upsilon_{k-1}}\tilde{m}_{k-1} \bmod \mathfrak{a}_{\mu+\upsilon}M$. Since also $m \equiv x_k^{\upsilon_k}m_k \bmod \mathfrak{a}_{\mu+\upsilon}M$, it follows that $x_1^{\nu_1}\ldots x_{k-1}^{\nu_{k-1}}\tilde{m}_{k-1} \in \mathfrak{a}_{\mu+\upsilon}M + x_k^{\upsilon_k}M$ and hence, by virtue of the pre-regularity, that

$\tilde{m}_{k-1} \in (x_1^{\mu_1},\ldots,x_{k-1}^{\mu_{k-1}},x_k^{\upsilon_k},x_{k+1}^{\mu_{k+1}+\nu_{k+1}},\ldots,x_n^{\mu_n+\upsilon_n})M$. Thus $x_1^{\upsilon_1}\ldots x_{k-1}^{\upsilon_{k-1}}\tilde{m}_{k-1} \in \mathfrak{a}_{\mu+\upsilon}M + x_1^{\upsilon_1}\ldots x_k^{\upsilon_k}M$ and the result follows.

5.2.8 COROLLARY. Situation as in the proposition. Then $\mathrm{Hom}_A(H^n_\mathfrak{a}(A), H^n_\mathfrak{a}(M)) \simeq \hat{M}$, the $\mathfrak{a}$-adic completion of M.

PROOF. $\mathrm{Hom}_A(H^n_\mathfrak{a}(A), H^n_\mathfrak{a}(M)) = \mathrm{Hom}_A(\varinjlim_t A/\mathfrak{a}_t, H^n_\mathfrak{a}(M))$, and since direct limits behave well on the contravariant side of the Hom-functor, this last module is isomorphic to $\varprojlim_t \mathrm{Hom}_A(A/\mathfrak{a}_t, H^n_\mathfrak{a}(M))$. In view of 5.2.7, inspection of the identification maps easily yields that our module is isomorphic to

$\varprojlim_t M/\mathfrak{a}_t M$, the limit taken over the canonical projection maps $M/\mathfrak{a}_{s+t}M \to M/\mathfrak{a}_t M$. This inverse limit equals $\hat{M}$ and the result follows.

5.2.9 COROLLARY. Situation as in 5.2.7. The extension h_1: $M/\mathfrak{a}M \to H^n_\mathfrak{a}(M)$ is essential.

PROOF. Since $H^n_\mathfrak{a}$ is the n-th derived functor of $L_\mathfrak{a}$, the local cohomology module $H^n_\mathfrak{a}(M)$ is the union of its ascending chain of submodules $Y_t = \{y \in H^n_\mathfrak{a}(M) \mid \mathfrak{a}^t y = 0\}$, $t = 1, \ldots$. If the submodule $Z \subset H^n_\mathfrak{a}(M)$ is not (0), it contains a $z \in Y_{t+1}$, but not in Y_t. Then $\mathfrak{a}^t Z \cap Y_1 \neq (0)$. But $Y_1 = \mathrm{im}(h_1)$ according to 5.2.7, which proves the corollary.

5.3 DEPTH

In this section we shall principally treat two notions of depth: the ordinary depth using regular sequences and the Ext-depth first introduced by Rees [Re 57] under another name. The first one has a more intuitive appeal, while the second possesses better formal properties. As Rees showed, they coincide for finitely generated modules over a noetherian ring. We shall not discuss depth in the most general case; it has been studied by Barger [Barg], Hochster [Ho 74] Northcott [No 76] and Alfonsi [Al]. We do point out, however, that the Ext-depth can be measured equally well by local cohomology functors. We finish with a result of D. Lazard which connects depth once again in a very concrete way with regular sequences - a recurring theme in this section.

5.3.1 DEFINITION. Let A be a ring, M an A-module and $\mathfrak{a} \subset A$ an ideal with $\mathfrak{a}M \neq M$. We define the depth of M in $\mathfrak{a}$ as the supremum of the lengths of all M-sequences contained in $\mathfrak{a}$. Notation $dp_A(\mathfrak{a},M)$, but if A is local and $\mathfrak{a} = \mathfrak{m}$, its maximal ideal, we just write dp M and speak of the depth of M.

Clearly $dp_A(\mathfrak{a},M)$ is a nonnegative integer or is ∞; it is rather obvious that the latter case cannot occur when A is noetherian, as we see by using standard methods as developed in Chapter 8 and 5.3.4. In fact, this is true for any finitely generated ideal in an arbitrary ring, as follows from 5.3.7 and 6.1.8. Notice that we have not defined a depth for "divisible" modules with $\mathfrak{a}M = M$. The reader should be warned that $dp_A(\mathfrak{a},M)$ is variously referred to by different authors as classical depth, chain depth, grade, classical grade or codimension of M in $\mathfrak{a}$, Bourbaki not having pontificated yet.

A maximal M-sequence in $\mathfrak{a}$ is one which cannot be lengthened, but there may exist longer M-sequences in $\mathfrak{a}$. Simple properties of regular sequences are reflected in

5.3.2 EXERCISE. Let $\mathfrak{a} \subset \mathfrak{b}$ be ideals in the ring A and M an A-module with $\mathfrak{b}M \neq M$. Let $x_1,\dots,x_n$ be an M-sequence in $\mathfrak{a}$. Then for $1 \leq i \leq n$ we have

(i) The sequence $x_i,\dots,x_n$ is regular on $M/(x_1,\dots,x_{i-1})M$ and it is maximal in $\mathfrak{a}$ iff the above sequence is maximal in $\mathfrak{a}$;

(ii) $dp_A(\mathfrak{a},M/(x_1,\dots,x_i)M) \leq dp_A(\mathfrak{a},M) - i$ with equality iff $dp_A(\mathfrak{a},M/(x_1,\dots,x_n)M) = dp_A(\mathfrak{a},M) - n$;

(iii) $dp_A(\mathfrak{a},M/(x_1,\dots,x_n)M) = 0$ iff $x_1,\dots,x_n$ is a maximal M-sequence in $\mathfrak{a}$;

(iv) $dp_A(\mathfrak{a},M) \leq dp_A(\mathfrak{b},M)$;

(v) If $\mathfrak{R}(\mathfrak{a}) = \mathfrak{R}(\mathfrak{b})$ then $dp_A(\mathfrak{a},M) = dp_A(\mathfrak{b},M)$ (use 5.1.3 d)).

We next describe the behaviour of depth under flat extensions.

5.3.3 PROPOSITION. Let $A \to B$ be a flat ring homomorphism $\mathfrak{a} \subset A$ an ideal. Suppose the A-module M has the property that $(B \otimes_A \mathfrak{a}).(B \otimes_A M) \neq B \otimes_A M$. Then $dp_A(\mathfrak{a}, M) \leq dp_B(B\mathfrak{a}, B \otimes_A M)$.

PROOF. First notice that $B\mathfrak{a}$ may be identified with $B \otimes_A \mathfrak{a}$ because of flatness and that the condition on M implies that $\mathfrak{a}M \neq M$, so that both depths are defined. The fact that injectivity of maps like $.x_i \colon M/(x_1, \ldots, x_{i-1})M \to M/(x_1, \ldots, x_{i-1})M$ is preserved by $B \otimes_A -$, readily allows one to conclude. Localizing being flat, this implies

5.3.4 COROLLARY. Let $\mathfrak{a}$ be an ideal in the ring A and M an A-module with $\mathfrak{a}M \neq M$. Then $dp_A(\mathfrak{a}, M) \leq \inf dp_{A_\mathfrak{p}}(\mathfrak{a}_\mathfrak{p}, M_\mathfrak{p})$ where the inf is taken over all primes $\mathfrak{p}$ in Supp $M/\mathfrak{a}M$.

The inequalities in 5.3.3 and 5.3.4 are in fact equalities in case A is noetherian, M finitely generated and, in 5.3.3, the ring extension $A \to B$ is faithfully flat. The easiest proof goes by showing that in this situation depth equals the Ext-depth, and the latter behaves properly. Before defining this new depth, we collect a few useful facts about the Ext-functors.

5.3.5 PROPOSITION. Let A be a ring, M and N modules.

(i) If E is an injective envelope of M and $M' = E/M$, then $\mathrm{Ext}^i_A(N,M) \simeq \mathrm{Ext}^{i-1}_A(N,M')$ for $i \geq 2$; and for $i = 1$ if N is cyclic and $\mathrm{Hom}_A(N,M) = 0$;

(ii) If $x_1,\ldots,x_n$ is an M-sequence contained in Ann N, then $\mathrm{Ext}^i_A(N,M) = 0$ for $0 \le i < n$ while $\mathrm{Ext}^n_A(N,M) \simeq \mathrm{Hom}_A(N,M/(x_1,\ldots,x_n)M)$;

(iii) If A is noetherian and N finitely generated, then $\mathrm{Ass}(\mathrm{Hom}_A(N,M)) = \mathrm{Supp}\, N \cap \mathrm{Ass}\, M$.

PROOF. (i) The module E being injective, $\mathrm{Ext}^i_A(N,E) = 0$ for $i \ge 1$ and the statement for $i \ge 2$ is clear. If N is cyclic, $\mathrm{Hom}_A(N,M) = 0$ implies $\mathrm{Hom}_A(N,E) = 0$. To see this, suppose $0 \ne f\colon N \to E$, then $f(v) \ne 0$ for a generator v of N. Hence $Af(v) \cap M \ne 0$ because of the envelope property of $M \subset E$. Therefore $0 \ne af \in \mathrm{Hom}_A(N,M)$ for a certain $a \in A$. Thus $\mathrm{Hom}_A(N,E) = 0$ and the result follows for $i = 1$.

(ii) Since the x_j's annihilate all $\mathrm{Ext}^i_A(N,-)$, the short exact sequence $0 \to M \overset{.x_1}{\to} M \to M/x_1M \to 0$ gives rise to short exact sequences in the Ext's. The proof follows by induction.

(iii) This is [Bo 61b, Ch. 4, §1, Prop. 10].

5.3.6 DEFINITION. Let $\mathfrak{a}$ be an ideal in a ring A and M an A-module. Then $\text{E-dp}_A(\mathfrak{a},M)$ is defined to be the least integer i for which $\mathrm{Ext}^i_A(A/\mathfrak{a},M) \ne 0$.

Thus E-dp is a nonnegative integer or, if such an i does not exist, it is ∞. Observe that this time we have not required $\mathfrak{a}M \ne M$ and that for instance $\text{E-dp}_A(\mathfrak{a},M) = \infty$ for the null module M and every ideal $\mathfrak{a}$. Comparing the first part of the next proposition with 5.3.2 (ii) shows up one of the advantages of the Ext-depth.

5.3.7 PROPOSITION. Let $\mathfrak{a}$ be an ideal in the ring A and M an A-module with $\mathfrak{a}M \ne M$.

(i) If $x_1,\ldots,x_n$ is an M-sequence in $\mathfrak{a}$, then $\text{E-dp}_A(\mathfrak{a},M/(x_1,\ldots,x_n)M) = \text{E-dp}_A(\mathfrak{a},M) - n$;

(ii) $dp_A(\alpha,M) \le E\text{-}dp_A(\alpha,M)$.

PROOF. (i) We get short exact sequences

$0 \to Ext_A^j(A/\alpha,M) \to Ext_A^j(A/\alpha,M/x_1M) \to Ext_A^{j+1}(A/\alpha,M) \to 0$ for $j \ge 0$, so $E\text{-}dp_A(\alpha,M) = \infty$ implies $E\text{-}dp_A(\alpha,M/x_1M) = \infty$ and if $E\text{-}dp_A(\alpha,M)$ is finite, then $E\text{-}dp_A(\alpha,M/x_1M)$ is one less. Induction finishes the proof.

(ii) This follows from (i), since an E-dp is never negative.

When are both depths equal? Here is a criterion:

5.3.8 PROPOSITION. Situation as above. The following are equivalent:

(i) $dp_A(\alpha,M) = E\text{-}dp_A(\alpha,M) = s < \infty$;

(ii) There exists an M-sequence $x_1,\ldots,x_s$ in α and a nonnull element $z \in M/(x_1,\ldots,x_s)M$ such that $\alpha z = 0$.

PROOF. (i) $\Rightarrow$ (ii): There exists a maximal M-sequence $x_1,\ldots,x_s$ in α; write $\bar{M} = M/(x_1,\ldots,x_s)M$. By virtue of 5.3.5 (ii), $Ext_A^s(A/\alpha,M) \simeq Hom_A(A/\alpha,\bar{M})$. The former is $\neq 0$ by assumption, hence so is the latter. Then (ii) follows.

(ii) $\Rightarrow$ (i): Use the same isomorphism in reverse and take into account that $Ext_A^i(A/\alpha,M) = 0$ for $0 \le i < s$, again by 5.3.5 (ii). The inequality 5.3.7 (ii) finishes the proof.

Clearly a sequence $x_1,\ldots,x_s$ as in (ii) is maximal and an arbitrary M-sequence $y_1,\ldots,y_n$ in α is maximal iff $n = s$ or, equivalently, iff there exists a $z' \neq 0$ in $M/(y_1,\ldots,y_n)M$ such that $\alpha z' = 0$.

5.3.9 COROLLARY. Let α be an ideal in the noetherian ring A and M a finitely generated A-module with $\mathfrak{a}M \neq M$. Then all maximal M-sequences in $\mathfrak{a}$ have the same finite length, to wit $dp_A(\mathfrak{a},M) = E\text{-}dp_A(\mathfrak{a},M)$.

PROOF. Let $x_1,\ldots,x_n$ be an M-sequence in $\mathfrak{a}$, then $(x_1) \subset (x_1,x_2) \subset \ldots \subset (x_1,\ldots,x_n)$ form a strictly ascending chain of ideals in the noetherian ring A. Thus there exist a maximal such sequence, say $x_1,\ldots x_s$. We need to show that there exists a $z \in M/(x_1,\ldots,x_s)M = \overline{M}$, $z \neq 0$, with $\mathfrak{a}z = 0$. Now all elements of $\mathfrak{a}$ are zero divisors on $\overline{M}$, therefore $\mathfrak{a}$ is contained in the union of the prime ideals in Ass $\overline{M}$. Since $\overline{M}$ is finitely generated, there are only finitely many of these, and $\mathfrak{a}$ is contained in a single $\mathfrak{p} \in$ Ass $\overline{M}$. There exists a monic $f\colon A/\mathfrak{p} \to \overline{M}$; if one writes $z = f(1 \bmod \mathfrak{p})$, then z has the desired property.

One cannot dispense with the condition that M be finitely generated, as the following example shows. Let k be a field and let $A = k[[X,Y]]$, the local ring of formal power series in two variables. Take $M = \oplus A/(a)$, where a ranges over all nonnull non-units of A, in other words, nonnull power series without constant term. Then dp M = 0 while E-dp M ≥ 1 as the reader may easily show. In fact, using 6.1.10, he or she will find that actually E-dp M = 1. However, even if M is not finitely generated in 5.3.9, we shall prove in 6.1.8 that $E\text{-}dp_A(\mathfrak{a},M)$ is always finite. It has also been shown, by means of a topological argument, that 5.3.9 remains valid for countably generated modules over a complete noetherian local ring [SVb, Th. 3.4].

The next device allows one to reduce many questions to the case of E-dp = 0.

5.3.10 LEMMA. Let A be a ring, $\mathfrak{a} \subset A$ an ideal and M an A-module with $E\text{-}dp_A(\mathfrak{a},M) \neq 0$. If E is an injective envelope of M and $M' = E/M$, then

$E\text{-}dp_A(\mathfrak{a},E) = \infty$ and $E\text{-}dp_A(\mathfrak{a},M') = E\text{-}dp_A(\mathfrak{a},M) - 1$.

PROOF. This is a consequence of 5.3.5 (i) where N is the cyclic module $A/\mathfrak{a}$.

5.3.11 PROPOSITION. Let $\mathfrak{a} \subset \mathfrak{b} \subset A$ be ideals and M an A-module. Then $E\text{-}dp_A(\mathfrak{a},M) \leq E\text{-}dp_A(\mathfrak{b},M)$.

PROOF. We may assume that $E\text{-}dp_A(\mathfrak{b},M) = s$. By iterated application of the lemma, we arrive at a module, say N, for which $E\text{-}dp_A(\mathfrak{b},N) = 0$ and it suffices to prove that $Hom_A(A/\mathfrak{a},N) \neq 0$. Compose a nonnull homomorphism $A/\mathfrak{b} \to N$ with the projection map $A/\mathfrak{a} \to A/\mathfrak{b}$ to obtain the desired nonnull map $A/\mathfrak{a} \to N$.

5.3.12 PROPOSITION. Let $\mathfrak{a}$ be an ideal in the noetherian ring A and let M be an A-module. If $A \to B$ is a flat ring homomorphism, then $E\text{-}dp_A(\mathfrak{a},M) \leq E\text{-}dp_B(B\mathfrak{a},B\otimes_A M)$; in case the homomorphism is faithfully flat, there is even equality.

PROOF. Under our assumptions the natural homomorphism $B\otimes_A Ext^i_A(A/\mathfrak{a},M) \to Ext^i_B(B\otimes_A A/\mathfrak{a},B\otimes_A M)$ is an isomorphism for all $i \geq 0$ [Bo 80, §6, Prop. 10 b)] and flatness also allows us to identify $B\otimes_A\mathfrak{a}$ with $B\mathfrak{a}$. The result follows right away.

5.3.13 COROLLARY. Let $\mathfrak{a}$ be an ideal in the noetherian ring A and let M be an A-module. Then $E\text{-}dp_A(\mathfrak{a},M) = \inf E\text{-}dp_{A_\mathfrak{p}}(\mathfrak{a}_\mathfrak{p},M_\mathfrak{p})$ where the inf is taken over all primes $\mathfrak{p}$ of A.

PROOF. By the proof above, $(\mathrm{Ext}^i_A(A/\mathfrak{a},M))_{\mathfrak{p}} \simeq \mathrm{Ext}^i_{A_{\mathfrak{p}}}(A_{\mathfrak{p}}/\mathfrak{a}_{\mathfrak{p}},M_{\mathfrak{p}})$ and since for any A-module N we know that $N \neq 0$ iff $N_{\mathfrak{p}} \neq 0$ for at least a single $\mathfrak{p}$, we have the corollary.

5.3.14 COROLLARY. In the situation above, $\text{E-dp}_A(\mathfrak{a},M) = \inf \text{E-dp}_{A_{\mathfrak{p}}} M_{\mathfrak{p}} = \inf\{i \mid \mu^i(\mathfrak{p},M) \neq 0\}$ where $\mathfrak{p}$ ranges over all primes containing $\mathfrak{a}$.

PROOF. Notice that in 5.3.13 one may restrict the inf to the primes $\mathfrak{p}$ containing $\mathfrak{a}$, since for the others $\text{E-dp}_{A_{\mathfrak{p}}}(\mathfrak{a}_{\mathfrak{p}},M_{\mathfrak{p}}) = \infty$. In view of 5.3.11 we have established that $\text{E-dp}_A(\mathfrak{a},M) \leq \inf \text{E-dp}_{A_{\mathfrak{p}}} M_{\mathfrak{p}}$ so we still need to prove the reverse inequality. Using 5.3.10 we can reduce to the case where $\text{E-dp}_A(\mathfrak{a},M) = 0$; since injective envelopes localize well in the case of noetherian rings, 3.3.2, we need only prove that for some $\mathfrak{p} \supset \mathfrak{a}$ there exists a nonnull homomorphism $f\colon k(\mathfrak{p}) \to M_{\mathfrak{p}}$, where $k(\mathfrak{p})$ is the residue class field of the local ring $A_{\mathfrak{p}}$. Now $\mathrm{Hom}_A(A/\mathfrak{a},M) \neq 0$ is a module over the noetherian ring A, therefore there exists a prime $\mathfrak{p} \in \mathrm{Ass}(\mathrm{Hom}_A(A/\mathfrak{a},M))$. According to 5.3.5 (iii) with $N = A/\mathfrak{a}$, this $\mathfrak{p}$ contains $\mathfrak{a}$ and there is an injection $A/\mathfrak{p} \to M$. Localizing the latter at $\mathfrak{p}$, we find the desired f, and the first equality is proved. The second is clear from 3.3.3.

We next discuss how the E-dp can be measured by using local cohomology in stead of the Ext-functors. Introduce the notation of 4.1.10 for the $H_{\mathfrak{a}}$.

5.3.15 PROPOSITION. Let $\mathfrak{a}$ be an ideal in the ring A, and M an A-module. Then $\text{E-dp}_A(\mathfrak{a},M) = h^-_{\mathfrak{a}}(M)$.

PROOF. First recall that $H^0_{\mathfrak{a}}(M) = \{m \in M\colon \mathfrak{a}^t m = 0$ for some $t \geq 1\}$ while $\mathrm{Ext}^0_A(A/\mathfrak{a},M) = \{m \in M \mid \mathfrak{a}m = 0\}$, so that if either of these is nonnull, so is the other. Hence $h^-_{\mathfrak{a}}(M) = 0$ iff $E\text{-dp}(\mathfrak{a},M) = 0$. When $0 < h^-_{\mathfrak{a}}(M) < \infty$ or $0 < E\text{-dp}(\mathfrak{a},M) < \infty$ we can use 4.1.10 resp. 5.3.10 to arrive at the case just treated and then argue in reverse. Remains the case where $h^-_{\mathfrak{a}}(M) = E\text{-dp}(\mathfrak{a},M) = \infty$.

5.3.16 COROLLARY. Let $\mathfrak{a}$ and $\mathfrak{b}$ be ideals in a ring A which define the same topology. Then $E\text{-dp}_A(\mathfrak{a},M) = E\text{-dp}_A(\mathfrak{b},M)$ for every A-module M.

Indeed, local cohomology functors $H^i_{\mathfrak{a}}(-)$ for noetherian rings only depend on the topology induced by the ideal $\mathfrak{a}$. Notice that 5.3.11 and 5.3.16 are the counterpart of 5.3.2 for Ext-depth.

In the last theorem of this chapter we return to regular sequences. First a lemma stating a connection with Ext's.

5.3.17 LEMMA. Let A be a noetherian ring, M and N finitely generated A-modules, and $\mathfrak{a} = \text{Ann } N$. Suppose that $\mathfrak{b}$ is an ideal containing $\mathfrak{a}$, that $\text{dp}_A(\mathfrak{a},M) = s$ and $\mathfrak{b}M \neq M$. Then $\text{dp}_A(\mathfrak{b},M) = s$ if and only if $\mathfrak{b} \subset \mathfrak{p}$ for some prime $\mathfrak{p} \in \text{Ass}(\text{Ext}^s_A(N,M))$.

PROOF. Let $x_1,\dots,x_s$ be a maximal M-sequence in $\mathfrak{a}$ and write $\bar{M} = M/(x_1,\dots,x_s)M$. Then by 5.3.5 (ii) and (iii) $\text{Ass}(\text{Ext}^s_A(N,M)) = \text{Supp } N \cap \text{Ass } \bar{M}$ and therefore consists of primes $\mathfrak{p} \in \text{Ass } \bar{M}$ which contain $\mathfrak{a}$. Now $\text{dp}_A(\mathfrak{b},\bar{M}) = \text{dp}_A(\mathfrak{b},M) - s$, so $\text{dp}_A(\mathfrak{b},M) = s$ precisely when $\mathfrak{b}$ is entirely made up of zero divisors on $\bar{M}$. But since $\text{Ext}^s_A(N,M)$ is a finitely generated module over the noetherian ring A, its Ass consists of only finitely many primes, so $\mathfrak{b}$ is contained in one of these if and only if every element of $\mathfrak{b}$ is a zero divisor on $\bar{M}$.

In order to state the theorem, let $(y_1,\ldots,y_n) \subset (x_1,\ldots,x_n)$ be ideals in a ring A and recall the vector notation $\underline{y} = T\underline{x}$ introduced in 5.1.14. The ideals are equal precisely when T is an n-square invertible matrix, i.e. det T is a unit in A. If $T = E_{ij}(a)$, the elementary matrix with 1's on the main diagonal, $a \in A$ in spot i,j, $i \neq j$, and 0's elsewhere, then $y_i = x_i + ax_j$ and $y_k = x_k$ for $k \neq i$. Such an elementary matrix has determinant 1, and the group which they generate, the elementary group E(n,A), consists of invertible matrices. The following extends a result of Lazard [La, Th. 1] to the module case.

5.3.18 THEOREM. Let M be a finitely generated module over a noetherian ring A and $(x_1,\ldots,x_n) = \mathfrak{a}$ be a proper ideal with $dp_A(\mathfrak{a},M) = s$. Then there exists a set of generators $y_1,\ldots,y_n$ of $\mathfrak{a}$ with $\underline{y} = T\underline{x}$ and $T \in E(n,A)$ such that all subsequences of s elements among the y's, in whichever order, form a regular sequence on M.

PROOF. By virtue of 5.1.3 e) it is enough to find a $T \in E(n,A)$ with $\underline{y} = T\underline{x}$ such that all subsequences of length $\leq s$ of $y_1,\ldots,y_n$, in the given order, are regular on M. In other words, for $m \leq n$, $k \leq s$ and $1 \leq i_1 < i_2 < \ldots < i_k \leq m$, the sequence $y_{i_1},\ldots,y_{i_k}$ should be regular on M. We make our induction on m since for $m = 0$ there is nothing to prove and for $m = n$ it is the statement we are after.

Now assume the statement has been proved for $0,1,\ldots,m-1$ with $\underline{z} = S\underline{x}$, $S \in E(n,A)$. Then if $1 \leq i_1 < i_2 < \ldots < i_k = m$, all subsequences of $z_{i_1},z_{i_2},\ldots,z_{i_{k-1}}$ of length r, $1 \leq r \leq k-1$ are regular on M. All these subsequences generate ideals of depth $r \leq k-1 < s$ on M. By the previous lemma, such an ideal of depth r on M is contained in only finitely many primes of depth r on M which are maximal with respect to this property.

Even if we vary r between 0 and k-1 and if we consider all possible chains of integers $i_1,\ldots,i_k$ with $1 \le i_1 < i_2 < \ldots < i_k = m$, we still have only finitely many primes, the depth of each of which is < s on M. Now $dp_A(\mathfrak{a},M) = s$ so $\mathfrak{a}$ cannot be contained in any of these by 5.3.2, whence it is not in their union, say D. Now $(z_1,\ldots,z_n) = \mathfrak{a}$, so $(z_m,\mathfrak{c}) \not\subset D$, where $\mathfrak{c}$ is the ideal $(z_1,\ldots,z_{m-1},z_{m+1},\ldots,z_n)$. By a well known trick first explicited by Davis [Ka 74, Th. 124], we can find $z'_m = z_m + \Sigma_{i\neq m}\, a_i z_i \notin D$ for suitable choices of the a_i's in A. The transformation S' which takes z_m to z'_m and leaves all the other z's invariant, is clearly in E(n,A). Then $\underline{z}' = S'S\underline{x}$ solves the induction step, and carrying on till m = n we find our $\underline{y} = T\underline{x}$.

We conclude this chapter with

5.3.19 EXERCISE. Let A be a noetherian ring and suppose the elements $x_1,\ldots,x_n$ generate an ideal $\mathfrak{a}$ which is contained in the Jacobson radical of A. Prove that, if M is a finitely generated module with $dp(\mathfrak{a},M) = n$, then the $x_1,\ldots,x_n$ form a regular sequence on M.

6 EXACTNESS OF COMPLEXES AND LINEAR EQUATIONS OVER RINGS

The Acyclicity Lemma made its surprise appearance in the joint thesis of Peskine and Szpiro [PS 73, Ch. I, Lemme 1.8] which injected new vigour into the entire field. It played an important role in their treatment of the Homological Conjectures - ostensibly the subject of this book. They considered noetherian local rings and their maximal ideals, using local cohomology. It was quickly realized that Ext-functors work equally well, and the lemma was extended in various directions, [No 76, Ch. 5, Th. 21], [Fo 77b, Lemma 1.3], [Sc 82, Kor. 2.3.2]. Here, however, we take a different tack, considering several specializations of the abstract Acyclicity Lemmas 1.1.1 and 1.1.2. This approach, due to A.-M. Simon, allows for greater flexibility as we shall see, and establishes that Ext-depth can also be measured by the Koszul complex, 6.1.6.

In section 6.2 we treat the Buchsbaum-Eisenbud criterion for a finite free complex to be exact, a tool which has proved its value in many a concrete situation. As a rather minor application, we show in section 6.3 how it affects the solution of systems of linear equations over rings. Since the ring is seldom in doubt, we usually drop the subscript A from notations like $\text{E-dp}_A(\alpha,M)$ and $\text{dp}_A(\alpha,M)$.

6.1 THE ACYCLICITY LEMMA AND A FEW CONSEQUENCES

By considering in the following noetherian local rings, their maximal ideals and finitely generated modules, the reader may recover the original Acyclicity Lemma's.

6.1.1 THEOREM. Let $\mathfrak{a}$ be an ideal in a ring A and

$$C_\bullet = 0 \to C_s \to C_{s-1} \to \dots \to C_0 \qquad (s \geq 1)$$

a complex of A-modules such that, for $1 \leq i \leq s$, $\text{E-dp}(\mathfrak{a},C_i) \geq i$ and either $H_i(C_\bullet) = 0$ or $\text{E-dp}(\mathfrak{a},H_i(C_\bullet)) = 0$. Then the complex $C_\bullet$ is exact.

PROOF. Immediate from 1.1.1 if one takes $F^i = \text{Ext}^i_A(A/\mathfrak{a},-)$, since then $f^- = \text{E-dp}(\mathfrak{a},-)$ by Definition 5.3.6.

6.1.2 COROLLARY. Let $\mathfrak{a}$ be an ideal in the ring A and

$$C_\bullet = 0 \to C_s \to C_{s-1} \to \dots \to C_0 \to 0$$

be a complex of A-modules such that, for $0 \leq i \leq s$, $\text{E-dp}(\mathfrak{a},C_i) > i$ and either $H_i(C_\bullet) = 0$ or $\text{E-dp}(\mathfrak{a},H_i(C_\bullet)) = 0$. Then the complex is exact.

PROOF. Follows from the theorem by raising the indices by one and calling the 0 on the right the new C_0.

The next two corollaries consider complexes of flat modules and more or less trivially generalize versions dealing with finite free complexes due to various authors. For a flat module F and an arbitrary module M over a noetherian ring A, we shall use that $\text{E-dp}(\mathfrak{a},F\otimes_A M) \geq \text{E-dp}(\mathfrak{a},M)$, which follows from the isomorphism $\text{Ext}^i_A(A/\mathfrak{a},F\otimes_A M) \simeq F\otimes_A \text{Ext}^i_A(A/\mathfrak{a},M)$

[Bo 80, §6, Prop. 7 c)].

6.1.3 COROLLARY. Let $\mathfrak{a}$ be an ideal in the noetherian ring A and

$$F_\bullet = 0 \to F_s \to F_{s-1} \to \dots \to F_0$$

be a complex of flat A-modules with $s \leq$ E-dp$(\mathfrak{a},A)$ and either $H_i(F_\bullet) = 0$ or E-dp$(\mathfrak{a},H_i(F_\bullet)) = 0$, $1 \leq i \leq s$). Then the complex is exact.

PROOF. Clear from 6.1.1 and the above.

The beauty of such acyclicity statements is that they also allow one to conclude the other way: the nonexactness of certain complexes implies an inequality involving Ext-depth. To avoid technical complications, we state the next result for a noetherian local ring, sufficient for its application in 13.1.1.

6.1.4 COROLLARY. Let $(A,\mathfrak{m},k)$ be a noetherian local ring and

$$F_\bullet = 0 \to F_s \to F_{s-1} \to \dots \to F_0 \to 0$$

be a nonexact complex of flat A-modules such that $\ell(H_i(F_\bullet)) < \infty$. Let M be an A-module such that $\mathfrak{m}M \neq M$. Then E-dp $M \leq s$.

PROOF. We may suppose $H_0(F_\bullet) \neq 0$ (if $H_0(F_\bullet) = 0$ then take $K_1 = \ker(F_1 \to F_0)$ and obtain a complex $\dots \to F_2 \to K_1 \to 0$ consisting of fewer modules). Consider the complex

(*) $$F_\bullet \otimes_A M = 0 \to F_s \otimes_A M \to F_{s-1} \otimes_A M \to \dots \to F_0 \otimes_A M \to 0$$

in which E-dp $H_i(F_\bullet \otimes_A M) = 0$ or $H_i(F_\bullet \otimes_A M) = 0$ since $\ell(H_i(F_\bullet)) < \infty$ implies that $H_i(F_\bullet \otimes M)$ is at most supported in $\mathfrak{m}$. We recall that E-dp$(F_i \otimes_A M) \geq$ E-dp M.

Because of the Acyclicity Lemma 6.1.2 it is now sufficient to prove that in the complex (*), $H_0(F_\bullet \otimes_A M) \neq 0$. But the right-exactness of the functor $-\otimes_A M$ yields $H_0(F_\bullet \otimes_A M) \simeq H_0(F_\bullet)\otimes_A M$. Since $\ell(H_0(F_\bullet)) < \infty$, $H_0(F_\bullet)$ is finitely generated, so $H_0(F_\bullet) \neq mH_0(F_\bullet)$. Now tensor $H_0(F_\bullet)\otimes_A M$ with k. The result is $(H_0(F_\bullet)\otimes_A M)\otimes_A k \simeq H_0(F_\bullet)/mH_0(F_\bullet)\otimes_k M/mM$, and this is a tensor product of nonnull vector spaces. Therefore $H_0(F_\bullet\otimes_A M) \neq 0$ which finishes the proof.

A more general result can be proved with the methods of Chapter 7.

Our next aim is to measure the Ext-depth in terms of the Koszul (co)homology of section 4.2. For $x = x_1,\dots,x_n$ elements in a ring A and an A-module M, we considered the descending resp. ascending Koszul complexes $K_\bullet(x,M)$ and $K^\bullet(x,M)$. For their homologies we proved in 4.2.8 that $H^i(x,M) \simeq H_{n-i}(x,M)$ for $0 \le i \le n$ and they are of course 0 outside of that range. Put $h_+(x,M)$ resp. $h_-(x,M)$ for the sup resp. the $\inf\{i \mid H_i(x,M) \neq 0\}$. Similarly define $h^+(x,M)$ and $h^-(x,M)$ using $H^i(x,M)$.

6.1.5 PROPOSITION. Let M be a module over a ring A and $x = x_1,\dots,x_n$ a sequence of elements in A. Then

(i) $h^+(x,M) = n - h_-(x,M)$, $h^-(x,M) = n - h_+(x,M)$ and if one of these invariants is finite, then so are all four of them;

(ii) If A is a noetherian local ring then also $h_+(x,M) = h^+(x,M^\vee)$ and $h^-(x,M) = h_-(x,M^\vee)$, where $-^\vee$ stands for the Matlis dual.

PROOF. (i) is clear from the isomorphisms recalled above. For (ii), it is enough to prove that $H_i(x,M)^\vee \simeq H^i(x,M^\vee)$ and $H^i(x,M)^\vee \simeq H_i(x,M^\vee)$ for all i, because the Matlis dual is faithful. This functor is also contravariant exact, so commutes with taking homology, thus we need only prove that $K_\bullet(x,M)^\vee \simeq K^\bullet(x,M^\vee)$ and $K^\bullet(x,M)^\vee \simeq K_\bullet(x,M^\vee)$. This in turn is clear from the identities $(N\otimes_A M)^\vee \simeq \mathrm{Hom}_A(N,M^\vee)$ and $\mathrm{Hom}_A(N,M)^\vee \simeq N\otimes_A M^\vee$, 3.4.14, which hold for finitely generated N.

6.1.6 THEOREM. Let $x = x_1,\dots,x_n$ generate an ideal α in a ring A and let M be an A-module. Then $\text{E-dp}(\alpha,M) = h^-(x,M)$.

PROOF. $\text{E-dp}(\alpha,M) \le h^-(x,M)$: We may assume that $h^-(x,M) = s < \infty$. Suppose that $\text{E-dp}(\alpha,M) \ge s+1$, and consider the truncated ascending Koszul complex $C^\bullet = 0 \to K^0(x,M) \to \dots \to K^s(x,M) \to K^{s+1}(x,M)$. Its chain modules C^i are finite direct sums of copies of M, and since the $\text{Ext}^i_A(A/\alpha,-)$ are additive functors, $\text{E-dp}(\alpha,C^i) = \text{E-dp}(\alpha,M) \ge s+1$ for all i. All the homology in $C^\bullet$ is annihilated by α, so either $H^i(C^\bullet) = 0$ or $\text{E-dp}(\alpha,H^i(C^\bullet)) = 0$. By virtue of 6.1.1 the complex $C^\bullet$ is exact, and we have a contradiction.

$h^-(x,M) \le \text{E-dp}(\alpha,M)$: Now assume that $\text{E-dp}(\alpha,M) = s < \infty$ and take a free resolution of the A-module A/α; let $L_\bullet$ be the associated free complex with $H_0(L_\bullet) = A/\alpha$. The chain modules of the complex $C^\bullet = \text{Hom}_A(L_\bullet,M)$ are direct products of copies of the module M, and since $K^\bullet(x,\Pi M) \simeq \Pi K^\bullet(x,M)$, we find that $h^-(x,C^i) = h^-(x,M)$ for all i. Now the truncated complex $C^\bullet_t = 0 \to C^0 \to \dots \to C^s \to C^{s+1}$ is not exact, and indeed its first nonnull homology is $H^s(C^\bullet_t) = H^s(C^\bullet) = \text{Ext}^s_A(A/\alpha,M)$. Also $H^0(x,H^s(C^\bullet_t)) \simeq \text{Hom}_A(A/\alpha,H^s(C^\bullet_t)) \simeq H^s(C^\bullet_t)$. In the proof of 4.3.4 we observed that the $H^i(x,-)$ are a right exact connected sequence of functors. In case $h^-(x,M) \ge s+1$, we are therefore in a position to put $F^i = H^i(x,-)$ in the abstract Acyclicity Lemma 1.1.1, to show that $C^\bullet_t$ is an exact complex. This contradiction implies that $h^-(x,M) \le s$, and finishes the proof of our theorem.

For infinitely generated ideals one can also relate the Ext-depth (= local cohomology depth, according to 5.3.15) to Koszul complexes, see [KM]. Yet another notion of depth, which uses polynomial extensions, was considered in that paper and previously in [Barg], [Ho 74] and [No 76].

We now introduce a notion dual to Ext-depth, the Tor-codepth. For any ideal $\mathfrak{a}$ in a ring A and an A-module M, define T-codp($\mathfrak{a}$,M) = $\inf\{i \mid \mathrm{Tor}_i^A(A/\mathfrak{a},M) \neq 0\}$. In case of a local ring (A,m,k) we just write T-codp for T-codp(m,-). Notice that for a finitely generated nonnull module M in this case, T-codp M = 0 by Nakayama's Lemma, but that for instance T-codp M = ∞ if we take A = $\mathbb{Z}_{(p)}$ for some prime p and M = $\mathbb{Q}$. We invite the reader to prove a dual result to 6.1.6, now using the dual abstract Acyclicity Lemma 1.1.2 rather that 1.1.1:

6.1.7 THEOREM. Let $x = x_1,\ldots,x_n$ generate an ideal $\mathfrak{a}$ in a ring A and let M be an A-module. Then T-codp($\mathfrak{a}$,M) = $h_-(x,M)$.

6.1.8 COROLLARY. E-dp($\mathfrak{a}$,M) < ∞ if and only if T-codp($\mathfrak{a}$,M) < ∞. In this case, E-dp($\mathfrak{a}$,M) + T-codp($\mathfrak{a}$,M) $\leq$ n, and equality is achieved precisely when $K^\bullet(x,M)$ has its homology concentrated in $h^-(x,M) = h^+(x,M)$ or, equivalently, $K_\bullet(x,M)$ in $h_-(x,M) = h_+(x,M)$.

PROOF. Follows right away from the equalities in 6.1.5, 6.1.6 and 6.1.7.

6.1.9 COROLLARY. If E-dp($\mathfrak{a}$,M) = n, then $M/\mathfrak{a}M \neq 0$ and $h_+(x,M) = h_-(x,M) = 0$. In particular, $x_1,\ldots,x_n$ is a pre-regular sequence on M.

PROOF. In view of the previous corollary, T-codp($\mathfrak{a}$,M) = $h_-(x,M) = 0$ so $M/\mathfrak{a}M \neq 0$. The last statement follows from 5.1.7, where condition (ii) implies condition (iv), and the latter was defined as pre-regularity in 5.1.13.

A few remarks are in order. Since local cohomology only depends on the topology defined by the ideal $\mathfrak{a}$, so does E-dp($\mathfrak{a}$,-), and hence $h^-(x,-)$ does not depend on the generators we have chosen for an ideal yielding the same topology. It is not hard to show that this is also the case for T-codp($\mathfrak{a}$,-) and $h_-(x,-)$.

In the important case of a noetherian local ring $(A,\mathfrak{m},k)$ we can circumvent this last point as follows. As a consequence of the Ext-Tor duality vis-à-vis the Matlis dual 3.4.14, and the faithfulness of this dual, one has T-codp M = E-dp $M^\vee$ and E-dp M = T-codp $M^\vee$. Thus T-codp M can equally well be measured using any $\mathfrak{m}$-primary ideal instead of $\mathfrak{m}$. For any two sets of generators $x = x_1,\ldots,x_n$ and $y = y_1,\ldots,y_m$ for two such $\mathfrak{m}$-primary ideals, we thus have $h^-(x,M) = h^-(y,M)$ and $h_-(x,M) = h_-(y,M)$ according to 6.1.6 and 6.1.7 (but of course not for h^+ and h_+, since these depend on n and m). It is a standard fact in the theory of noetherian local rings, which will be treated extensively in section 8.2, that the minimal number of elements required to generate such an $\mathfrak{m}$-primary ideal is equal to the Krull dimension of the ring, say d. Such systems $x = x_1,\ldots,x_d$ always exist and are called systems of parameters of A.

6.1.10 COROLLARY. Let A be a d-dimensional noetherian local ring and M an A-module. Then E-dp M is finite if and only if T-codp M is. In this case E-dp M + T-codp M $\leq$ d. Furthermore E-dp M = d if and only if $h_+(x,M) = 0$ for every system of parameters x and it is enough to require this for a single one.

This was proved by J. Bartijn in his thesis [Ba, Ch. IV, §4] using methods we shall encounter in Chapter 7. The present proof of the more general Corollary 6.1.8 and the whole approach to the so-called depth-sensitivity of the Koszul complex in 6.1.6, which exploits the abstract Acyclicity Lemma, is due to A.-M. Simon.

6.1.11 EXERCISES.

1. Use 4.2.9 and 6.1.6 to prove that Corollary 6.1.3 remains valid for a finitely generated ideal $\mathfrak{a}$ in a nonnoetherian ring.

2. Let M be a module over a noetherian local ring with E-dp $M < \infty$. Prove that E-dp $M \leq \dim M$.

6.2 THE BUCHSBAUM-EISENBUD CRITERION

In this section we shall treat the Buchsbaum-Eisenbud criterion, which is a criterion to determine the exactness of a complex of finitely generated free modules over a noetherian ring A by considering their rank and the rank of the boundary homomorphisms. In this case, acyclicity statements become quite specific; since they are the same by 5.3.9, we shall write dp for E-dp. ($\mathrm{dp}(\mathfrak{a},M)$ was only defined in 5.3.1 when $\mathfrak{a}M \neq M$; in case $\mathfrak{a}M = M$ here let us put $\mathrm{dp}(\mathfrak{a},M) = \infty$ since $\text{E-dp}(\mathfrak{a},M) = \infty$ in virtue of 5.3.14). In this section, we make the blanket assumption that A is noetherian and now introduce some terminology.

Let L_1 and L_2 be finitely generated free modules over A with fixed bases, and let $\phi: L_1 \to L_2$ be an A-linear map. We can represent ϕ by a matrix $\mathrm{mat}(\phi)$ of size $\mathrm{rk}\, L_2 \times \mathrm{rk}\, L_1$. We shall also deal with ideals of A generated by subdeterminants (minors) of $\mathrm{mat}(\phi)$.

DEFINITIONS.

(i) We define $I_k(\phi) \subset A$ as the ideal generated by the k×k-minors of $mat(\phi)$ and set by convention $I_0(\phi) = A$. Using some results from the theory of exterior algebras one can check that the definition of $I_k(\phi)$ does not depend on the basis chosen.

(ii) $rk(\phi) = \max\{k \mid I_k(\phi) \neq 0\}$.

(iii) $redrk(\phi)$ = reduced rank$(\phi) = \max\{k \mid dp_A(I_k(\phi),A) \geq 1\}$.

Of course $redrk(\phi) \leq rk(\phi)$. If $k = rk(\phi)$, we shall sometimes abbreviate $I_k(\phi)$ to $I(\phi)$.

The main theorem of this section is the following:

THEOREM. (Buchsbaum-Eisenbud [BE, Th. 1]) Let

$$L_\bullet = 0 \to L_s \stackrel{d_s}{\to} L_{s-1} \to \dots \stackrel{d_1}{\to} L_0$$

be a complex of finitely generated free A-modules. Then $L_\bullet$ is exact if and only if

(1) $dp_A(I(d_i),A) \geq i$ for $i = 1,\dots,s$;

(2) $rk(d_i) + rk(d_{i+1}) = rk\, L_i$ for $i = 1,\dots,s$.

Since the proof is rather long and elaborate, we treat separately an "if"-Theorem 6.2.3, and an "only if"-Theorem 6.2.6.

Before considering the "if"-theorem, we shall first prove a few lemmas:

6.2.1 LEMMA. Let L_1, L_2 be two finitely generated free A-modules and $\phi: L_1 \to L_2$ an A-linear map. Then $coker(\phi)$ is projective with stable rank (i.e. the rank is invariant with respect to localization at prime ideals), if and only if $I_k(\phi) = A$, where $k = rk(\phi)$.

PROOF. Since $I(\phi) = A$ if and only if $I(\phi)_{\mathfrak{p}} = A_{\mathfrak{p}}$ for all primes $\mathfrak{p}$, and also coker(ϕ) is projective with stable rank means that for each prime $\text{coker}(\phi)_{\mathfrak{p}} = A_{\mathfrak{p}}^n$ with fixed n, we may restrict ourselves to proving the lemma for a local ring A, replacing the statement "coker(ϕ) is projective with stable rank" by "coker(ϕ) is free".

First suppose coker(ϕ) is free. Then im(ϕ) is free too and the exact sequences

$$0 \to \ker(\phi) \to L_1 \to \text{im}(\phi) \to 0 \text{ and } 0 \to \text{im}(\phi) \to L_2 \to \text{coker}(\phi) \to 0$$

both split. So $L_1 \simeq \ker(\phi) \oplus \text{im}(\phi)$; $L_2 \simeq \text{im}(\phi) \oplus \text{coker}(\phi)$, and we can choose bases for L_1 and L_2 such that $\text{mat}(\phi) = \begin{pmatrix} I & 0 \\ 0 & 0 \end{pmatrix}$, where I is a k×k-identity submatrix. One easily sees now that $I_k(\phi) = A$, which proves one half of the lemma.

Now assume $I_k(\phi) = A$. Then ϕ has an invertible k×k submatrix Y, say $\text{mat}(\phi) = \begin{pmatrix} Y & * \\ * & * \end{pmatrix}$. After base change: $\text{mat}(\phi) = \begin{pmatrix} I & * \\ * & * \end{pmatrix}$. After a second base change: $\text{mat}(\phi) = \begin{pmatrix} I & 0 \\ 0 & X \end{pmatrix}$. But $\text{rk}(\phi) = k$, so X must be zero and coker(ϕ) is free and ($\text{rk}\, L_2 - k$) is its rank.

6.2.2 LEMMA. Let $L' \xrightarrow{\phi_1} L \xrightarrow{\phi_2} L''$ be a complex of finitely generated free A-modules with $I(\phi_1) = I(\phi_2) = A$. Then the complex is exact if and only if $\text{rk}(\phi_1) + \text{rk}(\phi_2) = \text{rk}\, L$.

PROOF. Since $I(\phi_1) = I(\phi_2) = A$, the integers $\text{rk}(\phi_1)$ and $\text{rk}(\phi_2)$ do not change under localization at a maximal ideal, so suppose again A is local. In the same way as in the proof of the previous lemma the statement $I(\phi_1) = I(\phi_2) = A$ implies that $\ker(\phi_1)$, $\text{im}(\phi_1)$, $\ker(\phi_2)$ and $\text{im}(\phi_2)$ are free A-modules. So one can choose bases for L', L and L" such that

$$\text{mat}(\phi_1) = \begin{pmatrix} I & 0 \\ 0 & 0 \end{pmatrix} \Big]\, \text{rk}\, L, \qquad \text{where rk } I = \text{rk}(\phi_1);$$

$$\text{mat}(\phi_2) = \underbrace{\begin{pmatrix} 0 & 0 \\ 0 & I \end{pmatrix}}_{\text{rk } L}, \qquad \text{where rk } I = \text{rk}(\phi_2).$$

Since $L' \to L \to L''$ is a complex, we have $\text{rk}(\phi_1) + \text{rk}(\phi_2) \le \text{rk } L$, and one easily sees that this complex is exact if and only if $\text{rk}(\phi_1) + \text{rk}(\phi_2) = \text{rk } L$.

We shall now prove the "if" part of the Buchsbaum-Eisenbud criterion.

6.2.3 THEOREM. Let $L_\bullet = 0 \to L_s \overset{d_s}{\to} L_{s-1} \overset{d_{s-1}}{\to} \dots \to L_0$ be a complex of finitely generated free A-modules satisfying

(i) $\text{dp}_A(I(d_i),A) \ge i$ for $i = 1,2,\dots,s$;

(ii) $\text{rk}(d_i) + \text{rk}(d_{i+1}) = \text{rk } L_i$ for $i = 1,2,\dots,s$.

Then $L_\bullet$ is exact.

PROOF. Since for all primes $\mathfrak{p}$ we have $\text{dp}_{A_\mathfrak{p}}(I(d_i)_\mathfrak{p},A_\mathfrak{p}) \ge \text{dp}_A(I(d_i),A) \ge i$ $(i = 1,2,\dots,s)$ (so $\text{rk}(d_{i_\mathfrak{p}}) = \text{rk}(d_i)$) and furthermore $H_i(L_\bullet) = 0$ if and only if $H_i(L_{\bullet\mathfrak{p}}) = 0$ for all primes $\mathfrak{p}$, we may suppose that A is local with maximal ideal $\mathfrak{m}$.

We use induction on $d = \dim A$ (i.e. the maximal length of a chain of prime ideals $\mathfrak{p}_0 \subsetneq \mathfrak{p}_1 \subsetneq \mathfrak{p}_2 \subsetneq \dots \subsetneq \mathfrak{p}_d = \mathfrak{m}$; the notion of dimension will be treated extensively in Chapter 8). If $d = 0$ then the maximal ideal $\mathfrak{m}$ is also minimal and contains only zerodivisors. Since $\text{dp}_A(I(d_i),A) \ge i > 0$ shows that $I(d_i)$ contains a unit, we see that $I(d_i) = A$ $(i = 1,\dots,s)$ and the result follows from Lemma 6.2.2.

Now assume the theorem has been proved for $\dim A = 0,1,\dots,d-1$. Then for every prime ideal $\mathfrak{p} \neq \mathfrak{m}$ the complex $L_{\bullet\mathfrak{p}}$ is exact, so $\text{Supp } H_i(L_\bullet) \subset \{\mathfrak{m}\}$ and $\text{dp } H_i(L_\bullet) = 0$ or $H_i(L_\bullet) = 0$ $(i = 1,\dots,s)$. Suppose $I(d_s) = \dots = I(d_{k+1}) = A$ and $I(d_k) \neq A$ (possibly $k = s$). Then by the Lemmas 6.2.1 and 6.2.2 the complex $0 \to L_s \to \dots \to L_k \to \text{coker}(d_{k+1}) \to 0$ is exact and

$L'_k = \operatorname{coker}(d_{k+1})$ is free, so we still have to show the exactness of the "tail" of $L_\bullet$, i.e. the complex $0 \to L'_k \xrightarrow{d'_k} L_{k-1} \xrightarrow{d_{k-1}} \dots \xrightarrow{d_1} L_0$ where d'_k is induced by d_k (note that the homology at L'_k equals $H_k(L_\bullet)$). In the latter complex we have $\operatorname{Supp} H_i(L_\bullet) \subset \{m\}$, so $\operatorname{dp} H_i(L_\bullet) = 0$ or $H_i(L_\bullet) = 0$ ($i = 1,\dots,k$) and furthermore $\operatorname{dp} L'_k = \operatorname{dp} A \geq \operatorname{dp}_A(I(d_k),A) \geq k$ (the first inequality reflects $I(d_k) \subset m$). The exactness of $0 \to L'_k \to L_{k-1} \to \dots \to L_0$ now follows from Corollary 6.1.3 with $\alpha = m$.

We shall next consider the converse statement to 6.2.3. Our treatment more or less follows Hochster [Ho 75a, Th. 6.4]. First some notation and two more lemmas. For each finitely generated A-module M define

$$V_i = V_i(M) = \{\mathfrak{p} \in \operatorname{Spec} A \mid \operatorname{pd}_{A_\mathfrak{p}} M_\mathfrak{p} \geq i\} \text{ and}$$

$$\mathfrak{A}_i = \mathfrak{A}_i(M) = \{a \in A \mid \operatorname{pd}_{A_a} M_a < i\}.$$

6.2.4 LEMMA.

(i) V_i is closed in Spec A (in the Zariski topology);

(ii) $V_i = \{\mathfrak{p} \in \operatorname{Spec} A \mid \mathfrak{A}_i \subset \mathfrak{p}\}$;

(iii) $\mathfrak{A}_i$ is a radical ideal, i.e. $\mathfrak{A}_i = \cap_{\mathfrak{p}\in V_i} \mathfrak{p}$.

PROOF. From the definition of V_i and $\mathfrak{A}_i$ follows that $V_0 = \operatorname{Supp} M$ and $\mathfrak{A}_0 = \{a \in A \mid \operatorname{pd}_{A_a} M_a < 0\} = \{a \in A \mid M_a = 0\} = \mathfrak{R}(\operatorname{Ann} M)$ so for $i = 0$ the result holds.

For $i > 0$ we shall give an inductive proof for (i), starting at V_1. Since $V_1 = V_0 \setminus \{\mathfrak{p} \in \operatorname{Spec} A \mid M_\mathfrak{p} \text{ is free over } A_\mathfrak{p}\}$ we need to prove that the set $W = \{\mathfrak{p} \in \operatorname{Spec} A \mid M_\mathfrak{p} \text{ is free over } A_\mathfrak{p}\}$ is open. Well, if for $\mathfrak{p} \in \operatorname{Spec} A$ the module $M_\mathfrak{p}$ is $A_\mathfrak{p}$-free, then there exists an integer n and a homomorphism $\phi: A^n \to M$ such that $\phi_\mathfrak{p}: A_\mathfrak{p}^n \to M_\mathfrak{p}$ is an isomorphism. So some element $a \in A\setminus\mathfrak{p}$ annihilates both $\ker(\phi)$ and $\operatorname{coker}(\phi)$, whence M_a is free over A_a and for each $\mathfrak{q} \in \operatorname{Spec} A$ such that $a \notin \mathfrak{q}$, the $A_{\mathfrak{q}}$-module $M_{\mathfrak{q}}$ is free over $A_{\mathfrak{q}}$. The set $\{\mathfrak{q} \in \operatorname{Spec} A \mid a \notin \mathfrak{q}\}$ is open, contains $\mathfrak{p}$ and is a subset of W. It is easy to see that W is a union of open sets, so open.

Now let $i > 1$. Consider a free presentation of M: $0 \to M' \to L \to M \to 0$. Using the long $\operatorname{Ext}_A(-,N)$ sequence and the fact that localization is an exact functor one easily deduces that $V_i(M) = V_{i-1}(M')$, whence the statement (i).

The statements (ii) and (iii) finally follow from the equivalences $a \in \cap_{\mathfrak{p}\in V_i}\mathfrak{p} \Leftrightarrow \operatorname{pd}_{A_{\mathfrak{q}}} M_{\mathfrak{q}} < i$ for all $\mathfrak{q}$ such that $a \notin \mathfrak{q}$ $\Leftrightarrow \operatorname{pd}_{A_a} M_a < i \Leftrightarrow a \in \mathfrak{A}_i(M)$.

6.2.5 LEMMA. In the above situation, if $\operatorname{pd}_A M < \infty$ then $\operatorname{dp}_A(\mathfrak{A}_i(M),A) \geq i$.

PROOF. We recall from 5.3.14 that $\operatorname{dp}_A(\mathfrak{a},A) = \inf\{\operatorname{dp}_{A_{\mathfrak{p}}} A_{\mathfrak{p}} \mid \mathfrak{a} \subset \mathfrak{p}\}$. Furthermore $\operatorname{pd}_{A_{\mathfrak{p}}} M_{\mathfrak{p}} + \operatorname{dp}_{A_{\mathfrak{p}}} M_{\mathfrak{p}} = \operatorname{dp}_{A_{\mathfrak{p}}} A_{\mathfrak{p}}$ for each $\mathfrak{p} \in \operatorname{Supp} M$, in case $\operatorname{pd}_A M < \infty$, a result that we shall prove in 7.1.5 but which most readers will probably recognize as the Auslander-Buchsbaum equality. So, if $\operatorname{dp}_A(\mathfrak{A}_i(M),A) < i$, there would exist a prime ideal $\mathfrak{p} \supset \mathfrak{A}_i$ such that $i > \operatorname{dp}_{A_{\mathfrak{p}}} A_{\mathfrak{p}} \geq \operatorname{pd}_{A_{\mathfrak{p}}} M_{\mathfrak{p}} \geq i$, contradiction.

6.2.6 THEOREM. (Buchsbaum-Eisenbud criterion, second part) Let

$$L_{\bullet} = 0 \to L_s \xrightarrow{d_s} L_{s-1} \xrightarrow{d_{s-1}} \dots \xrightarrow{d_1} L_0$$

be an exact complex of finitely generated free A-modules. Then

(1) $\operatorname{rk}(d_i) + \operatorname{rk}(d_{i+1}) = \operatorname{rk} L_i \quad (1 \leq i \leq s)$;

(2) $\operatorname{dp}_A(I(d_i),A) \geq i$.

PROOF. We shall first prove (1) by showing that

$$\operatorname{rk}(d_i) = \operatorname{rk} L_i - \operatorname{rk} L_{i+1} + \dots + (-1)^{s-i}\operatorname{rk} L_s.$$

Put $M = \operatorname{coker}(d_1)$, $S =$ the collection of nonzerodivisors in A and $B = S^{-1}A$

(the total ring of fractions). Note that $dp_{B_\mathfrak{p}} B_\mathfrak{p} = 0$ for each $\mathfrak{p} \in \mathrm{Spec}\, B$.

We write $d_i \otimes_A 1_B = \bar{d}_i$ $(i = 1,\dots,s)$. Since $L_\bullet$ is a free resolution of M, $pd_A M$ is finite, so for all $\mathfrak{p} \in \mathrm{Spec}\, B$, $pd_{B_\mathfrak{p}} M_\mathfrak{p} \le dp_{B_\mathfrak{p}} B_\mathfrak{p} = 0$ and

$$0 \to L_{s\mathfrak{p}} \xrightarrow{\bar{d}_{s\mathfrak{p}}} \dots \xrightarrow{\bar{d}_{1\mathfrak{p}}} L_{0\mathfrak{p}} \xrightarrow{\bar{d}_{0\mathfrak{p}}} M_\mathfrak{p} \to 0$$

is an exact complex of free $B_\mathfrak{p}$-modules which completely splits, so

$\mathrm{rk}\, M_\mathfrak{p} = \Sigma_{k=0}^{s}(-1)^k \mathrm{rk}\, L_k$ and

$\mathrm{rk}\, L_0 = \mathrm{rk}(\bar{d}_{1\mathfrak{p}}) + \mathrm{rk}(\bar{d}_{0\mathfrak{p}}) = \mathrm{rk}(\bar{d}_{1\mathfrak{p}}) + \mathrm{rk}\, M_\mathfrak{p} = \mathrm{rk}(\bar{d}_{1\mathfrak{p}}) + \Sigma_{k=0}^{s}(-1)^k \mathrm{rk}\, L_k$,

whence $\mathrm{rk}(\bar{d}_{1\mathfrak{p}}) + \Sigma_{k=1}^{s}(-1)^k \mathrm{rk}\, L_k = 0$.
From this we deduce:

a. $\mathrm{rk}(\bar{d}_1) = \Sigma_{k=1}^{s}(-1)^{k-1} \mathrm{rk}\, L_k$;

b. $S^{-1}M$ is projective with stable rank, so by Lemma 6.2.1
$I(\bar{d}_1) = S^{-1}I(d_1) = B$.

We can repeat this argument for $\mathrm{coker}(\bar{d}_i)$ for all $i \le s$, so
$\mathrm{rk}(\bar{d}_i) = \Sigma_{k=i}^{s}(-1)^{k-i} \mathrm{rk}\, L_k$ and $S^{-1}I(d_i) = B$. To show finally that $\mathrm{rk}(d_i) = \mathrm{rk}(\bar{d}_i)$, we put $\mathrm{rk}(\bar{d}_i) = r_i$. Then $S^{-1}I_{r_i}(d_i) = B$ whence $I_{r_i}(d_i) \ne 0$ and $S^{-1}I_{r_i+1}(d_i) = 0$, so $I_{r_i+1}(d_i) = 0$ since S contains no zerodivisors. Therefore (1) holds.

To prove (2) we again put $M = \mathrm{coker}(d_1)$ and $r_i = \mathrm{rk}(d_i)$. Then for each $\mathfrak{p} \in \mathrm{Spec}\, A$ we have the following equivalences for $i \ge 1$.

$\mathfrak{A}_i(M) \subset \mathfrak{p} \Leftrightarrow pd_{A_\mathfrak{p}} M_\mathfrak{p} \ge i \Leftrightarrow \ker(d_{i-2})_\mathfrak{p}$ is not free $\Leftrightarrow$

$\Leftrightarrow \mathrm{coker}(d_i)_\mathfrak{p}$ is not free (since $L_\bullet$ is exact) $\Leftrightarrow$

$\Leftrightarrow I_{r_i}(d_i).A_\mathfrak{p} \ne A_\mathfrak{p}$ (because of Lemma 6.2.1) $\Leftrightarrow$

$\Leftrightarrow I_{r_i}(d_i) \subset \mathfrak{p}$.

So $\mathfrak{A}_i(M) = \mathfrak{R}(I(d_i))$, whence $dp_A(\mathfrak{A}_i(M),A) = dp_A(I(d_i),A)$. From Lemma 6.2.5 it now follows that $dp_A(I(d_i),A) \ge i$.

Eagon and Northcott [EN, §7], see also [No 76, Ch. 6, Th. 15] proved a more general result for an arbitrary ring and involving an arbitrary module, using a version of depth which we have not treated in Chapter 5.

6.3 LINEAR EQUATIONS OVER RINGS

We shall give a nice proof of some useful facts in the theory of linear equations over noetherian rings, as indicated by D. Eisenbud.

For a noetherian ring A consider the following system of linear equations with unknowns $x_1,\ldots,x_n$, $\alpha_{ij} \in A$, $y_i \in A$:

$$\begin{aligned} \alpha_{11}x_1 + \ldots + \alpha_{1n}x_n &= y_1 \\ \alpha_{21}x_1 + \ldots + \alpha_{2n}x_n &= y_2 \\ \vdots \qquad\qquad & \quad \vdots \\ \alpha_{m1}x_1 + \ldots + \alpha_{mn}x_n &= y_m \end{aligned}$$

where $m \geq n$.

Let ϕ be the matrix (α_{ij}) and view ϕ as a linear map $A^n \to A^m$, A^n and A^m having standard bases. Then we can contract the above equations to $\phi(\underline{x}) = \underline{y}$ and wonder whether this equation is soluble. We shall give a criterion for this by using the results of the previous section. First we consider the homogeneous case, i.e. $\underline{y} = 0$.

6.3.1 PROPOSITION. The equation $\phi(\underline{x}) = 0 \in A^m$ has a nontrivial solution if and only if $\mathrm{redrk}(\phi) < n$ (in other words: ϕ is injective $\Leftrightarrow$ $\Leftrightarrow \mathrm{redrk}(\phi) = n$).

PROOF. Follows immediately from the equivalences: $\phi(\underline{x}) = 0$ has only the null-solution $\Leftrightarrow$ $0 \to A^n \xrightarrow{\phi} A^m$ is exact $\Leftrightarrow$ (Buchsbaum-Eisenbud criterion) $\mathrm{rk}(\phi) = n$ and $\mathrm{dp}(I_n(\phi),A) \geq 1 \Leftrightarrow \mathrm{redrk}(\phi) = n$.

If $y \neq 0$ the equation is more difficult to handle:

6.3.2 PROPOSITION. [Sharpe, Th.]. Let $y = (y_1,\dots,y_m) \in A^m$ and let ψ be the matrix

$$\psi = \begin{pmatrix} \alpha_{11} & \cdots & \alpha_{1n} & y_1 \\ \vdots & & \vdots & \vdots \\ \alpha_{m1} & \cdots & \alpha_{mn} & y_m \end{pmatrix}.$$

If $rk(\psi) = n$ and $dp(I_n(\phi),A) \geq 2$, then $\phi(\underline{x}) = \underline{y}$ has a unique solution in A^n.

PROOF. Consider the complex $L_\bullet = 0 \to A^n \xrightarrow{\phi} A^m \xrightarrow{\chi} \overset{n+1}{\Lambda} A^m$ where $\chi(a_1,\dots,a_m) = (a_1,\dots,a_m) \wedge (\alpha_{11},\dots,\alpha_{m1}) \wedge \dots \wedge (\alpha_{1n},\dots,\alpha_{mn})$ (recall that $m \geq n$). By calculating $\chi\circ\phi(e_i)$ (e_i is a basiselement of A^n) one sees that $L_\bullet$ is indeed a complex. From $rk(\psi) = n$ follows that $\chi(\underline{y}) = 0$, so if we can prove that $L_\bullet$ is exact, we are done, since in that case $\underline{y} \in im(\phi)$ and ϕ is injective.

For this proof we shall verify that the conditions of the Buchsbaum-Eisenbud criterion are satisfied, namely

1) $rk(\phi) = n$;

2) $rk(\phi) + rk(\chi) = m$ (so $rk(\chi) = m-n$);

3) $dp(I(\phi),A) = dp(I_n(\phi),A) \geq 2$;

4) $dp(I(\chi),A) = dp(I_{m-n}(\chi),A) \geq 1$.

Conditions 1) and 3) follow from the assumptions of the proposition, so it is sufficient for us to prove $rk(\chi) \leq m-n$ and after this $dp(I_{m-n}(\chi),A) \geq 1$.

The inequality $rk(\chi) \leq m-n$ follows from the following lemma:

6.3.3 LEMMA. Let $L' \xrightarrow{\phi} L \xrightarrow{\chi} L''$ be a complex of finitely generated free modules over a ring A, with $rk\ L = m$, $redrk(\phi) = n$ and $rk(\chi) = s$. Then

$s \leq m-n$.

PROOF. The proof will be given by elementary algebra. Since $\mathrm{redrk}(\phi) = n$, $I_n(\phi)$ must contain a nonzerodivisor. If $I_n(\phi) \subset \mathfrak{R}(\mathrm{Ann}\ I_s(\chi))$, then $I_s(\chi)$ must be zero, contradicting the assumption $\mathrm{rk}(\chi) = s$. So we have $I_n(\phi) \not\subset \mathfrak{R}(\mathrm{Ann}\ I_s(\chi))$, and $\mathrm{mat}(\phi)$ contains an $n\times n$-submatrix T such that the minor $\det T = \Delta \notin \mathfrak{R}(\mathrm{Ann}\ I_s(\chi))$.

After localizing at a prime ideal $\mathfrak{p} \supset \mathrm{Ann}\ I_s(\chi)$ such that $\Delta \notin \mathfrak{p}$ we have the situation $I_n(\phi) = A$ and $I_s(\chi) \neq (0)$. In the same way as in Lemma 6.2.1 we may suppose $\mathrm{mat}(\phi) = \begin{pmatrix} I & 0 \\ 0 & * \end{pmatrix} \Big]\, m$ (where I is an $n\times n$-identity submatrix). Since $\chi\circ\phi = 0$, the first n columns of $\mathrm{mat}(\chi)$ are zero, so $\mathrm{mat}(\chi)$ looks like $\begin{pmatrix} 0 & \cdots & 0 & \\ \vdots & & \vdots & * \\ 0 & \cdots & 0 & \end{pmatrix}$, and one easily sees now that $\mathrm{rk}(\chi) \leq m-n$.

To finish the proof of Proposition 6.3.2 we still have to show that $\mathrm{dp}(I_{m-n}(\chi),A) \geq 1$, and therefore we return to the complex $L_{\bullet} = 0 \to A^n \xrightarrow{\phi} A^m \xrightarrow{\chi} \overset{n+1}{\Lambda} A^m$, where $\{e_i \mid i = 1,\ldots,m\}$ is the standard basis for A^m and $\{e_{i_1}\wedge e_{i_2}\wedge\ldots\wedge e_{i_{n+1}} \mid 1 \leq i_1 < \ldots < i_{i_{n+1}} \leq m\}$ the same for $\overset{n+1}{\Lambda} A^m$.

Let T be an $n\times n$ submatrix of $\mathrm{mat}(\phi)$ with determinant Δ, say $\mathrm{mat}(\phi) = \begin{pmatrix} T \\ * \end{pmatrix}$. Consider then the submatrix of $\mathrm{mat}(\chi)$ of which the columns correspond to the basis elements e_i $(i=n+1,\ldots,m)$ of A^m and the rows to the basiselements $e_1\wedge\ldots\wedge e_n\wedge e_j$ $(j=n+1,\ldots,m)$ of $\overset{n+1}{\Lambda} A^m$. This submatrix upon inspection has the form $\begin{pmatrix} \pm\Delta & & 0 \\ & \ddots & \\ 0 & & \pm\Delta \end{pmatrix}$, and its determinant is $\pm\Delta^{m-n} \in I_{m-n}(\chi)$. In this way, the determinant of every $n\times n$-submatrix of $\mathrm{mat}(\phi)$ has its $(m-n)$-th power in $I_{m-n}(\chi)$, so there exists a positive integer N such that $(I_n(\phi))^N \subset I_{m-n}(\chi)$. Now the statement $\mathrm{dp}(I_{m-n}(\chi),A) \geq 1$ follows from the existence of a nonzerodivisor in $I_n(\phi)$. (In fact $(I_n(\phi))^N \subset I_{m-n}(\chi)$ shows that $\mathrm{dp}(I_{m-n}(\chi),A) \geq 2$.)

In the situation of Proposition 6.3.2, the reader may observe:

(1) If $rk(\psi) = n+1$, then $\phi(\underline{x}) = \underline{y}$ has no solution in A^n;

(2) If the equation $\phi(\underline{x}) = \underline{y}$ has a unique solution, then $rk(\psi) = n$ and $dp(I_n(\phi),A) \geq 1$.

These statements together with Proposition 6.3.2 imply the following slightly more general result:

6.3.4 PROPOSITION. In the situation of Proposition 6.3.2, if $redrk(\phi) = n$ (so $dp(I_n(\phi),A) \geq 1$) and $rk(\psi) = n$, then $\phi(\underline{x}) = \underline{y}$ has at most a single solution in A^n. If moreover $dp(I_n(\phi),A) \geq 2$, then such a solution does exist.

These considerations too extend to nonnoetherian rings if one uses another notion of depth, see [Sharpe].

7 COMPARING HOMOLOGICAL INVARIANTS

In this chapter we compare various homological invariants. All three sections have in common that certain results are proved by using double complexes of modules. In the first two, we are dealing with a particular double complex in distinct ways, Theorems 7.1.2 and 7.2.14. In the first case we get identities, the second leads to inequalities. Of course, double complexes have been employed more extensively in commutative algebra by such authors as Foxby [Fo 77a], [Fo 79], and P. Roberts [Ro 76], [Ro 80b], while one might also introduce the full machinery of derived categories as mentioned in section 1.2. We shall however restrict ourselves to a single type of down to earth complex as discussed in that section, freely using the Matlis dual to obtain pairs of results.

The content of 7.1 and 7.2 is in large part taken from Bartijn's thesis [Ba], which in turn owes much to the works of Foxby. Section 7.3 explains certain ideas of P. Roberts [Ro 76], who used a double complex to relate the annihilators of local cohomology modules to the exactness of finite free complexes. These results will play a key role in Chapter 11.

Observe that only homological notions figure in this chapter, like Ext-depth, various homological dimensions and grade. Though the notion is doubtless familiar to most readers, in this book we take the somewhat perverse view that Krull dimension is a more subtle invariant and only introduce it in the next chapter.

7.1 AUSLANDER-BUCHSBAUM - AND BASS IDENTITIES GENERALIZED

Throughout this section $(A,\mathfrak{m},k)$ is a noetherian local ring. We shall work with flat dimension fd and injective dimension id and we write $\mathrm{fd}^k M = \sup\{i \mid \mathrm{Tor}_i^A(k,M) \neq 0\}$ and $\mathrm{id}^k M = \sup\{i \mid \mathrm{Ext}_A^i(k,M) \neq 0\}$. Since we work over a fixed ring A, we have dropped the A from most notation. It is well known that for finitely generated modules $\mathrm{fd}^k M = \mathrm{fd}\ M = \mathrm{pd}\ M$ (the projective dimension) and $\mathrm{id}^k M = \mathrm{id}\ M$, e.g. 3.3.6 and [Ma 86, §19, Lemma 1], but $\mathrm{fd}_{\mathbb{Z}_{(p)}}^{\mathbb{F}_p} \mathbb{Q} = -\infty$, p a prime, while in a general local A, if the prime $\mathfrak{p}$ is not maximal, then $\mathrm{id}^k E(A/\mathfrak{p}) = -\infty$.

Recalling the Ext-Tor duality 3.4.14 stemming from the faithful Matlis dual $-^{\vee}$ the reader will have no trouble to prove

7.1.1 PROPOSITION. Let M be a module over a noetherian local ring $(A,\mathfrak{m},k)$. Then

(i) $\mathrm{fd}\ M = \mathrm{id}\ M^{\vee}$ and $\mathrm{id}\ M = \mathrm{fd}\ M^{\vee}$;

(ii) $\mathrm{fd}^k M = \mathrm{id}^k M^{\vee}$ and $\mathrm{id}^k M = \mathrm{fd}^k M^{\vee}$.

We have seen in 3.4.12 that if A is complete in its $\mathfrak{m}$-adic topology, the Matlis dual takes finitely generated modules to artinian ones and vice versa, so that in this case we find that for artinian modules $\text{E-dp}\ M = 0$ and $\text{T-codp}\ M < \infty$, while $\mathrm{fd}^k M = \mathrm{fd}\ M$ and $\mathrm{id}^k M = \mathrm{id}\ M$. These last two equalities have been extended to all modules which are complete in their $\mathfrak{m}$-adic topology over an arbitrary noetherian local ring $(A,\mathfrak{m},k)$ in [Si].

We can now formulate the theorem which will have several interesting identities as corollaries.

7.1.2 THEOREM. Let (A,m,k) be a noetherian local ring and M be a module with $\text{fd } M < \infty$. If N is a module such that $\text{Tor}_i^A(M,N) = 0$ for $i > 0$, then

$$\text{Ext}_A^n(k, M\otimes_A N) \simeq \oplus_{i+j=n} \text{Tor}_{-i}^A(k,M)\otimes_k \text{Ext}_A^j(k,N) \text{ for all } n.$$

Moreover, $\text{id } N < \infty$ implies $\text{id}(M\otimes_A N) < \infty$.

PROOF. First observe that the displayed isomorphism just means that the dimensions of two k-vectorspaces are equal, finite or not. Let $0 \to F_s \overset{d_s}{\to} \ldots \overset{d_1}{\to} F_0 \to M \to 0$ be a flat resolution of M and write $F^\bullet$ for the ascending complex $0 \to F^{-s} \overset{d^{-s}}{\to} \ldots \overset{d^{-1}}{\to} F^0 \to 0$ with $F^{-i} = F_i$ and $d^{-i} = d_i$. Let $E^\bullet$ be an injective resolution of N. The tensor product complex $I^\bullet = F^\bullet\otimes_A E^\bullet$ consists of injectives and has the form $0 \to I^{-s} \to \ldots \to I^0 \to I^1 \to \ldots$ where we shall write δ for the boundary operator. Since $H^i(I^\bullet) = H^i(F^\bullet\otimes_A N) = \text{Tor}_{-i}^A(M,N)$ for all i in view of 1.2.4 (in which C is our N), we find that the homology of $I^\bullet$ is concentrated in 0 and $H^0(I^\bullet) = M\otimes_A N$. Now $I'^\bullet = 0 \to I^{-s} \to \ldots \to \delta(I^{-1}) \to 0$ is a split exact subcomplex of $I^\bullet$, consisting of injectives. The quotient complex $I''^\bullet = I^\bullet/I'^\bullet = 0 \to I''^0 \to I^1 \to I^2 \to \ldots$ consists of injectives and $H^i(I''^\bullet) = H^i(I^\bullet)$ for all i, so that $0 \to M\otimes_A N \to I''^0 \to I^1 \to \ldots$ is an injective resolution of $M\otimes_A N$. If $E^\bullet$ is finite, this resolution is finite, which proves the last statement.

In any case, we can apply the functor $\text{Hom}_A(k,-)$ to the short exact sequence of complexes $0 \to I'^\bullet \to I^\bullet \to I''^\bullet \to 0$. From the long exact sequence of homology we obtain $H^n(\text{Hom}_A(k,I^\bullet)) = H^n(\text{Hom}_A(k,I''^\bullet)) = \text{Ext}_A^n(k,M\otimes_A N)$. On the other hand, by 1.2.4 we know that $H^n(\text{Hom}_A(k,I^\bullet)) \simeq \oplus_{i+j=n} H^i(k\otimes_A F^\bullet)\otimes_k \text{Ext}_A^j(k,N)$. But $H^i(k\otimes_A F^\bullet) = \text{Tor}_{-i}^A(k,M)$, so the theorem is proved.

7.1.3 COROLLARY. E-dp$(M\otimes_A N)$ is finite if and only if both T-codp M and E-dp N are finite. In this case E-dp N = $\mathrm{fd}^k M$ + E-dp$(M\otimes_A N)$ and $\mathrm{id}^k N$ = T-codp M + $\mathrm{id}^k(M\otimes_A N)$.

PROOF. E-dp$(M\otimes_A N) < \infty$ means that $\mathrm{Ext}^n_A(k, M\otimes_A N) \neq 0$ for some n. The isomorphism in the theorem involves k-vector spaces, so this condition just means that for some i and j the vector spaces on the right are not 0, which proves the first line. Reading off the inf resp. the sup of the $n|\ \mathrm{Ext}^n_A(k, M\otimes_A N) \neq 0$ then establishes the identities.

7.1.4 PROPOSITION. Let (A,m,k) be a noetherian local ring and M be a module with id $M < \infty$. Let N be a finitely generated module with $\mathrm{Ext}^i_A(N,M) = 0$ for $i > 0$. Then $\mathrm{fd}(\mathrm{Hom}_A(N,M)) < \infty$ if id $N < \infty$. Moreover, T-codp$(\mathrm{Hom}_A(N,M))$ is finite if and only if both E-dp M and E-dp N are. In this case E-dp N = $\mathrm{id}^k M$ + T-codp$(\mathrm{Hom}_A(N,M))$ and $\mathrm{id}^k N$ = E-dp M + $\mathrm{fd}^k(\mathrm{Hom}_A(N,M))$.

PROOF. If id $M < \infty$, the Matlis dual $M^\vee$ has fd $M^\vee < \infty$. Since N is finitely generated we know that $\mathrm{Tor}^A_i(N,M^\vee) \simeq \mathrm{Ext}^i_A(N,M)^\vee$, 3.4.14. In particular, $N\otimes_A M^\vee \simeq \mathrm{Hom}_A(N,M)^\vee$ and by 7.1.1., $\mathrm{id}(N\otimes_A M^\vee) = \mathrm{fd}(\mathrm{Hom}_A(N,M))$. We see that $M^\vee$ and N satisfy the conditions of 7.1.2 and 7.1.3. One then applies 7.1.1 to the result, bearing in mind that E-dp M = T-codp $M^\vee$ by 3.4.14. Notice that the condition E-dp $N < \infty$ only serves to avoid the null module.

We now put N = A and take M to be finitely generated. Feeding these into 7.1.3 and 7.1.4 we recover two well known identities.

7.1.5 COROLLARY. Let A be a noetherian local ring, and let $M \neq 0$ be a finitely generated module.

If $\text{pd } M < \infty$, then $\text{dp } A = \text{pd } M + \text{dp } M$ (Auslander-Buchsbaum);

If $\text{id } M < \infty$, then $\text{dp } A = \text{id } M$ (Bass).

Of course, there exist various other proofs of these important classical results, but the above argument, due to Bartijn [Ba, Ch. IV, §3], makes particularly clear in which sense these identities are dual. The identities have been generalized in different directions in [Fo 79] and [Av, Ths. 3.5 & 5.2].

Our next application of 7.1.2 is to modules M which have finite Bass numbers and finite Betti numbers. These modules include the finitely generated and the artinian modules, and their subcategory is closed under the Matlis dual, see 3.4.15. We can then form the power series $\beta(M,T) = \Sigma_{i=0}^{\infty} \beta_i(M)T^i$ and $\mu(M,T) = \Sigma_{j=0}^{\infty} \mu^j(M)T^j$, often referred to as Poincaré series.

7.1.6 COROLLARY. Let (A,m,k) be a noetherian local ring and M a module with finite Bass and Betti numbers. Let N be a module with finite Bass numbers.

(i) If $\text{fd } M < \infty$ and $\text{Tor}_i^A(M,N) = 0$ for $i > 0$, then $\mu(M\otimes_A N,T) = \beta(M,T^{-1})\mu(N,T)$ in $\mathbb{Z}[[T]][T^{-1}]$;

(ii) If $\text{id } M < \infty$ and $\text{Ext}_A^i(N,M) = 0$ for $i > 0$ and N is finitely generated, then $\beta(\text{Hom}_A(N,M),T) = \mu(M,T^{-1})\mu(N,T)$ in $\mathbb{Z}[[T]][T^{-1}]$.

PROOF. The first identity is proved by counting the (finite) dimensions of the k-vector spaces which occur in the identity of Theorem 7.1.2. Our conditions on Bass and Betti numbers ensure that the power series on the right are well defined with nonnegative integer coefficients; so are therefore the Poincaré series for $M\otimes_A N$. The second identity follows from the first, applied to the modules $M^\vee$ and N.

The corollary allows one to express Bass and Betti numbers of $M\otimes_A N$ resp. $\mathrm{Hom}_A(N,M)$ in terms of those of M and N. For N = A and M finitely generated these identities first appeared in Foxby [Fo 77a, Cor. 4.3].

7.2 EQUALITIES AND INEQUALITIES INVOLVING GRADE

We first wish to interpret some of the work in section 6.1 in a slightly different way. In keeping with the notation introduced in Proposition 1.1, we attach to two modules N and M over a ring A the invariants

$$\mathrm{tor}_-(N,M) = \inf\{i \mid \mathrm{Tor}_i^A(N,M) \neq 0\} \text{ and}$$

$$\mathrm{ext}^-(N,M) = \inf\{i \mid \mathrm{Ext}_A^i(N,M) \neq 0\}$$

which can range between 0 and ∞.

7.2.1 THEOREM. Let N and M be two modules over a noetherian ring A. If N is finitely generated with annihilator $\mathfrak{a}$, then $\mathrm{ext}^-(N,M) = \text{E-dp}(\mathfrak{a},M)$ and $\mathrm{tor}_-(N,M) = \text{T-codp}(\mathfrak{a},M)$.

PROOF. By 5.3.5 (iii) we know that $\mathrm{Ass}(\mathrm{Hom}_A(N,M)) = \mathrm{Supp}\, N \cap \mathrm{Ass}\, M = \mathrm{Supp}\, A/\mathfrak{a} \cap \mathrm{Ass}\, M = \mathrm{Ass}(\mathrm{Hom}_A(A/\mathfrak{a},M))$. Over a noetherian ring a module X is 0 if and only if $\mathrm{Ass}\, X = \emptyset$. It follows that $\mathrm{ext}^-(N,M) = 0$ if and only if $\text{E-dp}(\mathfrak{a},M) = 0$.

Suppose now that $\text{E-dp}(\mathfrak{a},M) = s < \mathrm{ext}^-(N,M)$. In order to invoke the Acyclicity Lemma, take a resolution $L_\bullet$ of $A/\mathfrak{a}$ consisting of finitely generated free modules L_i. In the complex $C^\bullet = \mathrm{Hom}_A(L_\bullet, M)$ the chain modules C^i are finite direct sums of copies of M, thus $\mathrm{ext}^-(N,C^i) = \mathrm{ext}^-(N,M) \geq s+1$. The modules $H^i(C^\bullet) = \mathrm{Ext}^i_A(A/\mathfrak{a},M)$ are annihilated by $\mathfrak{a}$, hence if $H^i(C^\bullet) \neq 0$, then $\text{E-dp}(\mathfrak{a},H^i(C^\bullet)) = 0$ so $\mathrm{ext}^-(N,H^i(C^\bullet)) = 0$. Applying Proposition 1.1.1 with $F^i = \mathrm{Ext}^i_A(N,-)$, we find that the complex $0 \to C^0 \to \ldots \to C^s \to C^{s+1}$ is exact, which contradicts the assumption that $H^s(C^\bullet) = \mathrm{Ext}^s_A(A/\mathfrak{a},M) \neq 0$. We leave it to the reader to prove that $\text{E-dp}(\mathfrak{a},M) \leq \mathrm{ext}^-(N,M)$ by taking a free resolution of N.

The statement about tor_- can be proved by using Proposition 1.1.2. It is quicker to observe that the primes $\mathfrak{p}$ in the support of $\mathrm{Tor}^A_i(N,M)$ are the ones for which $\mathrm{Tor}^{A_\mathfrak{p}}_i(N_\mathfrak{p},M_\mathfrak{p}) \neq 0$ or, by the faithfulness of the Matlis dual $-^\vee$ over the local noetherian ring $A_\mathfrak{p}$, for which $\mathrm{Tor}^{A_\mathfrak{p}}_i(N_\mathfrak{p},M_\mathfrak{p})^\vee \simeq \mathrm{Ext}^i_{A_\mathfrak{p}}(N_\mathfrak{p},M^\vee_\mathfrak{p}))$, see 3.4.14, is not 0. In this way we reduce to the result for $\mathrm{ext}^-(N,M)$ just proved, leaving further details to the reader.

7.2.2 COROLLARY. $\mathrm{ext}^-(N,M) < \infty$ if and only if $\mathrm{tor}_-(N,M) < \infty$. In this case $\mathrm{ext}^-(N,M) + \mathrm{tor}_-(N,M)$ is bounded above by the minimal number of generators of each ideal in the radical class of Ann N.

PROOF. Combine 6.1.8, 7.2.1 and the remarks following 6.1.9.

The first statement may fail when N is not finitely generated, e.g. for $A = \mathbb{Z}$, $N = \mathbb{Q}$, $M = \mathbb{Q}/\mathbb{Z}$. In case both modules are finitely generated, there is an interesting symmetry, which does not look all that obvious from the point of view of the Yoneda Ext's [St 78, Ch. 3, §1].

7.2.3 COROLLARY. For two finitely generated modules N and M over a noetherian ring A the following are equivalent:

(i) $N \otimes_A M = 0$;

(ii) $\mathrm{Ext}^i_A(N,M) = 0$ for all $i \geq 0$;

(iii) $\mathrm{Ext}^i_A(M,N) = 0$ for all $i \geq 0$.

PROOF. Since N and M are finitely generated, $\mathrm{Supp}\, N \cap \mathrm{Supp}\, M = \mathrm{Supp}(M \otimes_A N)$ so that clearly $N \otimes_A M = 0$ is equivalent with $\mathrm{Tor}^A_i(N,M) = 0$ for all $i \geq 0$. The result now follows from 7.2.2 if one takes into account the symmetry of N and M in the tensor product.

7.2.4 EXERCISE. Use localization, 5.3.5 and 5.3.9 to obtain a different proof of 7.2.3.

In order to treat infinitely generated modules, we follow Foxby [Fo 79, §2] in defining the small support of a module in terms of the notion of Tor-codepth.

7.2.5 DEFINITION. For a module N over a ring A put $\mathrm{supp}\, N = \{\mathfrak{p} \in \mathrm{Spec}\, A \mid \text{T-codp}_{A_\mathfrak{p}} N_\mathfrak{p} < \infty\}$.

If we write $k(\mathfrak{p}) = A_\mathfrak{p}/\mathfrak{p}A_\mathfrak{p}$, then one also has $\mathrm{supp}\, N = \{\mathfrak{p} \in \mathrm{Spec}\, A \mid \mathrm{tor}_-(k(\mathfrak{p}),N_\mathfrak{p}) < \infty\}$ while yet another way to describe this is to take a flat resolution $F_\bullet$ of N and to observe that $\mathfrak{p} \in \mathrm{supp}\, N$ if and only if $\mathfrak{p} \in \mathrm{Supp}(k(\mathfrak{p}) \otimes_{A_\mathfrak{p}} F_{\bullet\mathfrak{p}})$. Here the support of a complex $C_\bullet$ is as usual taken to mean $\mathrm{Supp}\, C_\bullet = \cup_i \mathrm{Supp}\, H_i(C_\bullet)$. Notice also that $\mathrm{supp}\, N = \mathrm{Supp}\, N$ for a finitely generated module N, and if moreover A is noetherian, then $\mathrm{supp}\, N = \mathrm{Supp}\, A/\mathfrak{a}$ with $\mathfrak{a} = \mathrm{Ann}\, N$. The reader will have no trouble in verifying that for an infinitely generated module these three sets may be distinct: consider the modules $\mathbb{Q}$ and $\mathbb{Q}/\mathbb{Z}$ over the ring $\mathbb{Z}$.

We generalize somewhat.

7.2.6 DEFINITION. Let $F_\bullet = \ldots \to F_i \to F_{i-1} \to \ldots \to F_0 \to 0$ be a complex of flat modules over a ring A. Then put $\operatorname{supp} F_\bullet = \operatorname{Supp}(k(\mathfrak{p}) \otimes_{A_\mathfrak{p}} F_{\bullet\mathfrak{p}})$.

7.2.7 LEMMA. Let A be a noetherian ring and $F_\bullet$ be as above. Then $\operatorname{supp} F_\bullet \subset \operatorname{Supp} F_\bullet$ and both sets have the same minimal elements. In particular $\operatorname{supp} N \subset \operatorname{Supp} N$ for any A-module N and both sets have the same minimal elements.

PROOF. Since the second statement follows from the first if one takes for $F_\bullet$ a flat resolution of N, let $F_\bullet$ be as above. If $\mathfrak{p} \notin \operatorname{Supp} F_\bullet$, the complex $F_{\bullet\mathfrak{p}}$ is a flat resolution of the null module, so $k(\mathfrak{p}) \otimes_{A_\mathfrak{p}} F_{\bullet\mathfrak{p}}$ is exact and $\mathfrak{p} \notin \operatorname{supp} F_\bullet$.

It is enough to show that every prime in $\operatorname{Supp} F_\bullet$ contains a prime in $\operatorname{supp} F_\bullet$, and we need only prove that if $F_\bullet$ is not exact over a local noetherian ring which we again call A, then $\operatorname{supp} F_\bullet \neq \emptyset$. The noetherian module A has a finite filtration with subquotients of the type $A/\mathfrak{p}$, $\mathfrak{p}$ a prime in A. Since $-\otimes_A F_i$ is exact for all $i \geq 0$, there must be such a prime with $A/\mathfrak{p} \otimes_A F_\bullet$ not exact. By the noetherian condition, there exists a prime $\mathfrak{q}$ maximal in this respect. For $x \in A \backslash \mathfrak{q}$, the prime ideals which occur in an appropriate filtration of $A/(\mathfrak{q},x)$ all properly contain $\mathfrak{q}$. Therefore $A/(\mathfrak{q},x) \otimes_A F_\bullet$ is an exact complex and multiplication by x yields an automorphism of $H_i(A/\mathfrak{q} \otimes_A F_\bullet)$ for all $i \geq 0$ in view of the long exact homology sequence induced by $.x$. It follows that $H_i(k(\mathfrak{q}) \otimes_{A_\mathfrak{q}} F_{\bullet\mathfrak{q}}) \simeq H_i(A/\mathfrak{q} \otimes_A F_\bullet)_\mathfrak{q} \simeq$ $\simeq H_i(A/\mathfrak{q} \otimes_A F_\bullet)$ and the latter is not 0 for some $i \geq 0$. Thus $\mathfrak{q} \in \operatorname{supp} F_\bullet$.

The advantage of the small support is that we are working with vector spaces, which shows up favourably in

7.2.8 PROPOSITION. Let A be a noetherian ring.

(i) For two flat complexes $F_\bullet$ and $G_\bullet$ as above

$$\operatorname{supp} F_\bullet \cap \operatorname{supp} G_\bullet = \operatorname{supp} F_\bullet \otimes_A G_\bullet;$$

(ii) For any A-module M there is an inclusion

$$\operatorname{supp} F_\bullet \cap \operatorname{supp} M \subset \operatorname{Supp} F_\bullet \otimes_A M,$$

and both sets have the same minimal elements;

(iii) For any two A-modules N and M there are inclusions

$$\operatorname{supp} N \cap \operatorname{supp} M \subset \cup_{i=0}^{\infty} \operatorname{Supp} \operatorname{Tor}_i^A(N,M) \subset \operatorname{Supp} N \cap \operatorname{Supp} M,$$

and the first two sets have the same minimal elements.

PROOF. (i) For $\mathfrak{p} \in \operatorname{Spec} A$ there are isomorphisms between complexes of vector spaces $(k(\mathfrak{p})\otimes_{A_\mathfrak{p}} F_{\bullet\mathfrak{p}})\otimes_{k(\mathfrak{p})}(k(\mathfrak{p})\otimes_{A_\mathfrak{p}} G_{\bullet\mathfrak{p}}) \simeq$ $\simeq k(\mathfrak{p})\otimes_{A_\mathfrak{p}}(F_{\bullet\mathfrak{p}}\otimes_{A_\mathfrak{p}} G_{\bullet\mathfrak{p}}) \simeq k(\mathfrak{p})\otimes_{A_\mathfrak{p}}(F_\bullet\otimes_A G_\bullet)_\mathfrak{p}$. The latter complex is exact if and only if at least one of the original complexes is in view of 1.2.3, so (i) is clear.

(ii) Take a flat resolution $G_\bullet$ of $M_\bullet$. Then $H_i(F_\bullet\otimes_A M) \simeq H_i(F_\bullet\otimes_A G_\bullet)$ for all $i \geq 0$ by 1.2.1. Since supp M is by definition supp $G_\bullet$, statement (ii) follows from (i) and 7.2.7.

(iii) If $F_\bullet$ and $G_\bullet$ are flat resolutions of N resp. M, then $H_i(F_\bullet\otimes_A G_\bullet) \simeq$ $\simeq \operatorname{Tor}_i^A(N,M)$ as a special case of the above, so (iii) is a particular instance of (ii) except for the last inclusion. But this is immediate because the Tor's localize well.

From this result we conclude that $\operatorname{supp} N \cap \operatorname{supp} M \neq \emptyset$ if and only if $\operatorname{tor}_-(N,M) < \infty$. Next we extend the classical notion of grade in the sense of Bartijn [Ba, Ch. IV, Def. 3.7].

7.2.9 DEFINITION. For two modules N and M over a ring A the grade of N in M is $gr_A(N,M) = \inf \text{E-dp}_{A_{\mathfrak{p}}} M_{\mathfrak{p}}$ where $\mathfrak{p}$ ranges over all the primes in supp N.

In the case of a noetherian ring, 6.1.10 tells us that $\text{E-dp}_{A_{\mathfrak{p}}} M_{\mathfrak{p}} < \infty$ if and only if $\text{T-codp}_{A_{\mathfrak{p}}} M_{\mathfrak{p}} < \infty$ and we obtain, in view of 7.2.5:

7.2.10 PROPOSITION. For modules N and M over a noetherian ring the following are equivalent:

(i) $gr(N,M) < \infty$;

(ii) $\text{supp } N \cap \text{supp } M \neq \emptyset$;

(iii) $tor_-(N,M) < \infty$.

Compare this with 7.2.1 and 7.2.2 where we needed the module N to be finitely generated. In that case supp N = Supp N = Supp $A/\mathfrak{a}$, where $\mathfrak{a}$ is Ann N, and we find, using 5.3.14, that $gr(N,M) = ext^-(N,M) = \text{E-dp}(\mathfrak{a},M)$.

Putting M = A, we write gr(N,A) = gr N, the grade of N as introduced by Rees [Re 57] for finitely generated modules. Among the several equivalent descriptions above, perhaps the most intuitive is to look at all the local rings $A_{\mathfrak{p}}$ with $\mathfrak{p} \in$ supp N and to take their minimal depth. From the Ext-description we obtain Rees' inequality:

7.2.11 COROLLARY. For every finitely generated module N over a noetherian ring, gr N ≤ pd N.

Such a module is called perfect if pd N < ∞ and gr N = pd N; for several purposes, these are easier to work with than arbitrary modules of finite projective dimension. Notice that if an ideal $\mathfrak{a} \subset A$ is generated by a regular sequence of length r, the module $A/\mathfrak{a}$ is perfect of grade r.

In order to extend Rees' inequality to infinitely generated modules, just as in 7.1.2, we require flat dimension rather than projective dimension, an insight first emphasized by Foxby. Just as with grade, however, we need to involve the small support.

7.2.12 DEFINITIONS. For a module N over a ring A put $\mathrm{fd}^+N = \sup \mathrm{fd}_{A_\mathfrak{p}}^{k(\mathfrak{p})} N_\mathfrak{p}$ and $\mathrm{fd}^-N = \inf \mathrm{fd}_{A_\mathfrak{p}}^{k(\mathfrak{p})} N_\mathfrak{p}$ where $\mathfrak{p}$ ranges over supp N.

Then $\mathrm{fd}^-N \le \mathrm{fd}^+N \le \mathrm{fd}\, N$ for $N \ne 0$. We first prove

7.2.13 PROPOSITION. Let A be a noetherian ring and $N \ne 0$ an A-module. If $\mathrm{fd}\, N < \infty$, then $\mathrm{fd}^+N = \mathrm{fd}\, N$. If in addition N is finitely generated, then $\mathrm{fd}^-N = \mathrm{gr}\, N$.

PROOF. Put $r = \mathrm{fd}\, N$, and take a prime $\mathfrak{q}$ which is maximal with respect to the condition $\mathrm{Tor}_r^A(N, A/\mathfrak{q}) \ne 0$. This is possible because Tor's commute with direct limits and a noetherian module can be filtered with subquotients of type $A/\mathfrak{p}$, where finitely many primes $\mathfrak{p}$ occur. We wish to show that $\mathrm{Tor}_r^A(N, A/\mathfrak{q})_\mathfrak{q} \ne 0$, because this module is $\mathrm{Tor}_r^{A_\mathfrak{q}}(N_\mathfrak{q}, k(\mathfrak{q}))$, which establishes $r \le \mathrm{fd}^+N$. Take $x \in A\setminus\mathfrak{q}$ and filter $A/(\mathfrak{q},x))$ as just described. For the primes $\mathfrak{p}$ occuring in the subquotients, we know that $\mathrm{Tor}_i^A(N, A/\mathfrak{p}) = 0$ for $i = r$ and $i = r+1$. This implies that $.x\colon A/\mathfrak{q} \to A/\mathfrak{q}$ induces an automorphism of $\mathrm{Tor}_r^A(N, A/\mathfrak{q})$, which proves that $\mathfrak{q} \in \mathrm{Supp}\, \mathrm{Tor}_r^A(N, A/\mathfrak{q})$.

Now suppose N is finitely generated. For every $\mathfrak{p} \in \mathrm{Supp}\, N = \mathrm{supp}\, N$ there is an identity $\mathrm{dp}_{A_\mathfrak{p}} A_\mathfrak{p} = \mathrm{pd}_{A_\mathfrak{p}} N_\mathfrak{p} + \mathrm{dp}_{A_\mathfrak{p}} M_\mathfrak{p}$ according to 7.1.5, hence $\mathrm{pd}_{A_\mathfrak{p}} N_\mathfrak{p} \le \mathrm{dp}_{A_\mathfrak{p}} A_\mathfrak{p}$. By recalling that $\mathrm{pd}_{A_\mathfrak{p}} N_\mathfrak{p} = \mathrm{fd}_{A_\mathfrak{p}} N_\mathfrak{p} = \mathrm{fd}_{A_\mathfrak{p}}^{k(\mathfrak{p})} N_\mathfrak{p}$ and taking the inf over $\mathfrak{p} \in \mathrm{Supp}\, N$ we obtain $\mathrm{fd}^-N \le \mathrm{gr}\, N$. To prove the reverse inequality, take a $\mathfrak{p} \in \mathrm{Supp}\, N$ such that $\mathrm{fd}_{A_\mathfrak{p}} N_\mathfrak{p} = \mathrm{fd}^-N$ and let $\mathfrak{q} \subset \mathfrak{p}$ be a minimal element in Supp N. Clearly $fd_{A_q}N_q = fd^{-}N$ while $dp_{A_q}N_q = 0$ so that $dp_{A_q}A_q = fd^{-}N$. Therefore $gr\ N \leq dp_{A_q}A_q = fd^{-}N$ and the proposition is proved.

In order to proceed, we need a result about flat complexes over local rings, which we shall use again in 13.1.4. If $F_{\bullet} = \ldots \to F_i \to F_{i-1} \to \ldots \to F_0 \to 0$ is a complex of flat modules over a ring A and M is an A-module, we shall write $f_+^M = \sup\{i \mid H_i(F_{\bullet}\otimes_A M) \neq 0\}$ and H for the f_+^M-th homology group of the complex $F_{\bullet}\otimes_A M$. Thus f_+^M is a natural number or $f_+^M = -\infty$.

7.2.14 THEOREM. Let $(A,\mathfrak{m},k)$ be a noetherian local ring and M an A-module. Let $F_{\bullet}$ be as above. Then

(i) If $F_{\bullet}\otimes_A k$ is not exact, then $F_{\bullet}\otimes_A M$ is not exact provided E-dp $M < \infty$;

(ii) Suppose that $F_i = 0$ for $i > s \geq 0$ and that $\mathfrak{m} \in \text{Ass}\ H$. Then $F_{\bullet}\otimes_A k$ is not exact and E-dp $M = f_+^k - f_+^M$.

PROOF. For (i) we observe that E-dp $M < \infty$ implies T-codp $M < \infty$ by 6.1.10, or in our present notation, that $\mathfrak{m} \in \text{supp}\ M$. Since also $\mathfrak{m} \in \text{supp}\ F_{\bullet}$, we know by 7.2.8 (ii) that $\mathfrak{m} \in \text{Supp}\ F_{\bullet}\otimes_A M$, which is what we need to prove.
(ii) We write $F^{\bullet}$ for the ascending complex $0 \to F^{-s} \to \ldots \to F^0 \to 0$ as in the proof of 7.1.2 and let $E^{\bullet}$ be an injective resolution of M. The tensor product complex $I^{\bullet} = F^{\bullet}\otimes_A E^{\bullet}$ consists of injectives and its homology is that of $F^{\bullet}\otimes_A M$, 1.2.4, in which C is our M. If we put $f = f_+^M$, the complex $0 \to I^{-s} \to \ldots \to I^{-f-1} \to B \to 0$ therefore is split injective where $B \subset I^{-f}$ are the boundaries. But then H is a submodule of I^{-f}, and since $\mathfrak{m} \in \text{Ass}\ H$, we find that the complex $\text{Hom}_A(k,I^{\bullet})$ is exact for $r < -f$ but has homology at $-f$. From 1.2.4 we read off $-f_+^k + \text{E-dp}\ M = -f_+^M$, which yields (ii).

7.2.15 DEFINITION. For two modules N and M over a ring A, we put $\mathrm{tor}_+(N,M) = \sup\{i \mid \mathrm{Tor}_i^A(N,M) \neq 0\}$.

Notice that $\mathrm{fd}^k N = \mathrm{tor}_+(N,k)$ in the case of a local ring $(A,\mathfrak{m},k)$. But we think of this as an invariant of the module N, while $\mathrm{tor}_+(N,M)$ is the relative flat dimension of N and M. Clearly $\mathrm{tor}_+(N,M) \leq \mathrm{fd}\, N$ and $\mathrm{fd}\, M$, and $\mathrm{tor}_+^{A_\mathfrak{p}}(N_\mathfrak{p},M_\mathfrak{p}) \leq \mathrm{tor}_+^A(N,M)$ for all primes $\mathfrak{p} \subset A$. Observe that $\mathrm{tor}_+(N,M) > -\infty$ is equivalent with $\mathrm{tor}_-(N,M) < \infty$, so, by 7.2.10, with $\mathrm{gr}(N,M) < \infty$.

In 7.1.3 we obtained two identities in case $\mathrm{fd}\, N < \infty$ and $\mathrm{tor}_+(N,M) = 0$ (there, N and M were interchanged). In case $\mathrm{tor}_+(N,M)$ is finite, we get two inequalities [Ba, Ch. IV, Th. 3.8], one of which is a variant of [Ho 74, Th. 1].

7.2.16 THEOREM. Let N and M be two modules over a noetherian ring A, with $\mathrm{fd}\, N < \infty$. Suppose $\mathrm{gr}(N,M) < \infty$. Then $\mathrm{fd}^- N \leq \mathrm{gr}(N,M) + \mathrm{tor}_+(N,M) \leq \mathrm{fd}\, N$.

PROOF. Our data imply that $\mathrm{tor}_+(N,M)$ is finite, so all four invariants in the theorem are nonnegative integers.

Take a flat resolution of N and let $F_\bullet$ be its nonaugmented complex. Put $\mathrm{tor}_+(N,M) = r \geq 0$ and take a prime $\mathfrak{p}$ which is minimal in the support of $\mathrm{Tor}_r^A(N,M)$. By localizing at $\mathfrak{p}$, we put ourselves in a position to invoke 7.2.14 (ii) and obtain that $\mathfrak{p} \in \mathrm{supp}\, N$ while $\mathrm{E\text{-}dp}_{A_\mathfrak{p}} M_\mathfrak{p} = \mathrm{fd}^{k(\mathfrak{p})}_{A_\mathfrak{p}} N_\mathfrak{p} - r$. Therefore $\mathrm{gr}(N,M) + \mathrm{tor}_+(N,M) \leq \mathrm{fd}^{k(\mathfrak{p})}_{A_\mathfrak{p}} N_\mathfrak{p} \leq \mathrm{fd}^+ N = \mathrm{fd}\, N$, the last equality is in 7.2.13.

For the other inequality, we take a prime $\mathfrak{p} \in \mathrm{supp}\, N$ for which $\mathrm{E\text{-}dp}_{A_\mathfrak{p}} M_\mathfrak{p} = \mathrm{gr}(N,M)$. Since $\mathrm{tor}_+^{A_\mathfrak{p}}(N_\mathfrak{p},M_\mathfrak{p}) \leq \mathrm{tor}_+^A(N,M)$, we are done if we can prove that $\mathrm{fd}^{k(\mathfrak{p})}_{A_\mathfrak{p}} N_\mathfrak{p} \leq \mathrm{E\text{-}dp}_{A_\mathfrak{p}} M_\mathfrak{p} + \mathrm{tor}_+^{A_\mathfrak{p}}(N_\mathfrak{p},M_\mathfrak{p})$.

We are thus reduced to the following question. If N and M are two modules over a local noetherian ring $(A,\mathfrak{m},k)$ such that fd N, $\mathrm{tor}_+(N,M)$, gr(N,M), T-codp N and E-dp M are all nonnegative integers, we need to prove that $\mathrm{fd}^k N \leq$ E-dp M + $\mathrm{tor}_+(N,M)$. We use induction on q = inf{$\mathrm{fd}^k N$, $\mathrm{tor}_+(N,M)$}.

The case q = 0 is clear: if $\mathrm{fd}^k N = 0$ the inequality is obvious, if $\mathrm{tor}_+(N,M) = 0$ it follows from the first identity in 7.1.3 (remember that M and N are interchanged). Assume the result has been proved for 0,1,...,q-1, and take an epimorphism of a free module onto N with kernel N'. Then $\mathrm{Tor}_i^A(N,M) \simeq \mathrm{Tor}_{i-1}^A(N',M)$ for $i \geq 2$. This shows that $\mathrm{tor}_+(N',M) = \mathrm{tor}_+(N,M) - 1$. Similarly for $\mathrm{fd}^k N$, and the theorem has been proved.

Observe that 7.2.16 just collapses to 7.2.11 if we take N finitely generated and M = A, in view of 7.2.13. On the other hand, for a perfect module N one obtains a version of Hochster's Tor-equality [Ho 75a, Cor. 6.9]:

7.2.17 COROLLARY. Let N be a perfect module over a noetherian ring A and suppose that M is a module with T-$\mathrm{codp}_{A_\mathfrak{p}} M_\mathfrak{p} < \infty$ for some $\mathfrak{p} \in$ Supp N. Then

gr(N,M) + $\mathrm{tor}_+(N,M)$ = pd N.

7.2.18 EXERCISE. Play around with the Matlis dual and look what you get.

7.3 ANNIHILATORS OF LOCAL COHOMOLOGY

We shall begin this section with a rather technical theorem which goes back to [Ro 76, Th. 1]. Our version is [Sc 82b, Cor. 1], see also

[Sc 82a, Satz 2.3.1], where it easily follows from an argument which uses a filtration on a spectral sequence. In fact, this spectral sequence is one of the finitely convergent spectral sequences which are associated with the double complex we shall describe below. Our proof is somewhat longer but is more elementary in the sense that it only uses diagram chasing. A corollary will serve in Chapter 10 to relate (the annihilators of) the local cohomology with dualizing complexes (when these exist). The next part of this section is devoted to a study of relations between local cohomology, Koszul complexes, Ext and Tor in case regular sequences are involved. It paves the way to the Local Duality Theorem in 10.2.1 and 10.2.7. We finish off by comparing lengths of certain modules, in which we use for real the spectral sequence alluded to above. This result too will be taken up in Chapter 10.

7.3.1 DEFINITION. If $\mathfrak{a}$ is an ideal in the noetherian ring A and if M is an A-module, let $\mathfrak{c}^i(\mathfrak{a},M)$ be the ideal Ann $H^i_{\mathfrak{a}}(M)$. If A is local with maximal ideal $\mathfrak{m}$, we shall write $\mathfrak{c}^i(M)$ for $\mathfrak{c}^i(\mathfrak{m},M)$. If moreover M = A, we use the notation $\mathfrak{c}^i$ for $\mathfrak{c}^i(A)$.

7.3.2 THEOREM. Let A be a noetherian ring, and $\mathfrak{a}$ an ideal in A. Let $C^\bullet = 0 \to C^0 \to \dots \to C^s \to 0$ be a complex of A-modules for which $\mathfrak{a} \subset \bigcap_{i=0}^s \mathfrak{R}(\mathrm{Ann}_A H^i(C^\bullet))$. Then the product ideal $\mathfrak{c}^0(\mathfrak{a},C^k)\dots\mathfrak{c}^k(\mathfrak{a},C^0)$ annihilates $H^k(C^\bullet)$ for all $0 \le k \le s$.

PROOF. Consider the double complex $C^\bullet \otimes_A K^\bullet$, where $K^\bullet = K^\bullet_\infty(x,A)$, for elements $x_1,\dots,x_n$ generating $\mathfrak{a}$, is the complex which we discussed in 4.3.2 and 4.3.5. We depict this complex as follows:

$$\begin{array}{ccccccccc}
 & & 0 & & 0 & & 0 & & \\
 & & \uparrow & & \uparrow & & \uparrow & & \\
0 & \longrightarrow & C^s \otimes_A K^0 & \longrightarrow & C^s \otimes_A K^1 & \longrightarrow \cdots \longrightarrow & C^s \otimes_A K^n & \longrightarrow & 0 \\
 & & \uparrow & & \uparrow & & \uparrow & & \\
 & & \vdots & & \vdots & & \vdots & & \\
 & & \uparrow & & \uparrow & & \uparrow & & \\
0 & \longrightarrow & C^1 \otimes_A K^0 & \longrightarrow & C^1 \otimes_A K^1 & \longrightarrow \cdots \longrightarrow & C^1 \otimes_A K^n & \longrightarrow & 0 \\
 & & \uparrow & & \uparrow & & \uparrow & & \\
0 & \longrightarrow & C^0 \otimes_A K^0 & \longrightarrow & C^0 \otimes_A K^1 & \longrightarrow \cdots \longrightarrow & C^0 \otimes_A K^n & \longrightarrow & 0 \\
 & & \uparrow & & \uparrow & & \uparrow & & \\
 & & 0 & & 0 & & 0 & &
\end{array}$$

Notice that:

1. The description of $K^\bullet$ in 4.3.5 and the fact that $\mathfrak{a} \subset \cap_{i=0}^{s} \mathfrak{R}(\mathrm{Ann}_A H^i(C^\bullet))$ yield that the vertical homology in the double complex vanishes, except perhaps in the first column: the complexes $C^\bullet \otimes_A K^i$ are exact for all $i \geq 1$.

2. Since $K^0 = A$, we find that $C^\bullet \otimes_A K^0$ is isomorphic with $C^\bullet$.

3. Corollary 4.3.6 yields that the horizontal homology $H^j(C^i \otimes_A K^\bullet)$ is in fact $H^j_{\mathfrak{a}}(C^i)$ and hence $\mathfrak{c}^j(\mathfrak{a}, C^i)$ is the annihilator of $H^j(C^i \otimes_A K^\bullet)$.

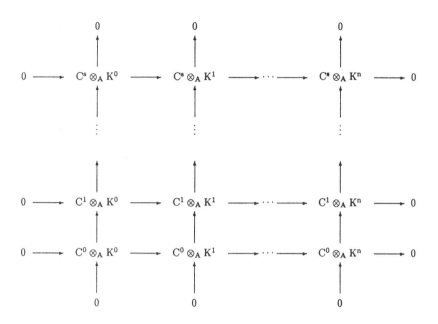

A sketch of proof now runs as follows: (i) we take a cycle in $C^\bullet \otimes_A K^0 \simeq C^\bullet$; (ii) with the use of 1., we descend to the bottom line; (iii)

then, climbing up against horizontal homology, we can use 3. to disturb the various elements effectively by boundaries; (iv) eventually we can draw the conclusion we want. The proof is visualized in the picture below.

Let us denote by d (resp. ∂) the horizontal (resp. vertical) differential of the double complex, where we shall use for convenience the convention $\partial d = d\partial$ (the reader may take his or her own favourite convention and adapt the proof). Let $u_k \in C^k \otimes_A K^0$ with $\partial u_k = 0$. Using 1., one easily finds $u_{k-1}, \ldots, u_0$, where $u_i \in C^i \otimes_A K^{k-i}$, $0 \leq i \leq k-1$, such that $du_{i+1} = \partial u_i$, $0 \leq i \leq k-1$:

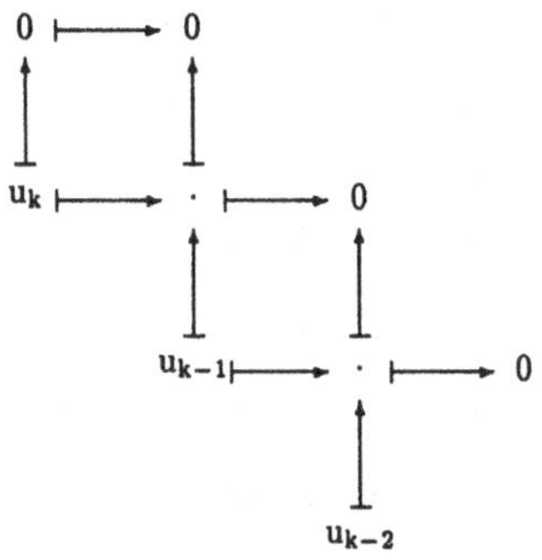

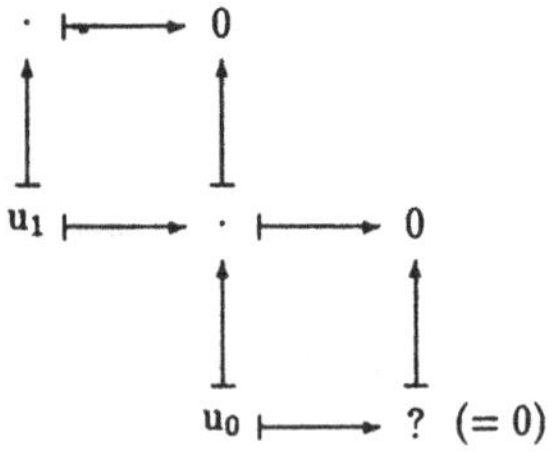

As is often the case, ascending is trickier than descending. We shall describe this process now. Let $c_i \in \mathfrak{c}^i(\mathfrak{a}, C^{k-i})$, $0 \leq i \leq k$. First, observe that ∂: $C^0 \otimes_A K^{k+1} \to C^1 \otimes_A K^{k+1}$ is monic and hence the question mark in the diagram is zero. It follows that u_0 is a horizontal cycle and therefore there exists a $w_0 \in C^0 \otimes_A K^{k-1}$ such that $dw_0 = c_k u_0$. We shall proceed by induction on i, $0 \le i \le k-1$, to find $w_i \in C^i \otimes_A K^{k-i-1}$ such that $d(\partial w_i - c_{k-i} \ldots c_k u_{i+1}) = 0$. For w_0, notice that $d(\partial w_0 - c_k u_1) = d\partial w_0 - c_k du_1 = \partial dw_0 - c_k \partial u_0 = \partial(c_k u_0) - c_k \partial u_0 = 0$. Assume now that we have constructed w_i for some $0 \le i < k-1$. Since $d(\partial w_i - c_{k-i} \ldots c_k u_{i+1}) = 0$, we can find w_{i+1} such that $dw_{i+1} = -c_{k-i-1}(\partial w_i - c_{k-i} \ldots c_k u_{i+1})$. It follows that $d(\partial w_{i+1} - c_{k-i-1} \ldots c_k u_{i+2}) = d\partial w_{i+1} - c_{k-i-1} \ldots c_k \partial u_{i+1} = d\partial w_{i+1} - \partial(dw_{i+1} + c_{k-i-1}\partial w_i) = -c_{k-i-1}\partial\partial w_i = 0$ and we are done.

We therefore have a $w_{k-1} \in C^{k-1} \otimes_A K^0$ such that $d(\partial w_{k-1} - c_1 \ldots c_k u_k) = 0$ and consequently $c_0(\partial w_{k-1} - c_1 \ldots c_k u_k) = 0$. We see that $c_0 \ldots c_k u_k = \partial(c_0 w_{k-1})$ is a boundary and hence zero in vertical homology.

7.3.3 COROLLARY. Let A be a noetherian local ring and let $L^\bullet = 0 \to L^0 \to \ldots \to L^s \to 0$ be a complex of finitely generated free A-modules with finite length homology. Then the ideal $\mathfrak{c}^0 \ldots \mathfrak{c}^k$ annihilates $H^k(L^\bullet)$ for all $0 \le k \le s$.

PROOF. Obviously, the theorem applies and we are done if we can prove that $\mathfrak{c}^i = \mathfrak{c}^i(L)$ for a finitely generated free module L. This is clear since local cohomology commutes with direct sums.

We next want to study the above complex $K^\bullet$ in case x is a regular sequence in A. Our approach is based on [Matl 74] and [Matl 78]. Let us use the notational conventions of Chapters 4 and 5. If $x = x_1, \ldots, x_n$ in A generate the ideal $\mathfrak{a}$, then $x(t)$ stands for the sequence $x_1^t, \ldots, x_n^t$ and $\mathfrak{a}_t$ for the ideal they generate.

7.3.4 THEOREM. Let $x = x_1, \ldots, x_n$ be a regular sequence in a noetherian ring A. Then $K_\bullet(x(t),A)$ is a free resolution of $A/\mathfrak{a}_t$ for every $t \geq 1$ and $K_\bullet^\infty(x,A)$ is a flat resolution of $H^n_\mathfrak{a}(A)$. Furthermore $\mathrm{Ext}^i_A(A/\mathfrak{a}_t,M) \simeq \mathrm{Tor}^A_{n-i}(A/\mathfrak{a}_t,M)$ for all i and every module M, while $H^i_\mathfrak{a}(M) \simeq \mathrm{Tor}^A_{n-i}(H^n_\mathfrak{a}(A),M)$. Finally $\mathrm{fd}\ H^n_\mathfrak{a}(A) = n$.

PROOF. For every $t \geq 1$ the sequence $x(t)$ is regular, so $H_i(x(t),A) = 0$ for $i \geq 1$, while $H_0(x(t),A) = A/\mathfrak{a}_t$, which proves the first statement. Self duality 4.2.8 of the Koszul complex then yields the isomorphisms between Ext's and Tor's. Now construct the complex $K_\bullet^\infty(x,A)$ as the direct limit of the $K_\bullet(x(t),A)$'s, see 4.3.1. Its chain modules are flat, and since taking direct limits is exact, its homology is concentrated in degree zero, where it is $H^n_\mathfrak{a}(A)$. Tensoring this flat resolution with M we get the complex $K_\bullet^\infty(x,M)$ where we read off homology. In view of 4.3.6 we obtain $H^i_\mathfrak{a}(M) \simeq \mathrm{Tor}^A_{n-i}(H^n_\mathfrak{a}(A),M)$. From its flat resolution we know that $\mathrm{fd}\ H^n_\mathfrak{a}(A) \leq n$, but taking $M = A/\mathfrak{a}$ in the above isomorphism we find $\mathrm{Tor}^A_n(H^n_\mathfrak{a}(A),A/\mathfrak{a}) \simeq H^0_\mathfrak{a}(A/\mathfrak{a}) = A/\mathfrak{a} \neq 0$, and we have proved the theorem.

7.3.5 COROLLARY. Let $\mathfrak{a}$ be an ideal in the noetherian ring A which is generated by a regular sequence in A of length n. Then $\mathfrak{c}^i(\mathfrak{a},A) = A$ for all $0 \leq i \leq n-1$, i.e. $H^0_\mathfrak{a}(A) = \ldots = H^{n-1}_\mathfrak{a}(A) = 0$.

PROOF. Immediate from the isomorphisms $H^i_\mathfrak{a}(A) \simeq \mathrm{Tor}^A_{n-i}(H^n_\mathfrak{a}(A),A)$.

7.3.6 COROLLARY. Let A be a noetherian local ring with maximal ideal m and let $L^\bullet = 0 \to L^0 \to \ldots \to L^s \to 0$ be a complex of finitely generated free A-modules with finite length homology. Then $H^i(L^\bullet) = 0$ for all $i < \mathrm{dp}\ A$.

PROOF. Let $n = \mathrm{dp}\ A$ and let $x = x_1,\ldots,x_n \in A$ be a regular sequence in A. If $\mathfrak{a}$ is the ideal generated by the x's, obviously

$\mathfrak{a} \subset \cap_{i=0}^{s} \mathfrak{R}(\mathrm{Ann}_A(H^i(L^\bullet))) = \mathfrak{m}$. We have already observed that if L is a finitely generated free A-module, then $\mathfrak{c}^i(\mathfrak{a},L) = \mathfrak{c}^i(\mathfrak{a},A)$. Therefore, by 7.3.3 and 7.3.5, we see that $A = \mathfrak{c}^0(\mathfrak{a},A)\ldots\mathfrak{c}^i(\mathfrak{a},A)$ annihilates $H^i(L^\bullet)$ for all $i \le n-1$. The result follows.

7.3.7 EXERCISES.

1. Show that in 7.3.2 and therefore also in 7.3.6 we may replace the last 0 in the complex by an arbitrary module and that the conclusion remains valid.

2. Obtain a stronger version of 7.3.6, with the condition that "either $H_i(L^\bullet) = 0$ or dp $H_i(L^\bullet) = 0$" replacing finite length for each i, as an immediate corollary of the Acyclicity Lemma 6.1.1.

3. Let $x_1,\ldots,x_r$ generate an ideal $\mathfrak{a}$ in a noetherian local ring A. Suppose dp $A_{\mathfrak{p}} \ge r$ for every prime $\mathfrak{p} \supset \mathfrak{a}$. Use 1. to prove that $H_i(x,A) = 0$ for $i \ge 1$ in the Koszul complex of $x = x_1,\ldots,x_r$.

We now return to the double complex $C^\bullet \otimes_A K^\bullet$ discussed in 7.3.2, replacing $C^\bullet$ by the complex $L^\bullet$ from 7.3.6. We shall not depict the double complex this time, but shall allow ourselves to use the theory of spectral sequences in a mild way, for which we refer e.g. to [CE, Ch. XIV]. The following result will be needed in 11.3.9. For clarity we write H_v for vertical and H_h for horizontal homology.

7.3.8 PROPOSITION. Let $(A,\mathfrak{m})$ and $L^\bullet$ be as in the preceding corollary. Then $\ell(H^t(L^\bullet)) \le \Sigma_{i+j=t}\, \ell(H_v^i(L^\bullet \otimes_A H_{\mathfrak{m}}^j(A)))$.

PROOF. Let $\mathfrak{a} = (x_1,\ldots,x_n)$ be an $\mathfrak{m}$-primary ideal and let $K^\bullet = K_\infty^\bullet(x,A)$ with $x = x_1,\ldots,x_n$ be the complex discussed in 7.3.2. The homology of the total complex $L^{\bullet}\otimes_A K^{\bullet}$ can be approximated by the terms E_r of the first and the second spectral sequence associated with it. The term E_2 of the second spectral sequence is given by $''E^{ji} = H_h^i(H_v^j(L^{\bullet}\otimes_A K^{\bullet}))$. We have seen in 7.3.2 that the columns of the double complex are exact for $j \geq 1$, so $''E_2^{ji} = 0$ for $j \geq 1$, which means that this spectral sequence degenerates and that the total homology $H(L^{\bullet}\otimes_A K^{\bullet})$ is measured by this term E_2: $H^t(L^{\bullet}\otimes_A K^{\bullet}) \simeq \simeq \oplus_{i+j=t}\ ''E_2^{ji} = {''E_2^{0t}} = H^t(L^{\bullet})$.

Now we calculate the term E_2 of the first spectral sequence: $'E_2^{ij} = H_v^i(H_h^j(L^{\bullet}\otimes_A K^{\bullet})) \simeq H_v^i(L^{\bullet}\otimes_A H_{\mathfrak{m}}^j(A))$ in view of the freeness of L^j and the fact that local cohomology only depends on the radical of the ideal $\mathfrak{a}$. Clearly if any of the terms $'E_2^{ij}$ has infinite length, the result is true, so assume not. The point of spectral sequences being that the i,j-th layer of $H^t(L^{\bullet}\otimes_A K^{\bullet})$ is successively approximated in a finite number of steps by subquotients of the $'E_t^{ij}$ with $i+j = t$, the inequality follows.

8 DIMENSION

In this chapter our rings will mostly be noetherian and the problems local. We begin with the basic theorem in noetherian dimension theory: the Principal Ideal Theorem of Krull, or in pithy German, the Hauptidealsatz. Later on in this chapter we discuss two natural generalizations, the Eisenbud-Evans-Bruns result on heights of order ideals, and the Homological Height Theorem of Hochster. The latter has several consequences which, in turn, suggest fresh questions. First though we introduce systems of parameters and develop some dimension theory. Parameters are compared to regular sequences or heights to depths, and this gives rise to a natural question concerning parameters: Hochster's Monomial Conjecture. In this work we use the theory developed in Chapter 5, sections 1 and 2. As part of the material in this chapter is well known and can be found in most books on commutative algebra, we state a number of standard facts in the form of excercises. At other points our treatment may present distinctive features.

8.1 KRULL'S HAUPTIDEALSATZ

We shall first introduce the Krull dimension.

8.1.1 DEFINITIONS. Let A be a ring and W a subset of Spec A. The Krull dimension of W (notation dim W) is defined to be the supremum of the lengths of all chains of prime ideals $\mathfrak{p}_0 \subsetneq \dots \subsetneq \mathfrak{p}_n$, where the $\mathfrak{p}_i$, $i = 0,\dots,n$, are in W and the length of such a chain is counted as n. If this supremum is finite it is of course achieved in a saturated chain: between two successive primes no third prime ideal can be inserted. In this way we can attach dimensions to rings and modules by putting dim A = dim(Spec A) and $\dim_A M = \dim(\mathrm{Supp}_A M)$ for the ring A and its module M. If the ring is not in doubt, we also write dim M and for M = (0) we set dim M = -1. The height of a prime ideal $\mathfrak{p}$ is defined as $\mathrm{ht}_A \mathfrak{p} = \dim A_{\mathfrak{p}}$; it is the supremum of the lengths of chains of prime ideals going down from $\mathfrak{p}$. Again we often write ht $\mathfrak{p}$. The height of an arbitrary ideal $\mathfrak{a}$ is then defined as $\mathrm{ht}\,\mathfrak{a} = \inf\{\mathrm{ht}\,\mathfrak{p} \mid \mathfrak{p} \in \mathrm{Spec}\,A,\ \mathfrak{p} \supset \mathfrak{a}\}$.

The following list of properties is easy to verify.

8.1.2 EXERCISE.

a) $\dim A = \sup\{\mathrm{ht}\,\mathfrak{p} \mid \mathfrak{p} \in \mathrm{Spec}\,A\} = \sup\{\mathrm{ht}\,\mathfrak{m} \mid \mathfrak{m} \text{ is a maximal ideal in } A\}$.

b) $\dim A/\mathfrak{a} = \sup\{\dim A/\mathfrak{p} \mid \mathfrak{p} \in \mathrm{Spec}\,A,\ \mathfrak{p} \supset \mathfrak{a}\}$ for every ideal $\mathfrak{a}$ in A.

c) $\mathrm{ht}\,\mathfrak{p} + \dim A/\mathfrak{p} \le \dim A$ for every prime ideal $\mathfrak{p}$ in A.

d) $\dim_A M = \sup\{\dim_{A_{\mathfrak{p}}} M_{\mathfrak{p}} \mid \mathfrak{p} \in \mathrm{Supp}_A M\}$ for every A-module $M \neq (0)$.

e) For A noetherian and M and N finitely generated A-modules, it is well known that $\mathrm{Supp}_A M = \{\mathfrak{p} \in \mathrm{Spec}\,A \mid \mathrm{Ann}\,M \subset \mathfrak{p}\}$ and $\mathrm{Supp}(M \otimes_A N) = \mathrm{Supp}\,M \cap \mathrm{Supp}\,N$. Hence $\dim M = \dim(A/\mathrm{Ann}\,M)$ and $\dim(M \otimes_A N) = \dim(A/(\mathrm{Ann}\,M + \mathrm{Ann}\,N))$. In particular, if $\mathfrak{a}$ is an ideal in A, then $\dim M/\mathfrak{a}M = \dim(A/(\mathfrak{a} + \mathrm{Ann}\,M))$.

We state Krull's celebrated Hauptidealsatz. The best known proof is the one by Rees [Re 56]; see also [Ka 74, Th. 142]. The argument was further shortened by Caruth [Ca]; it is his proof we shall present.

8.1.3 THEOREM. Let $x \neq 0$ be a non unit in a noetherian ring A. Then ht $\mathfrak{p} \leq 1$ for every prime ideal $\mathfrak{p}$ which is minimal over (x). This height equals 1 if x is a non zerodivisor.

PROOF. The last statement follows from the first since a non zerodivisor does not lie in any minimal prime.

Suppose that ht $\mathfrak{p} > 1$, then there exist prime ideals $\mathfrak{p}_0 \subsetneq \mathfrak{p}_1 \subsetneq \mathfrak{p}$ where $\mathfrak{p}$ is minimal over (x). After dividing out $\mathfrak{p}_0$ and localizing at $\mathfrak{p}$, we are reduced to a noetherian local domain A with maximal ideal $\mathfrak{m}$, a chain of prime ideals $(0) \subsetneq \mathfrak{q} \subsetneq \mathfrak{m}$, and containing an element $x \neq 0$ such that (x) is an $\mathfrak{m}$-primary ideal. Take $y \neq 0$ in $\mathfrak{q}$. For any $a \in ((y):x^n)$ there is a unique $b \in A$ such that $x^n a = by$. This $b \in ((x^n):y)$, and the correspondence $a \mapsto b$ is easily seen to describe an isomorphism $((y):x^n)/(y) \simeq ((x^n):y)/(x^n)$. Since (x) is an $\mathfrak{m}$-primary ideal, the module $A/(x^n)$ has finite length. Now the length function is additive on exact sequences, so comparing cokernel and kernel of the endomorphism $.y\colon A/(x^n) \to A/(x^n)$ we find that $\ell(A/(x^n,y)) = \ell(((x^n):y)/(x^n)) = \ell(((y):x^n)/(y))$. The ascending chain of ideals $(y):x^n$ stabilizes, hence $(x^n,y) = (x^{n+1},y)$ for n large enough. An equation $x^n = ax^{n+1}+by$ yields $x^n(1-ax) \in (y)$. But $1-ax$ is a unit, so $x^n \in (y)$. This implies $x \in \mathfrak{q}$, a contradiction which proves the theorem.

Recall that a chain of prime ideals $\mathfrak{p}_0 \subsetneq \cdots \subsetneq \mathfrak{p}_n$ in A is called saturated if no other primes can be inserted. An easy generalization of the above theorem reads as follows: in a noetherian ring, let $\mathfrak{p}$ be a minimal prime over the ideal $(x_1,\ldots,x_n)$. Then there exists a saturated chain of prime ideal $\mathfrak{p}_0 \subsetneq \cdots \subsetneq \mathfrak{p}_m$ with $\mathfrak{p}_m = \mathfrak{p}$, $\mathfrak{p}_0$ a minimal prime of A and $m \leq n$. But we can say more: the generalized Hauptidealsatz.

8.1.4 THEOREM. Let $x_1,\dots,x_n$ be elements of a noetherian ring A such that $(x_1,\dots,x_n) \neq A$. Then ht $\mathfrak{p} \le n$ for every prime ideal $\mathfrak{p}$ minimal over $(x_1,\dots,x_n)$. If $x_1,\dots,x_n$ is a regular sequence, then ht $\mathfrak{p} = n$.

PROOF. Proof by induction on n, n = 1 being 8.1.3. By localizing at $\mathfrak{p}$ we reduce to the following question: in a noetherian local ring $(A,\mathfrak{m})$ let elements $x_1,\dots,x_n$ generate an $\mathfrak{m}$-primary ideal, then dim A = ht $\mathfrak{m} \le n$. Suppose that ht $\mathfrak{m} > n$ and let $\mathfrak{p}$ be a submaximal prime ideal with ht $\mathfrak{p} > n-1$. This ideal does not contain $(x_1,\dots,x_n)$ which is $\mathfrak{m}$-primary, so suppose $x_1 \notin \mathfrak{p}$. Then $\mathfrak{p} + (x_1)$ is $\mathfrak{m}$-primary, hence there exist $a_i \in A$, $b_i \in \mathfrak{p}$, $i = 2,\dots,n$, and a positive integer t such that $x_i^t = a_ix_1 + b_i$. Now $(b_2,\dots,b_n) \subset \mathfrak{p}$ and ht $\mathfrak{p} > n-1$, so by induction there exists a prime $\mathfrak{q}$ with $(b_2,\dots,b_n) \subset \mathfrak{q} \subsetneq \mathfrak{p}$. Obviously $x_1, x_2^t,\dots,x_n^t$ are in $\mathfrak{q} + (x_1)$, so the latter ideal is $\mathfrak{m}$-primary. Applying 8.1.3 to the local ring $\bar{A} = A/\mathfrak{q}$ with maximal ideal $\bar{\mathfrak{m}}$, we see that dim $\bar{A} = \mathrm{ht}_{\bar{A}}\, \bar{\mathfrak{m}} \le 1$, which contradicts $\mathfrak{q} \subsetneq \mathfrak{p} \subsetneq \mathfrak{m}$.

Now assume $x_1,\dots,x_n$ is a regular sequence in our local ring $(A,\mathfrak{m})$ and ht $\mathfrak{m} \le n$ as we have just seen. By induction, all minimal primes over $(x_1,\dots,x_{n-1})$ have height n-1, but x_n is not in a single one of them. Therefore ht $\mathfrak{m} = n$.

An immediate consequence of the theorem is that in a noetherian ring, the height of a prime ideal (and hence of any ideal) is finite, i.e. there is a bound to the length of chains of prime ideals going down from any given prime ideal. Thus a noetherian local ring has finite dimension; examples show that this need not be true for a noetherian ring with infinitely many maximal ideals.

8.2 PARAMETERS AND DIMENSION

One may ask oneself whether for a prime ideal $\mathfrak{p}$ of height n there always exist elements $x_1,\ldots,x_n$ such that $\mathfrak{p}$ is minimal over $(x_1,\ldots,x_n)$. Of course, this is a local question and reduces to proving that in an n-dimensional local ring $(A,\mathfrak{m})$ there exist an $\mathfrak{m}$-primary ideal $(x_1,\ldots,x_n)$. This is true, but we can say more as we shall see.

8.2.1 DEFINITIONS. Let $(A,\mathfrak{m})$ be a noetherian local ring, M a finitely generated A-module. Put $d(M) = \inf\{n \mid$ there exist $x_1,\ldots,x_n \in \mathfrak{m}$ with $\operatorname{Supp} M/(x_1,\ldots,x_n)M = \{\mathfrak{m}\}\}$. A set of elements $x_1,\ldots,x_d$ with $d = d(M)$ such that $\operatorname{Supp} M/(x_1,\ldots,x_d)M = \{\mathfrak{m}\}$ is called a system of parameters of or for M; we usually abbreviate to s.o.p. for M.

One can also introduce s.o.p.'s in semi-local rings, as in [Se, Ch. III, B-3]. Another way of looking at parameters follows from the identity $\operatorname{Supp} M/\mathfrak{a}M = \operatorname{Supp}(A/\mathfrak{a} + \operatorname{Ann} M)$ for any ideal $\mathfrak{a}$, see 8.1.2 e). Thus a s.o.p. for M is a shortest sequence of elements $x_1,\ldots,x_d$ such that $(x_1,\ldots,x_d) + \operatorname{Ann} M$ is an $\mathfrak{m}$-primary ideal. Bearing this in mind, the reader will have no trouble with the useful

8.2.2 EXERCISE. For a finitely generated module M over a noetherian local ring A, the following are equivalent.

(i) $x_1,\ldots,x_d$ is a s.o.p. for M;

(ii) There exists $t \geq 1$ such that $x_1^t,\ldots,x_d^t$ is a s.o.p. for M;

(iii) For all $\nu_i \geq 1$, $i = 1,\ldots,d$, the sequence $x_1^{\nu_1},\ldots,x_d^{\nu_d}$ is a s.o.p. for M;

(iv) $x_1,\ldots,x_d$ is a s.o.p. for the A-module A/Ann M;

(v) The images $\bar{x}_1,\ldots,\bar{x}_d$ in A/Ann M are a s.o.p. in that ring.

We now connect parameters and dimension.

8.2.3 THEOREM. Let $M \neq (0)$ be a finitely generated module over a noetherian local ring $(A,\mathfrak{m})$. Then $\dim M = d(M)$.

PROOF. Put $\bar{A} = A/\mathrm{Ann}\, M$. We know that $\dim_A M = \dim \bar{A}$ and $d_A(M) = d_{\bar{A}}(\bar{A})$ by 8.1.2 e) and 8.2.2 respectively, so it is enough to prove the statement for $M = A$.

Let $n = \dim A$, and $d = d(A)$. Then if $x_1,\ldots,x_d$ is a s.o.p., the ideal $(x_1,\ldots,x_d)$ is $\mathfrak{m}$-primary, so $n = \mathrm{ht}\,\mathfrak{m} \leq d$ by 8.1.4. Conversely, we shall construct $x_1,\ldots,x_n \in \mathfrak{m}$ such that for $i = 0,\ldots,n$ all prime ideals $\mathfrak{p} \supset (x_1,\ldots,x_i)$ satisfy $\mathrm{ht}\,\mathfrak{p} \geq i$. For then $\mathfrak{m}$ is the only prime containing $(x_1,\ldots,x_n)$, which must then be $\mathfrak{m}$-primary. This will establish the other inequality $d \leq n$.

The statement being trivial for $i = 0$, assume we have constructed $x_1,\ldots,x_{i-1}$ with the above property. Let $\mathfrak{p}_1,\ldots,\mathfrak{p}_s$ be the prime ideals minimal over $(x_1,\ldots,x_{i-1})$, all of which have exactly height $i\text{-}1 < n$. Thus $\mathfrak{m}$ properly contains $\cup_{j=1}^{s} \mathfrak{p}_j$ and we can choose an $x_i \in \mathfrak{m}$ which is in none of the $\mathfrak{p}_1,\ldots,\mathfrak{p}_s$. Now any prime $\mathfrak{p} \supset (x_1,\ldots,x_i)$ contains a prime minimal over $(x_1,\ldots,x_{i-1})$, say $\mathfrak{p}_j$, and the inclusion is strict because $x_i \notin \mathfrak{p}_j$. Then $\mathrm{ht}\,\mathfrak{p} > \mathrm{ht}\,\mathfrak{p}_j = i\text{-}1$, which finishes the proof.

The above dimension theorem can also be proved without using 8.1.4. This generalized Hauptidealsatz, the reader should note, is in turn an immediate consequence of 8.2.3. Such an alternative approach brings in characteristic polynomials and can be found for instance in [AM, Ch. 11] and [Se, Ch. III, B]. It offers certain advantages, but we have opted for the "prime avoidance" argument in the above proof, which in one form or another is frequently used in dimension theory. It is worth observing that in this proof we have actually constructed a system of parameters $x = x_1,\ldots,x_d$ for the module M, with the property that every prime, minimal above $(x_1,\ldots,x_i)$, $1 \le i \le d$, has height i. We might call this a "strong" system of parameters, since not every s.o.p. has this property, as easy examples show [Ma 86, §14]. By exercising a little more care, we might even have achieved this extra condition for every permutation $y = y_1,\ldots,y_d$ of the x's, and we may call this a balanced strong s.o.p. In order to further familiarize himself with the notion of parameters, we invite the reader to prove the following fact, which we shall not need: every s.o.p. $x_1,\ldots,x_d$ has the property that, for $1 \le r \le d$, at least one minimal prime $\mathfrak{p}$ of $(x_1,\ldots,x_r)$ has height r, and $\dim A/\mathfrak{p} = d-r$. Another refinement of this kind is best described using the notation of 5.1.14. Let $x = x_1,\ldots,x_d$ be an arbitrary s.o.p. Then there exists a matrix $T \in E(d,A)$ such that $T\underline{x} = \underline{y}$ where y is a balanced strong s.o.p. The proof requires a rather simpler version of the induction used in the proof of Theorem 5.3.18.

In any case, a nonnull finitely generated module over a noetherian local ring possesses s.o.p.'s and their length is always equal to the dimension of the module. As a corollary, we obtain a further relation between dimension and s.o.p.'s.

8.2.4 COROLLARY. Let (A,m) be a noetherian local ring and $x_1,\ldots,x_n \in m$. If $M \neq (0)$ is a finitely generated module, then $\dim M/(x_1,\ldots,x_n)M + n \ge \dim M$. Equality holds if and only if $x_1,\ldots,x_n$ are part of a s.o.p. for M.

PROOF. Let $\overline{M} = M/(x_1,\ldots,x_n)M$ and let $y_1,\ldots,y_s$ be a s.o.p. for $\overline{M}$. Then m is the only element in the support of $\overline{M}/(y_1,\ldots,y_s)\overline{M} \simeq M/(x_1,\ldots,x_n,y_1,\ldots,y_s)M$ and hence $\dim M = d(M) \le n+s = n + \dim \overline{M}$. If equality holds, the x's and y's together form a s.o.p. for M. Conversely, if $x_1,\ldots,x_n$ is part of a s.o.p. $x_1,\ldots,x_n,x_{n+1},\ldots,x_d$ for M, then m is the only element in the support of $M/(x_1,\ldots,x_d)M \simeq \overline{M}/(x_{n+1},\ldots,x_d)\overline{M}$ and hence $d-n \ge \dim \overline{M}$. It follows that $\dim M = d \ge n + \dim \overline{M}$ and so there is equality.

Our next result concerns parameter ideals, i.e. ideals generated by systems of parameters. It will be used in 9.1.1 and is copied from [BS, Prop. 1.6] and [Ba, Ch. III, Lemma 3.4].

8.2.5 PROPOSITION. Let A be a d-dimensional noetherian local ring and $(x_1,\ldots,x_d)$ and $(y_1,\ldots,y_d)$ parameter ideals. Then there exists a parameter ideal $(z_1,\ldots,z_d)$ and positive integers $r_1,\ldots,r_d$ such that $(x_1^{r_1},\ldots,x_d^{r_d}) \subset$ $\subset (z_1,x_2^{r_2},\ldots,x_d^{r_d}) \subset \ldots \subset (z_1,\ldots,z_{d-1},x_d^{r_d}) \subset (z_1,\ldots,z_d) \subset (z_1,\ldots,z_{d-1},y_d)$ $\subset \ldots \subset (z_1,y_2,\ldots,y_d) \subset (y_1,\ldots,y_d)$ is a chain of parameter ideals.

PROOF. We first construct a s.o.p. $w_1,\ldots,w_d$ such that $w_1,\ldots,w_i,x_{i+1},\ldots,x_d$ and $w_1,\ldots,w_i,y_{i+1},\ldots,y_d$ are s.o.p.'s for $0 \le i \le d$. Since the case $i = 0$ is trivial, assume we have $w_1,\ldots,w_{i-1}$ with the desired property. The rings $A/(w_1,\ldots,w_{i-1},x_{i+1},\ldots,x_d)$ and $A/(w_1,\ldots,w_{i-1},y_{i+1},\ldots,y_d)$ are one-dimensional by 8.2.4, and therefore the minimal primes of $(w_1,\ldots,w_{i-1},x_{i+1},\ldots,x_d)$ and $(w_1,\ldots,w_{i-1},y_{i+1},\ldots,y_d)$ are all submaximal. Since there are finitely many of these, there exists a $w_i \in m$ which is not in any of them. This w_i fills our requirements, and induction finishes the construction of the w's.

Now we can find exponents s_i such that $w_1^{s_1} \in (y_1,\ldots,y_d)$ and successively $w_i^{s_i} \in (w_1^{s_1},\ldots,w_{i-1}^{s_{i-1}},y_i,\ldots,y_d)$, $i = 1,\ldots d$, since we are dealing exclusively with $\mathfrak{m}$-primary ideals. Setting $z_i = w_i^{s_i}$ we have constructed the chain going up from $(z_1,\ldots,z_d)$. Next observe that $(z_1,\ldots,z_i,x_{i+1}^{r_{i+1}},\ldots,x_d^{r_d})$ are also parameter ideals for $i = 1,\ldots,d$ and arbitrary positive exponents $r_{i+1},\ldots,r_d$. Choose an exponent r_d such $x_d^{r_d} \in (z_1,\ldots,z_d)$ and then successively r_i such that $x_i^{r_i} \in (z_1,\ldots,z_i,x_{i+1}^{r_{i+1}},\ldots,x_d^{r_d})$. This finishes the proof. Notice that we have again used prime avoidance.

We now investigate relations between the dimensions of a ring and certain extension rings, in particular integral and polynomial extensions. Although these facts are well known, we like to have them available in the form in which we shall use them.

8.2.6 PROPOSITION. Let $f: A \to B$ be a faithfully flat ring extension and let $\mathfrak{p} \subset A$ be a prime ideal. Then there exists a prime $\mathfrak{P} \subset B$ with $\mathfrak{p} = f^{-1}(\mathfrak{P})$. If moreover there is a prime $\mathfrak{q} \subset \mathfrak{p}$, then there exists a prime ideal $\mathfrak{Q} \subset \mathfrak{P}$ in B with $\mathfrak{q} = f^{-1}(\mathfrak{Q})$.

PROOF. If we can prove that $\mathfrak{P} \mapsto f^{-1}(\mathfrak{P})$ describes a surjective map from Spec B to Spec A, we are done for then we can find a prime $\mathfrak{P} \subset B$ with $\mathfrak{p} = f^{-1}(\mathfrak{P})$ and if we apply this surjectivity to the faithfully flat extension $\bar{f}: A_\mathfrak{p} \to B_\mathfrak{P}$, we see that there exists a prime $\bar{\mathfrak{Q}} \subset B_\mathfrak{P}$ with $f^{-1}(\bar{\mathfrak{Q}}) = \mathfrak{q}A_\mathfrak{p}$. Contracting $\bar{\mathfrak{Q}}$ to B gives us the prime $\mathfrak{Q}$ we wanted. For the surjectivity, notice that $f_\mathfrak{p}: A_\mathfrak{p} \to B_\mathfrak{p}$ is faithfully flat, hence $\mathfrak{p}B_\mathfrak{p} \neq B_\mathfrak{p}$, so we can find a maximal ideal $\mathfrak{M} \subset B_\mathfrak{p}$ with $\mathfrak{p}B_\mathfrak{p} \subset \mathfrak{M}$. It is clear that then $f_\mathfrak{p}^{-1}(\mathfrak{M}) = \mathfrak{p}A_\mathfrak{p}$, and contracting $\mathfrak{M}$ to B we get the prime $\mathfrak{P}$ with $f^{-1}(\mathfrak{P}) = \mathfrak{p}$.

8.2.7 THEOREM. Let $f: A \to B$ be a homomorphism of noetherian rings. Let $\mathfrak{P} \subset B$ be a prime ideal and $\mathfrak{p} = f^{-1}(\mathfrak{P})$. Then $ht_B\mathfrak{P} \le ht_A\mathfrak{p} + ht_{B/\mathfrak{p}B}\mathfrak{P}/\mathfrak{p}B$ and equality holds if f is a flat extension.

PROOF. Replacing A and B by $A_{\mathfrak{p}}$ and $B_{\mathfrak{P}}$, we may assume that f is a homomorphism of local rings $(A,m) \to (B,\mathfrak{M})$ and that $f^{-1}(\mathfrak{M}) = m$. If the original f is flat, the new one is even faithfully flat since the homomorphism is local. For the first statement we must prove that $\dim B \le \dim A + \dim B/mB$. Let $\alpha = (x_1,..,x_d)$ be a parameter ideal in A, then $m^t \subset \alpha \subset m$ for some $t > 0$, hence $m^tB \subset \alpha B \subset mB$. It follows that $\mathfrak{R}(\alpha B) = \mathfrak{R}(mB)$, where $\mathfrak{R}$ stands for the radical of an ideal, which implies that $\dim B/\alpha B = \dim B/mB$. Now let $\bar{y}_1,\dots,\bar{y}_s$ be a s.o.p. for $B/\alpha B$. Then $x_1,\dots,x_d,y_1,\dots,y_s$ generate an $\mathfrak{M}$-primary ideal in B so $\dim B \le d+s = \dim A + \dim B/mB$.

Assuming now that f is faithfully flat, let $s = \dim B/mB$ and let $\mathfrak{Q}_0 \subsetneq \cdots \subsetneq \mathfrak{Q}_s = \mathfrak{M}$ be a chain of prime ideals in B such that $mB \subset \mathfrak{Q}_0$. Then $m = f^{-1}(\mathfrak{M}) \supset f^{-1}(\mathfrak{Q}_i) \supset m$ and hence $f^{-1}(\mathfrak{Q}_i) = m$ for $i = 0,\dots,s$. If $\mathfrak{p}_0 \subsetneq \cdots \subsetneq \mathfrak{p}_d = m$ is a chain of prime ideals in A, we can apply 8.2.6 to find a chain of distinct prime ideals $\mathfrak{Q}_0 \supsetneq \mathfrak{P}_{d-1} \supsetneq \cdots \supsetneq \mathfrak{P}_0$ in B with $f^{-1}(\mathfrak{P}_i) = \mathfrak{p}_i$. Hence $\mathfrak{P}_0 \subsetneq \cdots \subsetneq \mathfrak{P}_{d-1} \subsetneq \mathfrak{Q}_0 \subsetneq \cdots \subsetneq \mathfrak{Q}_s$ is a chain of distinct prime ideals in B and it follows that $\dim B \ge d+s$, which finishes the proof.

8.2.8 COROLLARY. Let M be a finitely generated module over a noetherian local ring (A,m). Then $\dim_{\hat{A}}\hat{M} = \dim_A M$, where ^ denotes the m-adic completion.

PROOF. By 8.1.2 e) we know that $\dim_A M$ is the dimension of the ring $A/\mathrm{Ann}_A M$ and since completion is an exact functor on finitely generated modules, we have $\hat{A}/\mathrm{Ann}_{\hat{A}}\hat{M} = (A/\mathrm{Ann}_A M)\hat{}$. On the factor ring $A/\mathrm{Ann}_A M$ the

$\mathfrak{m}$-adic topology is of course the topology defined by its maximal ideal $\mathfrak{m}/\mathrm{Ann}_A M$. We are therefore reduced to proving the corollary in the case $M = A$, and there it follows from 8.2.7 since $A \to \hat{A}$ is faithfully flat and $\mathfrak{m}\hat{A} = \hat{\mathfrak{m}}$, see section 2.2.

8.2.9 PROPOSITION. Let $A \subset B$ be an integral extension of noetherian local rings and let $\mathfrak{P} \subset B$ be a prime ideal with $\mathfrak{p} = \mathfrak{P} \cap A$. Then $\mathrm{ht}_B\mathfrak{P} = \mathrm{ht}_A\mathfrak{p}$.

PROOF. Since the ring extension $A_\mathfrak{p} \subset B_\mathfrak{p}$ remains integral, [AM, Prop. 5.6] the ring $B_\mathfrak{p}/\mathfrak{p}B_\mathfrak{p}$ is 0-dimensional [AM, Cor. 5.8] so $\dim B_\mathfrak{P}/\mathfrak{p}B_\mathfrak{P} = 0$. Therefore 8.2.7 shows that $\mathrm{ht}_B\mathfrak{P} \le \mathrm{ht}_A\mathfrak{p}$. The reverse inequality follows because we have the Lying Over and Going Up theorems available [AM, Ths. 5.10 & 5.11].

8.2.10 COROLLARY. If $(A,\mathfrak{m}) \subset (B,\mathfrak{M})$ is a module-finite extension of noetherian local rings, then every s.o.p. of A is a s.o.p. for B.

PROOF. Let $x_1,\ldots,x_d$ be a s.o.p. of A, then $\ell_A(B/(x_1,\ldots,x_d)B) < \infty$ hence also $\ell_B(B/(x_1,\ldots x_d)B) < \infty$. Since $A \subset B$ is certainly integral, by the proposition $\dim A = \dim B$, so $x_1,\ldots,x_d$ is a s.o.p. for B.

We now lead up to the results on polynomial extensions which we are after, in a series of exercises.

8.2.11 EXERCISES.

1. Let $f\colon A \to B$ be a ring homomorphism, $\mathfrak{a}$ an ideal in A and $\mathfrak{A} = \mathfrak{a}B$.

a) Let S be a multiplicatively closed system in A. For primes $\mathfrak{P}$ in B we have

$\mathfrak{P} \cap f(S) = \emptyset$ and $\mathfrak{A} \subset \mathfrak{P}$ if and only if $f^{-1}(\mathfrak{P}) \cap S = \emptyset$ and $\mathfrak{a} \subset f^{-1}(\mathfrak{P})$.

b) Let $\mathfrak{p}$ be a fixed prime in A. There exists an inclusion preserving correspondence between primes $\mathfrak{P}$ in B with $f^{-1}(\mathfrak{P}) = \mathfrak{p}$ and the primes of the ring $B_{\mathfrak{p}}/\mathfrak{p}B_{\mathfrak{p}}$.

2. Let A be a principal ideal domain which is not a field. Then dim A = 1. In particular, a polynomial ring in a single indeterminate over a field has Krull dimension one.

3. When $\mathfrak{a}$ is an ideal in the ring A, we denote by $\mathfrak{a}[T]$ the ideal in the polynomial ring A[T] consisting of all polynomials in the indeterminate T whose coefficients lie in $\mathfrak{a}$.

a) $\mathfrak{a}[T] = \mathfrak{a}A[T]$ and $\mathfrak{a}[T] \cap A = \mathfrak{a}$.

b) $\mathfrak{a}$ is prime if and only if $\mathfrak{a}[T]$ is prime.

c) If the prime $\mathfrak{p} \subset A$ is minimal over $\mathfrak{a}$, then $\mathfrak{p}[T] \subset A[T]$ is minimal over $\mathfrak{a}[T]$. Hint: if the prime $\mathfrak{P}$ in A[T] satisfies $\mathfrak{a}[T] \subset \mathfrak{P} \subset \mathfrak{p}[T]$, prove that $\mathfrak{P} \cap A = \mathfrak{p}$ and draw the desired conclusion.

d) For a prime $\mathfrak{p}$ in A let W be the set of primes $\mathfrak{P}$ in A[T] with $\mathfrak{P} \cap A = \mathfrak{p}$. Prove that dim W = 1 and that $\mathfrak{p}[T]$ is the unique minimal prime in W. Hint: use 1., 2. and the identification $A[T]_{\mathfrak{p}}/\mathfrak{p}A[T]_{\mathfrak{p}} \simeq (A_{\mathfrak{p}}/\mathfrak{p}A_{\mathfrak{p}})[T]$.

8.2.12 THEOREM. Let A be a noetherian ring and A[T] a polynomial extension. Then

(i) $ht_A\mathfrak{p} = ht_{A[T]}\mathfrak{p}[T]$ for every prime $\mathfrak{p} \subset A$;

(ii) If $\mathfrak{P}$ is a prime in A[T] with $\mathfrak{P} \cap A = \mathfrak{p}$ and $\mathfrak{P} \supsetneq \mathfrak{p}[T]$, then $ht_{A[T]}\mathfrak{P} = ht_A\mathfrak{p} + 1$, and such a $\mathfrak{P}$ always exists;

(iii) dim A[T] = dim A + 1;

(iv) If $x_1, \ldots, x_n \in A$ generate an ideal of height n, then $ht_{A[T]}\mathfrak{P} \geq n+1$ for every prime $\mathfrak{P} \subset A[T]$ which contains $x_1, \ldots, x_n$ and the indeterminate T.

PROOF. Since $\mathfrak{p}[T] = \mathfrak{p}A[T]$, (i) is a direct consequence of 8.2.7 since $A \to A[T]$ is flat. For (ii), use (i) and 8.2.11, 3 d). Now also (iii) is clear so let us prove (iv). Put $\mathfrak{p} = \mathfrak{P} \cap A$. Then $\mathfrak{p}$ contains $(x_1,..,x_n)$ and $\mathfrak{p}[T] \subsetneq \mathfrak{P}$ since $T \notin \mathfrak{p}[T]$. Hence $ht_{A[T]}\mathfrak{P} > ht_{A[T]}\mathfrak{p}[T] = ht_A\mathfrak{p} \geq n$.

8.3 PARAMETERS AND REGULAR SEQUENCES

In this section we further investigate parameters and compare them with regular sequences. In 8.1.4 we saw that regular sequences certainly behave like parameters, and here we explore the connection.

8.3.1 PROPOSITION. Let $x_1,\ldots,x_n$ generate an ideal of height n in a noetherian ring A. Let $f(X_1,\ldots,X_n) \in A[X_1,\ldots,X_n]$ be a homogeneous polynomial with $f(x_1,\ldots,x_n) = 0$. Now consider the ideal I in the polynomial ring $B = A[X_1,\ldots,X_n,T]$ generated by the polynomials $X_iT - x_i$, $i = 1,\ldots,n$. Then $f(X_1,\ldots,X_n)$, as an element of B, is in $\mathfrak{R}_B(I)$.

PROOF. Put $Y_i = X_iT - x_i$. If deg f = s, then $T^sf(X_1,\ldots,X_n) = f(Y_1+x_1,\ldots,Y_n+x_n) = g(Y_1,\ldots,Y_n)$, where g is a polynomial in the Y_i's with constant term $g(0,\ldots,0) = f(x_1,\ldots,x_n) = 0$. Hence $T^sf(X_1,\ldots,X_n) \in I$. Suppose $\mathfrak{P}$ is a prime ideal in B containing I. If $T \in \mathfrak{P}$, then the relations $X_iT - x_i \in I$ imply that the $x_i \in \mathfrak{P}$. Hence $ht_B\mathfrak{P} \geq n+1$ by 8.2.12 (iv). But this contradicts 8.1.4 since I is generated by n elements. Therefore $T \notin \mathfrak{P}$ and the membership $T^sf(X_1,\ldots,X_n) \in I$ implies that $f(X_1,\ldots,X_n) \in \mathfrak{P}$. Since $\mathfrak{P} \supset I$ is arbitrary, this proves our contention.

8.3.2 COROLLARY. Let $x_1,\ldots,x_n$ be elements of a noetherian ring A which generate an ideal $\mathfrak{a}$ of height n. Then $\mathfrak{b} \subset \mathfrak{R}(\mathfrak{a})$, where $\mathfrak{b}$ is the ideal $\mathfrak{b}_x$ associated to the x's before 5.1.11.

PROOF. Let $\mathfrak{p}$ be a prime ideal containing $\mathfrak{a}$, and write $\mathfrak{P}$ for the prime ideal $\mathfrak{p}B + (T)$ in the polynomial ring B. Since $\mathfrak{a} \subset \mathfrak{p}$, the polynomials $X_iT - x_i$ are all in $\mathfrak{P}$, hence $I \subset \mathfrak{P}$. Now let $f(X_1,\ldots,X_n) \in A[X_1,\ldots,X_n]$ be homogeneous with $f(x_1,\ldots,x_n) = 0$. In view of 5.1.10 it is enough to show that the coefficients of f are in $\mathfrak{p}$. But 8.3.1 tells us that $f \in \mathfrak{P} \cap A[X_1,\ldots,X_n] = \mathfrak{p}[X_1,\ldots,X_n]$, which says precisely that.

8.3.3 COROLLARY. Suppose the noetherian ring A contains a field k and the elements $x_1,\ldots,x_n$ generate an ideal $\mathfrak{a}$ of height n in A. Then these elements are algebraically independent over k.

PROOF. Let $f(X_1,\ldots,X_n) \in k[X_1,\ldots,X_n]$ be a nonnull polynomial with $f(x_1,\ldots,x_n) = 0$, and let f_s be its homogeneous component of lowest degree. Clearly $f_s(x_1,\ldots,x_n) \in \mathfrak{a}^{s+1}$. Considered as a polynomial in $A[X_1,\ldots,X_n]$, this f_s has its coefficients in $\mathfrak{b}$ by 5.1.10 with $M = A$, and therefore in $\mathfrak{R}(\mathfrak{a})$ by the previous corollary. The latter ideal is proper, its intersection with k is therefore (0). Hence $f_s = 0$, contradicting our assumption.

Actually, the converse of 8.3.2 is also true: $\mathfrak{b} \subset \mathfrak{R}(\mathfrak{a})$ implies ht $\mathfrak{a} = n$, as was proved in [Val, Cor. 2.9]. The important local case of 8.3.2 is known as the analytic independence of all systems of parameters and goes back to Zariski. The present proof is adapted from Davis [Da 67].

8.3.4 COROLLARY. Let $x = x_1,\dots,x_d$ be a s.o.p. in the d-dimensional local noetherian ring $(A,\mathfrak{m})$. Then $\mathfrak{b}_x \subset \mathfrak{m}$.

It is natural to ask whether the larger ideal $\mathfrak{q}_x$ of 5.1.10 is also contained in the maximal ideal. This is Hochster's Monomial Conjecture [Ho 73b, Conj. 1]. It is tantamount to requiring, in the notation of 5.1.10: for any given s.o.p. x none of the homomorphisms $.x^t\colon A/\mathfrak{a} \to A/\mathfrak{a}_{t+1}$ is null, $t \geq 1$. Corollary 5.1.16 allows us to state this for more general determinantal maps as introduced in 5.1.14.

8.3.5 MONOMIAL CONJECTURE. In a d-dimensional noetherian local ring A, suppose s.o.p.'s $x = x_1,\dots,x_d$ and $y = y_1,\dots,y_d$ generate ideals $\mathfrak{a}_x \supset \mathfrak{a}_y$. If $T\underline{x} = \underline{y}$ and $\Delta = \det T$, then the homomorphism $.\Delta\colon A/\mathfrak{a}_x \to A/\mathfrak{a}_y$ is nonnull.

Indeed, since $\mathfrak{R}(\mathfrak{a}_y) = \mathfrak{m}$, 5.1.16 with $M = A$ tells us that $(\mathfrak{a}_y:\Delta) \subset \mathfrak{q}_x$. By hypothesis $\mathfrak{q}_x \subset \mathfrak{m}$, so that $.\Delta\colon A/\mathfrak{a}_x \to A/\mathfrak{a}_y$ is not null. As a consequence of 8.3.4 and 5.1.14, the reader may prove that if $\mathfrak{a}_x = \mathfrak{a}_y$, then $.\Delta$ is actually an isomorphism.

Since $A/\mathfrak{a} = \hat{A}/\hat{\mathfrak{a}}$ for an $\mathfrak{m}$-primary ideal $\mathfrak{a}$, the Monomial Conjecture is true for A if it is true for its $\mathfrak{m}$-adic completion $\hat{A}$. The conjecture was proved by Hochster for equicharacteristic rings [Ho 73b, Prop. 3] and when $d \leq 2$. He has also analyzed the situation in detail in [Ho 83]. We shall return to the matter in section 10.3. Here we merely invite the reader to try his hand at the two-dimensional case, i.e. to show that $x_1^t x_2^t \notin (x_1^{t+1}, x_2^{t+1})$ for any given s.o.p. x_1, x_2 and $t \geq 1$; it will alert him to some of the difficulties involved.

Besides the analytic independence 8.3.4 there exists another motivation for the Monomial Conjecture. In 4.3.4 we have seen that the direct limit of the directed system of maps $.x^t\colon A/\mathfrak{a}_s \to A/\mathfrak{a}_{t+s}$ is $H^d_{\mathfrak{a}}(A)$. Since local cohomology only depends on the radical of the ideal involved, $H^d_{\mathfrak{a}}(A)$ is just $H^d_{\mathfrak{m}}(A)$. In Corollary 10.2.2 we shall prove a result of Grothendieck which asserts that $H^d_{\mathfrak{m}}(M) \neq 0$ for any d-dimensional finitely generated module M, so $H^d_{\mathfrak{m}}(A) \neq 0$. This implies that for s large enough all canonical maps $A/\mathfrak{a}_s \to H^d_{\mathfrak{m}}(A)$ have nonzero image, and therefore all the maps $.x^t\colon A/\mathfrak{a}_s \to A/\mathfrak{a}_{s+t}$ are nonnull. In particular, all maps $.x^{st}\colon A/\mathfrak{a}_s \to A/\mathfrak{a}_{s(t+1)}$ are nonnull, in other words, the Monomial Conjecture is true for the s.o.p. $x_1^s,\ldots,x_d^s$. The question then remains whether it is true for all systems of parameters.

Two remarks are in order. Firstly, it is easy to see that the statement is not valid for arbitrary d-dimensional modules [Ho 73b, Remark 6]. Secondly, the integer s can be uniformly chosen for all s.o.p.'s; this we shall see in 11.2.4.

We next wish to examine systems of parameters more closely. To this end let us recall that a reduced ring is one without nonzero nilpotent elements and that a ring A is called unmixed when Ass A consists entirely of minimal primes. An ideal $\mathfrak{a}$ is called unmixed if ht $\mathfrak{a}$ = ht $\mathfrak{p}$ for every $\mathfrak{p} \in$ Ass $A/\mathfrak{a}$; in particular, $A/\mathfrak{a}$ is then unmixed.

8.3.6 LEMMA. A reduced noetherian ring is unmixed. Moreover an element in a minimal prime ideal is always annihilated by an element outside of that ideal.

PROOF. If ab = 0 for two nonzero elements, then, b not being contained in all minimal primes, the element a has to be in at least one. Thus Ass A consists of minimal primes. Now let a be in the minimal prime $\mathfrak{p}$. Then $A_{\mathfrak{p}}$ is an artinian local ring whose maximal ideal $\mathfrak{p}A_{\mathfrak{p}}$ consists of nilpotents. Thus $(a/1)^t = 0$ for some $t \geq 1$. Hence $sa^t = 0$ for some $s \in A\setminus\mathfrak{p}$.

Then $(sa)^t = 0$ so $sa = 0$ which finishes the proof.

8.3.7 PROPOSITION. Let A be a noetherian ring and suppose the ideal $(x_1,\dots,x_n) \neq A$. Then $\mathrm{ht}(x_1,\dots,x_r) = r$ for every r, $0 \le r \le n$, precisely if for every $y \in A$, a relation $x_{r+1}y \in \mathfrak{R}(x_1,\dots,x_r)$ implies $y \in \mathfrak{R}(x_1,\dots,x_r)$, $r = 0,\dots,n-1$.

PROOF. Let $\mathfrak{p} \supset (x_1,\dots,x_r)$ be a prime ideal minimal with respect to this property. Then ht $\mathfrak{p} \le r$ by 8.1.4 but $\mathrm{ht}(x_1,\dots,x_{r+1}) = r+1$, so if $x_{r+1}y \in \mathfrak{R}(x_1,\dots,x_r)$, then $y \in \mathfrak{p}$. This holds for all such $\mathfrak{p}$, therefore $y \in \mathfrak{R}(x_1,\dots,x_r)$.

Conversely we shall show that all minimal primes of $(x_1,\dots,x_r)$ have height r, $0 \le r \le n$. Since this is obvious for $r = 0$, suppose it has been proved for $0,1,\dots,r < n$. Assume then that ht $\mathfrak{p} \le r$ for some minimal prime $\mathfrak{p} \supset (x_1,\dots,x_{r+1})$. Then $\mathfrak{p} \supset (x_1,\dots,x_r)$ is minimal (if not, there would be another prime $\mathfrak{q}$ with $\mathfrak{p} \supsetneq \mathfrak{q} \supset (x_1,\dots,x_r)$ with ht $\mathfrak{q} = r$ and this would imply ht $\mathfrak{p} > r$). Now $x_{r+1} \in \mathfrak{p}$, and we can apply 8.3.6 to the reduced ring $A/\mathfrak{R}(x_1,\dots,x_r)$. This yields a $y \notin \mathfrak{p}$ such that $x_{r+1}y \in \mathfrak{R}(x_1,\dots,x_r)$, which contradicts our assumption. Hence ht $\mathfrak{p} = r+1$ which finishes the proof.

It is worth comparing the above characterization of strong parameters with the definition of a regular sequence in 5.1.1; in the latter the radical signs are just dropped, the relationships hold "on the nose". Similarly, if we replace the inclusion $\mathfrak{b} \subset \mathfrak{R}(\mathfrak{a})$ in 8.3.2 by the stronger condition $\mathfrak{b} \subset \mathfrak{a}$, then the sequence $x_1,\dots,x_n$ is A-pre-regular by 5.1.12, and if A is local noetherian, then this means regular by 5.2.5, and the converse is also true. It is therefore small wonder that one of the central and recurrent themes in local algebra consists in comparing systems of parameters with regular sequences, in other words, dimensions with depths. We offer a few results in this vein, which will illustrate the point. First of all, Proposition 8.3.7 shows that a regular sequence $x_1,\ldots,x_n$ in a noetherian local ring A is certainly part of a strong system of parameters. For if $x_{r+1}y \in \mathfrak{R}(x_1,\ldots,x_r)$, then $(x_{r+1}y)^t \in (x_1,\ldots,x_r)$ for some $t \geq 1$. Since $x_1,\ldots,x_r,x_{r+1}^t$ is a regular sequence by 5.1.3 d), it follows that $y^t \in (x_1,\ldots,x_r)$ whence $y \in \mathfrak{R}(x_1,\ldots,x_r)$, which fulfils the criterion of 8.3.7. A module version reads

8.3.8 PROPOSITION. Let $M \neq 0$ be a finitely generated module over a noetherian local ring A. Any regular sequence on M can be extended to a s.o.p. of M. In particular, dp M $\leq$ dim M.

PROOF. In view of 8.2.4 we need to prove that $\dim M/(x_1,\ldots,x_n)M + n = \dim M$ for our regular sequence $x_1,\ldots,x_n$. Now $\dim M/x_1M = \dim M - 1$ when x_1 is not in any minimal prime of Supp M, which is certainly true if x_1 is a non zerodivisor on M, i.e. x_1 avoids all primes in Ass M. Carry on by induction.

One can actually prove a somewhat sharper inequality. To this end, for any prime $\mathfrak{p}$, we write $\mathrm{ht}^-\mathfrak{p}$ for the shortest length of a saturated chain of prime ideals descending from $\mathfrak{p}$ right down to a minimal prime in the ring; this is Kaplansky's notion of "little rank" and the second statement of 8.3.9 below is [Ka 74, Th. 138]. We can extend the notion to an arbitrary ideal $\mathfrak{a}$ by putting $\mathrm{ht}^-\mathfrak{a} = \inf\{\mathrm{ht}^-\mathfrak{p} \mid \mathfrak{p} \supset \mathfrak{a}\}$ and modules M by putting $\dim^- M$ is the shortest length of a saturated chain of primes connecting the maximal ideal $\mathfrak{m}$ with a minimal element of Supp M. Of course $\mathrm{ht}^-\mathfrak{p} \leq \mathrm{ht}\,\mathfrak{p}$ and $\dim^- M \leq \dim M$. Write $(A,\mathfrak{m},k)$ for our local ring.

8.3.9 PROPOSITION. In the situation of 8.3.8 $\operatorname{dp} M \leq \dim^{-} A/\mathfrak{p}$ for every $\mathfrak{p} \in \operatorname{Ass} M$, so in particular $\operatorname{dp} M \leq \dim^{-} M$. Also $\operatorname{gr} M \leq \operatorname{ht}^{-}\mathfrak{p}$ for every $\mathfrak{p} \in \operatorname{Supp} M$.

PROOF. If $\mathfrak{p} \in \operatorname{Ass} M$, then $A/\mathfrak{p} \subset M$ so $k(\mathfrak{p}) \subset M_{\mathfrak{p}}$. Thus $\mu^0(\mathfrak{p},M) \neq 0$, by 3.3.3. If $\dim^{-} M = r$, iterated use of 3.3.4 tells us that $\mu^r(\mathfrak{m},M) \neq 0$, so $\operatorname{dp} M \leq r$ by 5.3.14. For the second inequality, recall that $\operatorname{gr} M \leq \operatorname{dp} A_{\mathfrak{p}}$ for every $\mathfrak{p} \in \operatorname{Supp} M$ by definition, 7.2.9. So we need only prove that for a noetherian local ring $(A,\mathfrak{m})$ always $\operatorname{dp} A \leq \operatorname{ht}^{-}\mathfrak{m}$. But this is a weak form of the first inequality, so that the proof has been furnished.

We close this section with another pretty and useful result, which compares parameters and regular sequences in a particular case.

8.3.10 THEOREM. Let $(A,\mathfrak{m},k)$ be a noetherian local ring and $(x_1,\dots,x_n) = \mathfrak{p}$ a prime ideal of height n. Then $y = y_1,\dots,y_n$ is a regular A-sequence for every minimal set of generators y of $\mathfrak{p}$ and $(y_1,\dots,y_r)$ is a prime ideal of height r for $0 \leq r \leq n$.

PROOF. A minimal set of generators of $\mathfrak{p}$ consists of $\dim_k \mathfrak{p}/\mathfrak{m}\mathfrak{p}$ elements according to Nakayama; by 8.1.4 this number is n. It is enough therefore to prove the result for the x's. The ideal $\mathfrak{b}$ attached to the x's, is contained in $\mathfrak{R}(\mathfrak{p})$, 8.3.2. But $\mathfrak{p}$ is prime, so $\mathfrak{R}(\mathfrak{p}) = \mathfrak{p}$. Theorem 5.1.12 now tells us that A is x-pre-regular, so in view of 5.2.5 x-regular, and that the map $\beta_A\colon (A/\mathfrak{p})[X_1,\dots,X_n] \to G_{\mathfrak{p}}(A)$ is an isomorphism. The left hand ring is an integral domain, so is therefore the associated graded $G_{\mathfrak{p}}(A)$. Using the fact that in our local ring the intersection of the powers of $\mathfrak{p}$ is the null ideal, 2.1.4, one easily sees that A is an integral domain.

This takes care of the case $r = 0$.

We contend that the situation is similar for the factor rings $\bar{A} = A/(x_1,\ldots,x_r)$, $1 \le r \le n$. The images $\bar{x}_{r+1},\ldots,\bar{x}_n$ in $\bar{A}$ form a regular sequence and generate a prime ideal $\bar{\mathfrak{p}} \subset \bar{A}$. According to 8.1.4 we know $\mathrm{ht}_{\bar{A}}\bar{\mathfrak{p}} = n-r$. The first part of the proof tells us that $\bar{A}$ is an integral domain or that $(x_1,\ldots,x_r)$ is a prime ideal in A. The statement about its height is again a consequence of 8.1.4.

This result is due to Davis, cf. [Da 78], see also [HSV, Kap. I, Satz 3.16.7]. The condition "local" is essential in the theorem; there are counter examples even in two-dimensional integrally closed noetherian domains [He].

8.4 EXTENSIONS OF THE HAUPTIDEALSATZ

We continue this chapter with two later extensions of Krull's important Theorem 8.1.4. We shall need the concept of a catenary noetherian ring. This means that all saturated chains of prime ideals between any two given $\mathfrak{p} \subset \mathfrak{q}$ have the same length; this is therefore an invariant depending only on $\mathfrak{p}$ and $\mathfrak{q}$, and if A is local, this is simply $\dim A/\mathfrak{p} - \dim A/\mathfrak{q}$. A ring is called universally catenary if all its finitely generated ring extensions are catenary. These rings are not so special as one might believe; in section 9.3 we shall see that all complete noetherian local rings are universally catenary, a fact we shall use in the proof of 8.4.1.

To present the first generalization, write $M^* = \mathrm{Hom}_A(M,A)$ for each A-module M and consider the bilinear pairing $\tau\colon M^*\otimes_A M \to A$ given by

$\tau(f\otimes m) = f(m)$. The order ideal $\mathfrak{o}_m = \{f(m) \in A \mid f \in M^*\} \subset A$ reflects properties of the element $m \in M$. In case $M = A^n$, a free module of rank n, an element $m \in M$ is given by its coordinates $x_1,\ldots,x_n$ on a standard basis of A^n and 8.1.4 says that ht $\mathfrak{p} \le$ rk M for each minimal prime $\mathfrak{p}$ of $\mathfrak{o}_m$, provided $m \in \mathfrak{m}M$ for some maximal ideal $\mathfrak{m}$. In order to exploit this point of view, we need to extend the notion of rank. So let rk $M = \mathrm{rk}_{A_{(0)}} M_{(0)}$ in case A is an integral domain, and if not define rk $M = \max \mathrm{rk}_{A_\mathfrak{p}/\mathfrak{p}A_\mathfrak{p}}(M_\mathfrak{p}/\mathfrak{p}M_\mathfrak{p})$ where $\mathfrak{p}$ ranges over the minimal primes of A. Clearly, if M is finitely generated, its rank is a nonnegative integer. By localizing at $\mathfrak{m}$ we can reduce to a local ring:

8.4.1 THEOREM. If M is a finitely generated module over a noetherian local ring $(A,\mathfrak{m})$ and $m \in \mathfrak{m}M$, then ht $\mathfrak{p} \le$ rk $M_\mathfrak{p}$ for each minimal prime $\mathfrak{p}$ of $\mathfrak{o}_m$.

This was proved by Eisenbud and Evans for equicharacteristic rings [EE, Th. 1.1], and by Bruns [Br, Th. 1] in general. We shall follow the latter's proof.

PROOF. First observe that, since M is finitely presented, $B\otimes_A \mathfrak{o}_m$ is the order ideal of the element $1\otimes m \in B\otimes_A M$ in the ring B whenever B is a faithfully flat ring extension of A. Taking for B the $\mathfrak{m}$-adic completion of A, tensoring with which also preserves both height, 8.2.8, and rank, we may restrict ourselves to complete local rings. Now divide out by a minimal prime $\mathfrak{p}$ of A such that $\mathrm{ht}_{A/\mathfrak{p}}(\mathfrak{o}_m+\mathfrak{p})/\mathfrak{p} = \mathrm{ht}\ \mathfrak{o}_m$. Tensoring the pairing $\tau\colon M^*\otimes_A M \to A$ with $A/\mathfrak{p}$, we find that $(\mathfrak{o}_m+\mathfrak{p})/\mathfrak{p}$ is contained in the order ideal of m mod $\mathfrak{p}M$ in $A/\mathfrak{p}$ while $\mathrm{rk}_{A/\mathfrak{p}} M/\mathfrak{p}M \le \mathrm{rk}_A M$. Thus we may carry out our proof for a complete local domain which we again call A. Such a ring is universally catenary, cf. section 9.3.

Since $m \in \mathfrak{m}M$, write $m = \Sigma_{i=1}^{n} x_i m_i$ with $x_i \in \mathfrak{m}$, $m_i \in M$, $i = 1,\ldots,n$. In the polynomial ring $A[X_1,\ldots,X_n]$ the ideal $(\mathfrak{m},X_1,\ldots,X_n)$ is maximal; we localize at it to obtain a faithfully flat extension of local domains $A \to B$. The images of the elements $x_1+X_1,\ldots,x_n+X_n$ in B generate an ideal $\mathfrak{q} \subset B$. Since $B/\mathfrak{q} \simeq A$, the ideal $\mathfrak{q}$ is prime in B. Let K be the field of fractions of A, then K is contained in $B_{\mathfrak{q}}$ because $A \cap \mathfrak{q} = (0)$ and $M' = (B\otimes_A M)_{\mathfrak{q}} = B_{\mathfrak{q}}\otimes_K K\otimes_A M$ is a free $B_{\mathfrak{q}}$-module of finite rank. Consider the element $v = (x_1+X_1)m_1 + \ldots + (x_n+X_n)m_n \in B\otimes_A M$. Then its order ideal in B satisfies $\mathrm{ht}_B \mathfrak{o}_v = \mathrm{ht}_{B_{\mathfrak{q}}}(B_{\mathfrak{q}}\otimes_B \mathfrak{o}_v) \le \mathrm{rk}_{B_{\mathfrak{q}}} M' = \mathrm{rk}_A M$, where the inequality is given by our interpretation of 8.1.4 in terms of an order ideal of a free module and the left hand equality stems from 8.2.12 (i).

The local domain B is catenary, therefore there exists a prime ideal $\mathfrak{P} \subset B$ which contains $\mathfrak{o}_v$ as well as all the indeterminates $X_1,\ldots,X_n$, and such that $\mathrm{ht}_B\mathfrak{P} \le \mathrm{rk}_A M + n$. Now $B\otimes_A \mathfrak{o}_m \subset \mathfrak{o}_v$ because any $f \in M^*$ extends to an element $1\otimes f \in (B\otimes_A M)^*$. Therefore $\mathfrak{P}$ contains a minimal prime $\mathfrak{P}_0$ of $B\otimes_A \mathfrak{o}_m$. But all minimal primes of $B\otimes_A \mathfrak{o}_m$ are extended from minimal primes of $\mathfrak{o}_m$ by 8.2.11. Therefore $\mathfrak{P}_0 \subset \mathfrak{P}_0 + (X_1) \subset \ldots \subset \mathfrak{P}_0 + (X_1,\ldots,X_n)$ is a strictly increasing chain of prime ideals in B which are contained in $\mathfrak{P}$, so $\mathrm{ht}_A \mathfrak{o}_m = \mathrm{ht}_B(B\otimes_A \mathfrak{o}_m) \le \mathrm{ht}_B \mathfrak{P}_0 \le \mathrm{rk}_A M$. This finishes the proof.

8.4.2 EXERCISE. In the situation of the above theorem, prove that $\mathrm{ht}\ f(M) \le \mathrm{rk}\ M$ for each $f \in \mathfrak{m}M^*$.

The second extension of the Hauptidealsatz is Hochster's Homological Height Theorem [Ho 73a, Conj. (A)]:

8.4.3 THEOREM. Suppose $A \to B$ is a homomorphism of noetherian rings and M is a finitely generated nonnull A-module with annihilator $\mathfrak{a}$. Suppose $\mathfrak{P} \subset B$ is a prime ideal minimal above $\mathfrak{a}B$. Then $\mathrm{ht}_B\mathfrak{P} \le \mathrm{pd}_A M$.

To show that this implies 8.1.4, take $A = \mathbb{Z}[X_1,\ldots,X_n]$, $\mathfrak{a} = (X_1,\ldots,X_n)$, $M = A/\mathfrak{a}$. By Hilbert's Syzygies Theorem or e.g. 5.1.9, $\mathrm{pd}_A M = n$. A homomorphism of A into any given ring B is determined by specifying the images $x_1,\ldots,x_n \in B$ of $X_1,\ldots,X_n$. So we recover Theorem 8.1.4: if $\mathfrak{P} \subset B$ is a minimal prime of $\mathfrak{a}B$, then $\mathrm{ht}_B\mathfrak{P} \le n$.

Another consequence of the Homological Height is Rees' inequality 7.2.11: take $A = B$ and $\mathfrak{P}$ a minimal prime of the ideal $\mathfrak{a} \subset A$. Then we know that $\mathrm{gr}\ M = \mathrm{dp}(\mathfrak{a},A) \le \mathrm{ht}\ \mathfrak{P}$, 7.2.9, and we obtain $\mathrm{gr}\ M \le \mathrm{ht}\ \mathfrak{P} \le \mathrm{pd}\ M$.

The Homological Height Theorem was proved by Hochster in equal characteristic and conjectured in general. Recently P. Roberts has settled the remaining case. We shall discuss this in 13.1.3. Here, we take the proof as read and investigate several consequences. The first is the celebrated Intersection Theorem, first proved by Peskine-Szpiro [PS 73, Ch. II, Th. 2.1] when A contains a field, at least in most interesting cases.

8.4.4 THEOREM. (Intersection Theorem) Let $M \neq 0$ and N be finitely generated modules over a noetherian local ring $(A,\mathfrak{m},k)$. Then

$$\dim N \le \mathrm{pd}\ M + \dim(M\otimes_A N).$$

PROOF. Put $\mathfrak{a} = \mathrm{Ann}\ M$ and $\mathfrak{b} = \mathrm{Ann}\ N$. Then $\mathrm{Supp}(M\otimes_A N) = \mathrm{Supp}\ M \cap \mathrm{Supp}\ N = \mathrm{Supp}\ A/(\mathfrak{a}+\mathfrak{b})$, see 8.1.2 e). We proceed by induction on $\dim(M\otimes_A N)$.

If $\dim(M\otimes_A N) = 0$, the ideal $\mathfrak{a}+\mathfrak{b}$ is $\mathfrak{m}$-primary and 8.4.3 with $B = A/\mathfrak{b}$ yields $\dim N = \dim A/\mathfrak{b} = \mathrm{ht}_B \mathfrak{m}/\mathfrak{b} \le \mathrm{pd}_A M$.

If $\dim(M\otimes_A N) > 0$, take an element $x \in A$ which is part of a s.o.p. for both N and $M\otimes_A N$, see the beginning of the next section. Since $M\otimes_A N/xN \simeq (M\otimes_A N)/x(M\otimes_A N)$, we find $\dim N = \dim N/xN + 1 \le \mathrm{pd}\ M + \dim(M\otimes_A N/xN)$

$+ 1 = \text{pd } M + \dim(M \otimes_A N)$.

8.4.5 COROLLARY. Let $M \neq 0$ be a finitely generated module over a local noetherian ring A. If $\text{pd } M < \infty$, then

(i) $\dim A \leq \dim M + \text{pd } M$ (H.-B. Foxby);

(ii) $\dim A - \text{dp } A \leq \dim M - \text{dp } M$ (D. Buijs).

PROOF. Both statements are equivalent in view of the identity $\text{dp } A = \text{dp } M + \text{pd } M$ (7.1.5), while (i) is just 8.4.4 with $N = A$.

Buijs' formulation is particularly suggestive. It says that among finitely generated modules of finite projective dimension, the defect between dimension and depth is minimal for the ring itself. We elaborate on this theme in the next chapter.

8.5 DIMENSION CONJECTURES

Throughout this section we work with finitely generated modules M and N over a noetherian local ring $(A,\mathfrak{m},k)$. The standing reference is the joint thesis of Peskine and Szpiro [PS 73].

Suppose both M and N have positive dimension, then it is easy to find an element $x \in \mathfrak{m}$ which is part of a s.o.p. for M and part of a s.o.p. for N. We need to avoid the minimal primes of Supp M and those of Supp N and these are finite in number. Since $\mathfrak{m}$ is not among them by the dimension condition, there exist such an x by prime avoidance. Also $\dim M/xM = \dim M - 1$ and $\dim N/xN = \dim N - 1$ in view of 8.2.4, so we can continue until either dimension gives out. Similar considerations work for regular sequences on modules of positive depth. This time we need to avoid the primes in Ass M and in Ass N and find an $x \in \mathfrak{m}$ such that dp $M/xM = $ dp $M - 1$ and dp $N/xN = $ dp $N - 1$, cf. 5.3.7 and 5.3.9. We can continue to construct a common regular sequence on M and on N until either depth gives out.

Of course this does not mean that an arbitrary s.o.p. resp. regular sequence for M can always be extended to a s.o.p. resp. regular sequence for N. We illustrate the point by a banal example. Put $A = k[[X,Y]]/(XY)$ and $M = A/(y) = k[[x]]$ where x and y are the images of X resp. Y in A. Clearly x is both a parameter and a non zerodivisor on M, but not in A. On the other hand, x+y is both a common s.o.p. and a non zerodivisor on M and A.

For which modules can every s.o.p. on M be extended to a s.o.p. of A? Same question for regular sequences? M. Auslander conjectured that the second property is true whenever M has finite projective dimension, while Peskine-Szpiro did the same for the first property. Auslander's conjecture is now a theorem, and we shall treat it in 13.1.10. Here we shall discuss the parameter version and tie it in with Theorem 8.4.4.

8.5.1 PROPOSITION. For a finitely generated module $M \neq 0$ over a noetherian local ring A the two following statements are equivalent:

(i) Every s.o.p. of M can be extended to a s.o.p. of A;

(ii) $\dim M + \dim N \leq \dim A + \dim(M \otimes_A N)$ for every finitely generated module N.

PROOF. Clearly $\dim(M \otimes_A N) \leq \dim N$, so we can find a s.o.p. $x_1, \ldots, x_r$ of $M \otimes_A N$ which extends to a s.o.p. of N. If $\mathfrak{a} = (x_1, \ldots, x_r)$, then $\dim N = \dim N/\mathfrak{a}N + r$ by 8.2.4 and the associativity of the tensor product tells us that $M\otimes_A N/\mathfrak{a}N = (M\otimes_A N)/\mathfrak{a}(M\otimes_A N)$ which has finite length. If we put $\mathfrak{b} = \text{Ann } N/\mathfrak{a}N$, this implies $\ell(M\otimes_A A/\mathfrak{b}) < \infty$. Thus there exists a s.o.p. $y_1,\ldots,y_s$ of M which is contained in $\mathfrak{b}$, where $s = \dim M$. Extend this, according to (i), to a s.o.p. $y_1,\ldots,y_s,y_{s+1},\ldots,y_d$ of A, with $d = \dim A$. We also see that $\dim A/(y_1,\ldots,y_s) \geq \dim A/\mathfrak{b} = \dim N/\mathfrak{a}N$, so $d-s \geq \dim N/\mathfrak{a}N = \dim N - r$. The inequality (ii) follows.

Conversely, let $y_1,\ldots,y_s$ be a s.o.p. for M, and put $\mathfrak{c} = (y_1,\ldots,y_s)$. Since $M/\mathfrak{c}M = M\otimes_A A/\mathfrak{c}$ has finite length, condition (ii) says that $\dim M + \dim A/\mathfrak{c} \leq \dim A$. If $z_1,\ldots,z_t$ is a s.o.p. for $A/\mathfrak{c}$, then $(y_1,\ldots,y_s,z_1,\ldots,z_t)$ is an $\mathfrak{m}$-primary ideal so the inequality tells us that $s+t = d$ and we have extended to a s.o.p. of A, finishing the proof.

Having observed this, Peskine-Szpiro went on to relate version (ii) above with a strengthening of 8.4.4 which replaces projective dimension by grade. We need a general inequality involving grade.

8.5.2 PROPOSITION. Let M and N be finitely generated, nonnull modules over a noetherian local ring A. Then

$$\text{dp } N \leq \text{gr}(M,N) + \dim(M\otimes_A N) \leq \dim N.$$

PROOF. By definition of grade, and the fact that N is finitely generated, $\text{gr}(M,N) \leq \text{E-dp}_{A_\mathfrak{p}} N_\mathfrak{p} = \text{dp}_{A_\mathfrak{p}} N_\mathfrak{p} \leq \dim^-_{A_\mathfrak{p}} N_\mathfrak{p}$ where $\mathfrak{p}$ ranges over $\text{Supp } M \cap \text{Supp } N = \text{Supp}(M\otimes_A N)$ in view of 8.3.9. Since $\dim A/\mathfrak{p} + \dim_{A_\mathfrak{p}} N_\mathfrak{p} \leq \dim N$ for every such $\mathfrak{p}$, the right hand inequality follows.

For the left hand one, we induce on gr(M,N). If this is 0, some prime $\mathfrak{p}$ of Supp M is in Ass N. Then $\text{dp } N \leq \dim^- A/\mathfrak{p} \leq \dim(M\otimes_A N)$ by 8.3.9. If $\text{gr}(M,N) > 0$, take an $x \in \text{Ann } M$ which is regular on N. Then $\text{gr}(M,N/xN) = \text{gr}(M,N) - 1$ and $\text{dp } N/xN = \text{dp } N - 1$. The result follows by induction.

A slightly different formulation of the next result is contained in [PS 73, Th II 0.10].

8.5.3 THEOREM. For a finitely generated module $M \neq 0$ over a noetherian local ring A statements (C) and (D) are jointly equivalent to (S):

(C) $\operatorname{gr} M + \dim M = \dim A$;

(D) $\dim M + \dim N \leq \dim A + \dim(M \otimes_A N)$ for every finitely generated A-module N;

(S) $\dim N \leq \operatorname{gr} M + \dim(M \otimes_A N)$ for every finitely generated A-module N.

PROOF. Assuming (C) and (D), one rewrites dim A in (D), using the identity (C), to obtain (S). Conversely, assume (S). Add dim M on both sides of this inequality and use the right hand inequality of 8.5.2 with $N = A$ to obtain (D). To obtain (C), compare (S) and the right hand inequality of 8.5.2, both with $N = A$.

8.5.4 CONJECTURES. All three statements in the preceding theorem are true when $\operatorname{pd} M < \infty$.

For (C), this was conjectured by M. Auslander and it is sometimes called the Codimension Conjecture. The other ones are due to Peskine and Szpiro; one might call (D) the Dimension Conjecture while (S) is sometimes referred to as Strong Intersection. It is indeed stronger than the Intersection Theorem 8.4.4: $\operatorname{gr} M \leq \operatorname{pd} M$ by 7.2.11. This means that for perfect modules (S), and hence the two others, are true. Notice that (C) implies 8.4.5. in view of 7.2.11.

The statements have been proved in the graded case by Peskine-Szpiro in a later note [PS 74, Th. 2], and also a few very special cases are known. But in general these conjectures rate among the most important open questions in local algebra, cf. [Sz]. Peskine-Szpiro took these conjectures to assert that the support of a finitely generated module of finite projective dimension sits in Spec A in a particularly nice way. For the Dimension Conjecture this shows clearly in its parameter version 8.5.1. Statement (C) too can be viewed in this light, as we shall see in 9.1.7.

Finitely generated modules of finite injective dimension enjoy similar, though distinct privileges. Whereas 8.5.4 conjectures that in 8.5.2 (for $N = A$) the largest possible value is reached by $\operatorname{gr} M + \dim M$ when $\operatorname{pd} M < \infty$, the smallest bound is obtained when $\operatorname{id} M < \infty$ [PS 73, Ch. I, Cor. 4.9]:

8.5.5 PROPOSITION. Let M be a finitely generated nonnull module of finite injective dimension over a noetherian local ring $(A,\mathfrak{m},k)$. Then

$$\operatorname{id} M = \operatorname{dp} A = \operatorname{gr} M + \dim M.$$

PROOF. We need to prove that $\operatorname{dp} A \geq \operatorname{gr} M + \dim M$, since $\operatorname{dp} A = \operatorname{id} M$ by 7.1.5. Take $\mathfrak{q} \in \operatorname{Supp} M$ with $\dim A/\mathfrak{q} = \dim M$. Since $\operatorname{dp} A_{\mathfrak{p}} \geq \operatorname{gr} M$ for every $\mathfrak{p} \in \operatorname{Supp} M$, it is enough to prove that $\operatorname{dp} A \geq \operatorname{dp} A_{\mathfrak{p}} + \dim A/\mathfrak{p}$ for every such prime. So let $\operatorname{dp} A_{\mathfrak{p}} = s$. Since $\operatorname{id}_{A_{\mathfrak{p}}} M_{\mathfrak{p}} < \infty$, this integer is s. Theorem 3.3.3 and Corollary 3.3.6 imply that $\mu^s(\mathfrak{p},M) \neq 0$, and then also that $\mu^{s+t}(\mathfrak{m},M) \neq 0$ where $t = \dim A/\mathfrak{p}$ in view of 3.3.4. Thus $\operatorname{dp} A = \operatorname{id} M \geq \operatorname{dp} A_{\mathfrak{q}} + \dim A/\mathfrak{q}$ in particular, which proves the result.

This is a convenient place to record an observation of Bass [Bas, Lemma 3.5] concerning modules of infinite injective dimension.

8.5.6 PROPOSITION. Let M be a finitely generated module of infinite injective dimension over a d-dimensional noetherian local ring (A,m,k). Then $\mu^i(M) \neq 0$ for $i \geq d$.

PROOF. If $d = 0$, then A is artinian with unique prime ideal m, and $\mu^i(M) = 0$ would imply $id_A M < i$, so proceed by induction on d. In case $id_{A_\mathfrak{p}} M_\mathfrak{p} = \infty$ for some nonmaximal prime $\mathfrak{p}$, then $\mu^j(\mathfrak{p},M) = \mu^j(\mathfrak{p}A_\mathfrak{p},M_\mathfrak{p}) \neq 0$ for $j \geq \dim A_\mathfrak{p} = ht\ \mathfrak{p}$ by 3.3.2 and the induction hypothesis. Then $\mu^i(M) = \mu^i(m,M) \neq 0$ for $i \geq j + \dim^- A/\mathfrak{p}$ in virtue of 3.3.4, so certainly for $i \geq d$. In the other case, $d > \dim A_\mathfrak{p} \geq dp\ A_\mathfrak{p} = id_{A_\mathfrak{p}} M_\mathfrak{p}$ for every $\mathfrak{p} \neq m$, which means that $\mu^i(\mathfrak{p},M) = 0$ for $i \geq d$. The result follows, since otherwise a minimal injective resolution of the A-module M would turn out to be finite.

9 COHEN-MACAULAY MODULES AND REGULAR RINGS

In this chapter we investigate what happens when parameters actually form regular sequences. The spade work has been done in previous chapters, and we are quickly led to the notions in the title. Taking our cue through pre-regularity, we discuss Cohen-Macaulay properties in the first section, referring to the literature for some of the more familiar ones. In 9.2 the main properties of regular local rings are derived, which in section 9.3 lead us to the Cohen structure theorems for complete noetherian local rings. This is a natural place to introduce the Direct Summand Conjecture.

9.1 COHEN-MACAULAY MODULES AND RINGS

9.1.1 THEOREM. Let $(A,\mathfrak{m})$ be a d-dimensional noetherian local ring and $x = x_1,\ldots,x_d$ a s.o.p. If M is an x-pre-regular module then its $\mathfrak{m}$-adic completion $\hat{M}$ is y-regular for every s.o.p. y and z-regular for every pre-regular sequence $z = z_1,\ldots,z_m$ on M.

PROOF. Since $\mathfrak{a} = (x_1,\ldots,x_d)$ is a parameter ideal, the $\mathfrak{a}$-adic and $\mathfrak{m}$-adic topologies coincide. By Theorem 5.2.3 the module $\hat{M}$ is x-regular and whenever $\hat{M}$ is regular for some s.o.p. $v = v_1,\ldots,v_d$, then it is regular for every permutation of the v's. Now let $r_1,\ldots,r_d$ be positive integers and $z_1,\ldots,z_d$ be parameters such that there exists a chain of parameter ideals as in 8.2.5 connecting $(x_1^{r_1},\ldots,x_d^{r_d})$ and $(y_1,\ldots,y_d)$ step by step. It suffices to prove that if $(v_1,\ldots,v_d) \subset (v_1,\ldots,v_{d-1},w_d)$ are parameter ideals and $\hat{M}$ is v-regular, then w_d is a non zerodivisor on $\hat{M}/(v_1,\ldots,v_{d-1})\hat{M}$. But this is easily seen to be true for any module.

The statement on z-regularity is clear from 5.2.6.

In order to discuss the significance of this result, we introduce the notions of Cohen-Macaulay and pre-Cohen-Macaulay modules, usually abbreviated to CM resp. pre-CM. Over the noetherian local ring $(A,\mathfrak{m})$ let us suppose that $\mathfrak{m}M \neq M$. Then $\text{dp } M \leq \text{E-dp } M \leq \dim M \leq \dim A$ by 5.3.7, 6.1.10 and 6.1.11, while the first inequality is an equality in case M is finitely generated, 5.3.9. A finitely generated module with $\text{dp } M = \dim M$ is customarily called a Cohen-Macaulay module, and A is called a Cohen-Macaulay ring if $\text{dp } A = \dim A$.

9.1.2 DEFINITIONS. Let $x = x_1,\ldots,x_d$ be a s.o.p. in a noetherian local ring A. An A-module M is called a Big Cohen-Macaulay module w.r.t. x if it is x-regular; it is called x-pre-Cohen-Macaulay if it is x-pre-regular. A Big CM-module is called balanced if it is CM with respect to all s.o.p.'s.

A few remarks are in order. The designation "Big" signifies that the module has maximal depth, i.e. $\text{dp } M = \dim A$, but also that the module need not be finitely generated. For pre-CM modules we omit this adjective, since there is no usage before [BS, §1].

The next observations interpret previous results in terms of the new terminology. Theorem 5.1.7 says that a Big CM module is always pre-CM, while Theorem 9.1.1 now reads: the $\mathfrak{m}$-adic completion of a pre-CM module is a balanced Big CM module, so for a complete module the notions pre-CM and balanced Big CM imply one another. This is also true for finitely generated modules by Corollary 5.2.5, but not in general, as shown in Example 5.1.2 3). Also notice that a pre-CM module w.r.t. a single s.o.p. is pre-CM for every s.o.p. (so is "balanced", if you wish) by 5.2.6 and the fact that the completion is balanced.

Though we do not go into details here, let us just mention that the conditions in Theorem 5.1.7 between Big CM and pre-CM are all distinct, as can be shown by examples [Ba, Ch. V, §4]. For instance, a module M over a noetherian local ring A satisfies E-dp M = dim A precisely when $H_p(x,M) = 0$ for all s.o.p.'s x and all $p \geq 1$, and it is enough to require this for a single s.o.p. This result of Bartijn's follows from 6.1.10. By the above, however, all of these conditions coincide for finitely generated modules.

Cohen-Macaulay rings and modules have many pleasant properties, well documented in [Se, Ch. IV, B],[Ka 74, Ch. 3, §1] and [Ma 86, Ch. 6, §16]. Part of these properties have been extended to balanced Big CM modules by Sharp [Sh 81], Foxby (unpublished) and Zarzuela [Za 87] and [Za 88] or to complete Big CM-modules [Si]. We mention here that a CM ring is universally catenary [Ma 86, Th. 3.1] and that the noetherian local ring $(A,\mathfrak{m})$ is CM if and only if its $\mathfrak{m}$-adic completion $\hat{A}$ is. This last fact is easily seen by noticing that we need only prove that the ring is pre-CM w.r.t. some s.o.p. $x = x_1,\ldots,x_d$, and, by 5.1.12, it suffices to show that the maps $.x^t$: $A/\mathfrak{a} \to A/\mathfrak{a}_{t+1}$ are injective for $t \geq 1$, where we have used notation standard in that chapter. But $\mathfrak{a}$ and $\mathfrak{a}_{t+1}$ are $\mathfrak{m}$-primary ideals, hence $\hat{A}/\hat{A}\mathfrak{a} \simeq$

$A/\mathfrak{a}$ and $\hat{A}/\hat{A}\mathfrak{a}_{t+1} \simeq A/\mathfrak{a}_{t+1}$, 2.2.5, so that we are looking at the same map for the completion.

9.1.3 PROPOSITION. If a noetherian local ring (A,m) possesses a pre-CM module, then it satisfies the Monomial Conjecture.

PROOF. We wish to show that for a given s.o.p. $x = x_1,\dots,x_d$ none of the maps $.x^t\colon A/\mathfrak{a} \to A/\mathfrak{a}_{t+1}$ is null, $t \geq 1$, where the notation is as in 5.1.7. If M is a pre-CM module, it is pre-CM w.r.t. x. Tensoring the above maps with $- \otimes_A 1_M$ yields the maps $.x^t\colon M/\mathfrak{a}M \to M/\mathfrak{a}_{t+1}M$ which by pre-regularity are injective. But pre-regularity also implies that $M/\mathfrak{a}M \neq 0$, so the original maps cannot be null.

Notice that for an m-primary ideal $\mathfrak{a}$ the condition $M \neq \mathfrak{a}M$ is equivalent to $M \neq mM$ for any module M. Other consequences of the existence of pre-CM-modules will be treated in Chapter 13.

A few useful properties of Cohen-Macaulay rings are codified in

9.1.4 PROPOSITION. Let $\mathfrak{a}$ be a parameter ideal in a d-dimensional Cohen-Macaulay ring (A,m). Then $\mathrm{Ext}^i_A(A/\mathfrak{a},M) \simeq \mathrm{Tor}^A_{d-i}(A/\mathfrak{a},M)$ for every A-module M and all i. Moreover fd $H^d_m(A) = d$ and $H^i_m(M) \simeq \mathrm{Tor}^A_{d-i}(H^d_m(A),M)$.

PROOF. This is a direct consequence of Theorem 7.3.4, since a parameter ideal is generated by a regular sequence of length d, while $H^i_{\mathfrak{a}}(M) = H^i_m(M)$ for all i and all M.

We now state a number of results on finitely generated modules of finite projective dimension stemming from the inequality 8.4.5 (ii) and which use the notion of Cohen-Macaulay rings and modules.

9.1.5 PROPOSITION. Let M be a finitely generated module over a noetherian local ring (A,m,k) with pd M < ∞. Then the rings $A_{\mathfrak{p}}$ are Cohen-Macaulay for all primes $\mathfrak{p}$ which are minimal in Supp M. Moreover gr M = ht $\mathfrak{q}$ for at least one such prime $\mathfrak{q}$.

PROOF. If $\mathfrak{p}$ is minimal in Supp M, then dim $M_{\mathfrak{p}} = 0$ so dp $M_{\mathfrak{p}} = 0$. From 8.4.5 (ii), we conclude that $A_{\mathfrak{p}}$ is CM, since $\mathrm{pd}_{A_{\mathfrak{p}}} M_{\mathfrak{p}} \le \mathrm{pd}_A M < \infty$. Furthermore gr M = $\inf\{\mathrm{fd}^{k(\mathfrak{p})}_{A_{\mathfrak{p}}} M_{\mathfrak{p}} \mid \mathfrak{p} \in \mathrm{Supp}\, M\}$ by 7.2.13. But $\mathrm{fd}^{k(\mathfrak{p})}_{A_{\mathfrak{p}}} M_{\mathfrak{p}} = \mathrm{pd}_{A_{\mathfrak{p}}} M_{\mathfrak{p}}$ so the inequality just recalled shows that this infimum must be achieved for a prime $\mathfrak{q}$ minimal in Supp M. Thus gr M = $\mathrm{pd}_{A_{\mathfrak{q}}} M_{\mathfrak{q}}$. Since $\mathrm{dp}_{A_{\mathfrak{q}}} M_{\mathfrak{q}} = 0$, we have $\mathrm{pd}_{A_{\mathfrak{q}}} M_{\mathfrak{q}} = \mathrm{dp}\, A_{\mathfrak{q}}$ by 7.1.5, and we already know that the ring $A_{\mathfrak{q}}$ is Cohen-Macaulay, so gr M = dp $A_{\mathfrak{q}}$ = ht $\mathfrak{q}$.

9.1.6 COROLLARY. If $\mathfrak{a}$ = Ann M, then ht $\mathfrak{a}$ = gr M.

This allows one to reformulate the Codimension Conjecture 8.5.3 (C), cf. [Fo 79, Th. 5.5].

9.1.7 CONJECTURE. In the corollary, ht $\mathfrak{a}$ + dim $A/\mathfrak{a}$ = dim A.

This rather explains the name codimension conjecture and yet again illustrates the theme emphasized in section 8.5: the support of a finitely generated module of finite projective dimension is embedded in Spec A in a particularly nice way. It also shows that the Codimension Conjecture is true for a ring in which all maximal chains of prime ideals have length equal to the dimension of the ring, e.g. a catenary noetherian local domain. More generally, rings which satisfy the identity 9.1.7. for all ideals $\mathfrak{a}$ have been dubbed taut-level in [Ra]. We show that a Cohen-Macaulay ring is taut-level and that, a fortiori, the Codimension Conjecture is true for such a ring.

9.1.8 PROPOSITION. Let A be a Cohen-Macaulay local ring. Then $\text{ht } \mathfrak{a} + \dim A/\mathfrak{a} = \dim A$ for every ideal $\mathfrak{a}$ and if $\mathfrak{a} = \text{Ann } M$ for a finitely generated module M, then $\text{ht } \mathfrak{a} = \text{gr } M$.

PROOF. Since always $\text{gr } M \leq \text{ht } \mathfrak{a}$ and $\text{ht } \mathfrak{a} + \dim A/\mathfrak{a} \leq \dim A$ by the definitions, the inequality 8.5.2 with $N = A$ implies $\text{dp } A \leq \text{gr } M + \dim A/\mathfrak{a} \leq \text{ht } \mathfrak{a} + \dim A/\mathfrak{a} \leq \dim A$. If A is Cohen-Macaulay, these are equalities.

We next prove a result which combines observations of Peskine-Szpiro and Macaulay and tells us when strong parameters actually form a regular sequence.

9.1.9 THEOREM. Let $x_1,\ldots,x_r$ be elements in a noetherian local ring A which generate an ideal $\mathfrak{a}$ of height r. Then the following are equivalent:

(i) $\text{pd } A/\mathfrak{a} < \infty$;

(ii) $x_1,\ldots,x_r$ are a regular sequence;

(iii) $\text{pd } A/\mathfrak{a}^t < \infty$ for all $t \geq 1$.

PROOF. If one assumes (ii), then $\mathfrak{a}^t$ is generated by all monomials in $x_1,\ldots,x_r$ of degree t and $\text{pd } A/\mathfrak{a}^t < \infty$ by Theorem 5.1.8, in view of Corollary 5.2.5.
Since (iii) obviously implies (i), let us suppose $\text{pd } A/\mathfrak{a} < \infty$. According to 9.1.6 we have $\text{gr } A/\mathfrak{a} = \text{ht } \mathfrak{a} = r$, while $\text{gr } A/\mathfrak{a} = \text{E-dp}(\mathfrak{a},A) = \text{dp}(\mathfrak{a},A)$ by the remarks following 7.2.10. Now conclude by Exercise 5.3.19 that $x_1,\ldots,x_r$ is

A-regular. The reader is invited to provide an alternative proof using 5.2.5 and 6.1.6 or Exercise 3 of 7.3.7.

It is clear that the ideals considered in this theorem are perfect, i.e. the cyclic modules $A/\mathfrak{a}^n$ are perfect. A special case arises when $r = \dim A$: if $\operatorname{pd} A/\mathfrak{a} < \infty$ for a parameter ideal $\mathfrak{a}$, then A is Cohen-Macaulay, and vice versa. In fact we can characterize CM rings as follows.

9.1.10 PROPOSITION. A noetherian local ring is Cohen-Macaulay if and only if it possesses a module which has finite length and finite projective dimension.

PROOF. The "if" part is a corollary either of 8.4.5 (ii) or of 9.1.5. In fact the former shows that it is enough for the ring to possess a CM-module of finite projective dimension, and this module is then perfect.

This proposition and the stronger statement just mentioned can be conceived as dual versions of a question of Bass, which we shall deal with in 13.1.6. We end the section with a series of statements in the form of an exercise. Let us for brevity refer to a finitely generated module of finite projective dimension which satisfies an equality in 8.4.5 (ii) as a Buijs module.

9.1.11 EXERCISE. We consider finitely generated modules M over a noetherian local ring A.

a) If M is perfect, then $\operatorname{dp}(\mathfrak{p},A) = \operatorname{pd} M$ for every $\mathfrak{p} \in \operatorname{Ass} M$.

b) A perfect module is a Buijs module; a Buijs module M satisfies (D) of 8.5.3 and is perfect iff it satisfies (C).

c) If A is Cohen-Macaulay, perfect modules, Buijs modules and Cohen-Macaulay modules of finite projective dimension are all the same.

d) If M is Cohen-Macaulay, with annihilator $\mathfrak{a}$, then $B = A/\mathfrak{a}$ is both taut-level and catenary, and if $\mathfrak{b}$ is an ideal in B, then $dp(\mathfrak{b},M) = ht\,\mathfrak{b}$.

9.2 REGULAR LOCAL RINGS

Cohen-Macaulay local rings were characterized as those rings in which each s.o.p. forms a regular sequence, and it is enough to know this for a single s.o.p. What if the maximal ideal itself is generated by a s.o.p.?

9.2.1 THEOREM. Let $(A,\mathfrak{m})$ be a d-dimensional local ring and suppose that $\mathfrak{m} = (x_1,\dots,x_d)$. Then every permutation $y = y_1,\dots,y_d$ of the x's forms a regular sequence and $(y_1,\dots,y_r)$ is a prime ideal of height r for $0 \le r \le d$.

This is just the case $\mathfrak{p} = \mathfrak{m}$ of 8.3.10. A local ring as above is called regular; it is a special kind of Cohen-Macaulay ring and also a domain (case $r = 0$ in the theorem). In order to prove the main theorem characterizing regular rings, we need the following observation which appears in Grothendieck [Gro 64, Ch. 0, Lemme 17.3.1.3] but is probably due to Kaplansky.

9.2.2 LEMMA. In a noetherian local ring $(A,\mathfrak{m},k)$ let $x \in \mathfrak{m}\setminus\mathfrak{m}^2$. Then $\mathfrak{m}/(x)$ is a direct summand of $\mathfrak{m}/x\mathfrak{m}$ as an A-module.

PROOF. Let $x,y_1,\dots,y_r$ be a minimal set of generators of $\mathfrak{m}$, which is obtained by lifting a k-basis of the vector space $\mathfrak{m}/\mathfrak{m}^2$. Put

$\mathfrak{a} = (y_1, \ldots, y_r)$. Then $\mathfrak{a} + (x) = \mathfrak{m}$ while $\mathfrak{a} \cap (x) \subset x\mathfrak{m}$; the latter inclusion holds because $yx \in \mathfrak{a}$ cannot be true for a unit y. Therefore the composition

$$\mathfrak{m}/(x) = (\mathfrak{a} + (x))/(x) \simeq \mathfrak{a}/(\mathfrak{a} \cap (x)) \to \mathfrak{m}/x\mathfrak{m} \to \mathfrak{m}/(x)$$

is the identity and we have the desired splitting.

Recall that the global dimension of a ring is defined to be $\sup\{i \mid \mathrm{Ext}^i_A(M,N) \neq 0\}$ over all couples of modules M and N and that for a local noetherian ring $(A,\mathfrak{m},k)$ we know that gl dim A = $\mathrm{pd}_A k$, see for instance [Se, Ch. IV, C] or [Ma 86, §19, Lemma 1]. We are now in a position to prove a well known characterization of regular local rings due to Auslander-Buchsbaum and to Serre [Se, Ch. IV, Th. 9],[Ka 74, Ch. 3, §3] or [Ma 86, Ch. 7]. The usual procedure is to concentrate on $\mathrm{pd}_A k$. Here, as an alternative, we feature $\mathrm{id}_A \mathfrak{m}$, following an exercise in Kaplansky [Ka 74, Ch. 4, Exc. 6].

9.2.3 THEOREM. For a d-dimensional noetherian local ring $(A,\mathfrak{m},k)$ the following statements are equivalent:

(i) A is regular;

(ii) $\dim_k \mathfrak{m}/\mathfrak{m}^2 = d$;

(iii) $\mathfrak{m}$ can be generated by a regular sequence;

(iv) The map $\beta_A: k[X_1, \ldots, X_d] \to G_{\mathfrak{m}}(A)$ of 5.1.12 is an isomorphism, where $x_1, \ldots x_d$ generate $\mathfrak{m}$;

(v) gl dim A $< \infty$;

(vi) $\mathrm{id}_A \mathfrak{m} < \infty$.

PROOF. (i) $\Leftrightarrow$ (ii): Since a basis of the k-vectorspace $\mathfrak{m}/\mathfrak{m}^2$ lifts to a set of generators of $\mathfrak{m}$, the condition $\dim_k \mathfrak{m}/\mathfrak{m}^2 = d$ just means that $\mathfrak{m}$ can be generated by a set of parameters.

(i) $\Rightarrow$ (iii): Is proven in 9.2.1.

(iii) ⇒ (iv): Since this regular sequence is certainly a s.o.p., we know that the sequence has length d. A regular sequence is in turn pre-regular, and the proof that β_A is an isomorphism is furnished by Theorem 5.1.12.

(iv) ⇒ (ii): The images $\beta_A(X_i)$, $i = 1,\ldots,d$, form a k-basis of m/m^2.

(iii) ⇒ (v): The A-module k arises from A by dividing out a regular sequence of length d; hence $\mathrm{pd}_A k = d$. By the result recalled above, then gl dim A = d.

(v) ⇒ (vi): A fortiori.

(vi) ⇒ (iii): We first observe that $\mathrm{id}_A m = \mathrm{dp}\, A = s$ by 7.1.5, and we induce on this common integer and show that it is d.

Case s = 0: Let $x \in m$ and consider the ideal (x) in A. The injection $(x) \to m$ extends to a map $f: A \to m$, given by $1 \mapsto y$ for some $y \in m$. Then $f(x) = xy = x$ and $x(1-y) = 0$. However, 1-y is a unit, so x = 0. Thus A = k, and a field is a 0-dimensional regular local ring.

Case $s \geq 1$: Since dp $A = s \geq 1$, there exists a non zerodivisor in $m \backslash m^2$. For if not, then $m \backslash m^2 \subset \cup\, \mathfrak{p}_i$ where the finitely many $\mathfrak{p}_i$'s are the elements of Ass A. Then $m \subset m^2 \cup \mathfrak{p}_i$, and since $m \not\subset m^2$, then $m \subset \mathfrak{p}_i$ for some i [Bo 61a, Ch. 2, §1, Prop. 2]. Therefore $m \in$ Ass A which would mean dp A = 0.

Let x be a non zerodivisor in $m \backslash m^2$. Consider the maximal ideal $m/(x)$ in the factor ring $A/(x)$. By Lemma 9.2.2 it is a direct summand of m/xm, while $\mathrm{id}_{A/(x)} m/xm = s-1$ in view of 3.3.6 (ii). Thus $\mathrm{id}_{A/(x)} m/(x) \leq s-1$ but, because dp $A/(x) = s-1$, we must have equality. By the induction hypothesis, this maximal ideal $m/(x)$ is generated by a regular sequence $\bar{y}_1,\ldots,\bar{y}_{s-1}$ in the local ring $A/(x)$. Clearly m is generated by the regular sequence $x, y_1, \ldots, y_{s-1}$ in A, so s = d.

Once this result has been established, there are many facts about regular local rings which can be deduced. We refer to [Se, Ch. IV, D] and to [Ma 86, Ch. 7]. Here we only mention one of the crowning results in Serre's influential course. He showed that the Dimension Conjecture (D) in 8.5.4 is true for all finitely generated modules over a regular local ring [Se, Ch. V, Th. 3]. Since the Codimension Conjecture (C) is true even over a Cohen-Macaulay ring, 9.1.7, this means that Conjectures 8.5.4 are all satisfied for every finitely generated module over a regular local ring. Now such a module has finite projective dimension by Theorem 9.2.3, so it was imagined that this was its distinguishing feature. Thus Serre's achievement naturally led to Conjectures 8.5.4 and a number of related ones regarding finitely generated modules of finite projective dimension, which are certainly not satisfied for arbitrary modules over a noetherian local ring.

9.3 COMPLETE LOCAL RINGS AND THE DIRECT SUMMAND CONJECTURE

We first observe that if $\hat{A}$ is the $\mathfrak{m}$-adic completion of a noetherian local ring $(A,\mathfrak{m})$, then $\hat{A}$ is regular if and only if A is. This follows from any of the characterizations in 9.2.3. It is clear that if K is a field, the power series ring $K[[X_1,\ldots,X_d]]$ with $\mathfrak{m} = (X_1,\ldots,X_d)$ is d-dimensional regular local, complete in its $\mathfrak{m}$-adic topology.

Now suppose that $(A,\mathfrak{m},k)$ is a complete noetherian local ring which is equicharacteristic. In 2.3.3 and 2.3.5 we obtained I.S. Cohen's result that A contains a field of representatives or, as some authors write, a coefficient field. We now prove Cohen's two important structure theorems for such rings.

9.3.1 THEOREM. Let $(A,\mathfrak{m})$ be an equicharacteristic local ring, complete in its m-adic topology. Then A is a homomorphic image of a power series ring $K[[X_1,\ldots,X_n]]$ over a field K. If A is d-dimensional regular, we can choose $n = d$ and $K[[X_1,\ldots,X_d]] \simeq A$.

PROOF. Let K be a field of representatives in A, then $K + m = A$ and $K \cap m = (0)$. Let $x_1,\ldots,x_n$ be a minimal set of generators of m, then we can map $K[X_1,\ldots,X_n]$ to A by choosing the identity on K and sending X_i to x_i, $i = 1,\ldots,n$. This map is continuous w.r.t. the $(X_1,\ldots,X_n)$-adic topology in $K[X_1,\ldots,X_n]$ and the m-adic topology in A. Since A is complete, the map factors through the $(X_1,\ldots,X_n)$-adic completion of the polynomial ring so we obtain a continuous ring homomorphism ϕ: $K[[X_1,\ldots,X_n]] \to A$ which is clearly surjective.

In case A is regular, $n = d$ and A is an integral domain. Thus $\ker \phi$ is a prime ideal, and comparing dimensions shows that it is (0), which finishes the proof.

9.3.2 COROLLARY. An equicharacteristic complete noetherian local ring is universally catenary.

PROOF. It is the homomorphic image of a Cohen-Macaulay ring.

9.3.3 THEOREM. Let (A,m) be a d-dimensional equicharacteristic noetherian local ring, complete in its m-adic topology. Then there exists a field K and a subring $K[[X_1,\ldots,X_d]]$ of A such that A is a finitely generated module over this subring.

PROOF. Let K be a field of representatives in A, and $x = x_1,\ldots x_d$ a s.o.p. in A. As in the proof of 9.3.1 we obtain a continuous ring homomorphism $K[[X_1,\ldots,X_d]] \to A$ which is not surjective this time. It is injective, however: For suppose that $f = \Sigma f_i$, with f_i homogeneous of total degree i in $X_1,\ldots,X_d$, is a power series such that $f(x_1,\ldots,x_d) = 0 \in A$. If f_s is its lowest degree nonzero term, then $f_s(x_1,\ldots,x_d) \in \mathfrak{a}^{s+1}$, where $\mathfrak{a} = (x_1,\ldots,x_d)$. Write $f_s = \Sigma_{|\upsilon|=s} a_\upsilon X^\upsilon$ in the notation of 5.1.10 with $a_\upsilon \in K \subset A$, then that result with $M = A$ tells us that all the coefficients a_υ are in the ideal $\mathfrak{b}_x \subset A$. But according to 8.3.2 this $\mathfrak{b}_x$ is contained in $\mathfrak{R}(\mathfrak{a})$, and since x is a s.o.p., $\mathfrak{R}(\mathfrak{a}) = \mathfrak{m}$. So $a_\upsilon \in K \cap \mathfrak{m} = (0)$ and $f_s = 0$, so $f = 0$. This argument, by the way, explains the name "analytic independence" for property 8.3.2.

It remains to show that A is a finitely generated $K[[X_1,\ldots,X_d]]$-module. Identify X_i and x_i, $i = 1,\ldots,d$. Since $\mathfrak{m}^n \subset \mathfrak{a}$ for some $n \geq 1$, one easily sees that there exists a finite dimensional K-vectorspace $V \subset A$ such that $A = V \oplus \mathfrak{a}$. But then $A = V \oplus \mathfrak{a}A = V \oplus \mathfrak{a}V \oplus \mathfrak{a}^2 = V \oplus \mathfrak{a}V \oplus \mathfrak{a}^2V \oplus \mathfrak{a}^3 = \ldots$. Now use that A is complete in its $\mathfrak{m}$-adic topology, which coincides with the $\mathfrak{a}$-adic topology. Therefore every element of A is a uniquely determined limit of a sequence of elements in $K[X_1,\ldots,X_d]V$, so $A = K[[X_1,\ldots,X_d]]V$, which means that A is a finitely generated $K[[X_1,\ldots,X_d]]$-module.

In the mixed characteristic case, there exist variants of these results, originally also due to Cohen, which are a bit harder to state and to prove. We do not do so, chiefly because they are not needed for the main lines of argument in this book. We refer the reader to [Bo 83, Ch. IX], [Na, Ch. 3] or [Ma 86, Ch. 10]. Here we only mention that 9.3.2 is also true in the mixed characteristic case: a complete noetherian local ring is always universally catenary, being a homomorphic image of a complete regular local ring.

We have seen that over a field a complete noetherian local ring is a module-finite extension of a regular subring. This is one of the motivations for Hochster's [Ho 73b]

9.3.4 DIRECT SUMMAND CONJECTURE. Let R be a regular local ring and A a module-finite extension ring of R. Then R is a direct summand of A as an R-module.

This conjecture is related to the Monomial Conjecture 8.3.5 as we shall see presently. Like that one, it is enough to prove the Direct Summand Conjecture for the $\mathfrak{m}$-adic completion $\hat{R}$ of R w.r.t. to its maximal ideal $\mathfrak{m}$. For a splitting of $R \to A$ is tantamount to the map $\mathrm{Hom}_R(A,R) \to \mathrm{Hom}_R(R,R)$ being surjective. Tensoring with the faithfully flat R-module $\hat{R}$ does not change surjectivity. But $\hat{R}$ being a flat and A a finitely presented R-module, this map is $\mathrm{Hom}_{\hat{R}}(\hat{R}\otimes_R A, \hat{R}\otimes_R R) \to \mathrm{Hom}_{\hat{R}}(\hat{R}\otimes_R R, \hat{R}\otimes_R R)$ [Bo 61a, Ch. I, Prop. 11] or rather $\mathrm{Hom}_{\hat{R}}(\hat{A},\hat{R}) \to \mathrm{Hom}_{\hat{R}}(\hat{R},\hat{R})$ and this map is surjective precisely when $\hat{R} \to \hat{A}$ splits.

In this chapter we prove the Direct Summand Conjecture in equal characteristic 0, where in fact the following far more general statement is true. More general, because a regular local ring is certainly an integrally closed domain [Se, Ch. IV, Th. 9, Cor. 3], which in equal characteristic 0 contains the rationals.

9.3.5 PROPOSITION. Suppose R is an integrally closed integral domain which contains the rationals. If A is an extension ring which is finitely generated as an R-module, then this extension splits.

PROOF. We may divide out from A a minimal prime $\mathfrak{p}$ to obtain an integral domain $A/\mathfrak{p}$; for if the composition $R \to A \to A/\mathfrak{p}$ splits, so does $R \to A$. Hence we may as well assume A is a domain. A similar argument allows us to enlarge A if necessary so that its field of fractions L is a Galois extension of K, the field of fractions of R, and that A is the integral closure of R in L. If $[L : K] = n$, the map $1/n.\mathrm{Tr}: L \to K$ furnishes the splitting, for $\mathrm{Tr}(x) = nx$ for every $x \in K$, and the trace of an integral element in L is an integral element in K. Since R is integrally closed, this means that $1/n.\mathrm{Tr}$ maps A to R as required.

In positive characteristic the Direct Summand Conjecture is also true, but is a deeper statement, as we shall see in Proposition 10.3.5. Hochster has given an example to show that mere integral closure does not suffice [Ho 73b, Ex. 1]. Like most conjectures, this one can be globalized, but we shall not do so. In mixed characteristic, its validity is unknown, although there are proofs for certain types of extensions [Ro 80a], [HM], [Ko 86]. The Direct Summand Conjecture is discussed in great detail in [Ho 73b] and [Ho 83]. We include an exercise which shows that integrally closed is at least necessary [Ho 73b, p. 27].

9.3.6 EXERCISE. If A is a domain which is a direct summand of every module-finite extension, then A is integrally closed.

We conclude this chapter with

9.3.7 PROPOSITION. Suppose the Direct Summand Conjecture holds in equal characteristic. So does then the Monomial Conjecture.

PROOF. We only need prove the Monomial Conjecture for a complete local ring A, so let $X = X_1,\ldots,X_d$ be a s.o.p., whence A is a module-finite extension of $K[[X_1,\ldots,X_d]]$ as in 9.3.3. By assumption there exists a splitting $s: A \to K[[X_1,\ldots,X_d]]$ of that injection. If $X_1^t\ldots X_d^t =$

$\Sigma_{i=1}^{d} a_i X_i^{t+1}$ with $a_i \in A$ is a relation in A, then $X_1^t \ldots X_d^t = \Sigma_{i=1}^{d} s(a_i) X_i^{t+1}$ is true in the power series ring. But X, being a regular sequence in that ring, is certainly pre-regular by 5.1.7, so this equation is precluded by 5.1.12 or, more quickly, conclude by 9.1.3.

10 GORENSTEIN RINGS, LOCAL DUALITY, AND THE DIRECT SUMMAND CONJECTURE

A Gorenstein ring is a particularly nice Cohen-Macaulay ring and, as we shall see, complete intersections are always Gorenstein. In the first section we give various characterizations of Gorenstein rings. The second contains part of the local duality theory for homomorphic images of Gorenstein rings. In the last section of the chapter, we shall use some of these results to prove the equivalence of the Monomial Conjecture and the Direct Summand Conjecture in equal characteristic as well as the Monomial Conjecture itself in characteristic $p > 0$.

10.1 GORENSTEIN RINGS

10.1.1 DEFINITION. Let A be a noetherian local ring. Then A is said to be Gorenstein if it has finite injective dimension over itself, i.e. $\operatorname{id} A < \infty$.

We first mention a few elementary properties.

10.1.2 PROPOSITION. Let (A,m,k) be a noetherian local ring. Then

(i) The m-adic completion $\hat{A}$ of A is Gorenstein if and only if A is Gorenstein;

(ii) A is Gorenstein if and only if for every non zerodivisor $x \in m$ on A, the ring A/(x) is Gorenstein and a single one suffices;

(iii) If A is Gorenstein, also all localizations $A_{\mathfrak{p}}$ at a prime ideal $\mathfrak{p}$ are Gorenstein.

PROOF. Since $\mathrm{Ext}^i_A(A/m,A)\hat{} \simeq \mathrm{Ext}^i_{\hat{A}}(\hat{A}/\hat{m},\hat{A})$ [Bo 80, §6, Prop. 10 b)] and, since $\mathrm{id}^k_A A = \mathrm{id}_A A$ and $\mathrm{id}^k_{\hat{A}}\hat{A} = \mathrm{id}_{\hat{A}}\hat{A}$, the equivalence in (i) follows from the faithful exactness of the m-adic completion functor on finitely generated modules. Furthermore, (ii) is a consequence of 3.3.6 with M = A. As to (iii), notice that $\mathrm{Ext}^i_A(A/\mathfrak{p},A)_{\mathfrak{p}} \simeq \mathrm{Ext}^i_{A_{\mathfrak{p}}}(A_{\mathfrak{p}}/\mathfrak{p}A_{\mathfrak{p}},A_{\mathfrak{p}})$ (see 3.3.2) and the result follows easily.

We next locate Gorenstein rings in between two other types of ring already discussed.

10.1.3 PROPOSITION. A complete intersection is Gorenstein. A Gorenstein ring is Cohen-Macaulay.

PROOF. Let A be a complete intersection, i.e. $A = R/(x_1,\ldots,x_n)$, where R is a regular local ring and $x_1,\ldots,x_n$ is a regular sequence in R. Since R is regular, gl dim $R < \infty$, so R is certainly Gorenstein. The assertion that A is Gorenstein follows by iterated use of 10.1.2 (ii). Now let A be Gorenstein. Then 8.5.5 with M = A implies dp A = dim A since gr A is obviously 0.

In order to give a first characterization of Gorenstein rings, we need a lemma.

10.1.4 LEMMA. Let $f(T)$ be a polynomial in one variable over $\mathbb{Z}$ and assume that its coefficients are all nonnegative. In case $f(T).f(T^{-1}) = 1$ in $\mathbb{Z}[T,T^{-1}]$, then $f(T) = T^r$ for some $r \geq 0$.

PROOF. If r denotes the degree of $f(T)$, the element $g(T) = T^r f(T^{-1})$ of $\mathbb{Z}[T,T^{-1}]$ is in fact in $\mathbb{Z}[T]$. Hence we have a relation $f(T).g(T) = T^r$ in $\mathbb{Z}[T]$. This means that $f(T) = \pm T^s$ for some $s \leq r$. But since r is the degree of $f(T)$ and since $f(T)$ has no negative coefficients, we necessarily must have $f(T) = T^r$.

Recall the power series $\mu(A,T) = \Sigma_{i=0}^{\infty} \mu^i(A)T^i$ of 7.1.6.

10.1.5 THEOREM. Let A be a noetherian local ring of dimension d. Then A is Gorenstein if and only if $\mu(A,T) = T^d$.

PROOF. If A is Gorenstein, 7.1.6. yields that $\mu(A,T).\mu(A,T^{-1}) = \beta(A,T)$. Since $\text{id } A < \infty$, $\mu(A,T)$ is obviously in $\mathbb{Z}[T]$. Clearly, $\beta(A,T) = 1$ so $\mu(A,T) = T^r$ for some $r \geq 0$ by 10.1.4. But A is Gorenstein, so $\text{id}^k A = \text{id } A = \text{dp } A = \dim A = d$ and we find $\mu^d(A) \neq 0$. Hence $r = d$. Conversely, if $\mu(A,T) = T^d$, or more generally if $\mu(A,T)$ is a polynomial, obviously $\text{id } A < \infty$ and we are done.

10.1.6 COROLLARY. Let A be a Gorenstein ring. Then $\mu^i(\mathfrak{p},A) = \delta_{i,\text{ht}\mathfrak{p}}$ for all prime ideals $\mathfrak{p}$. Specifically, a minimal injective resolution of A is

$$0 \to A \to \oplus_{\text{ht}\mathfrak{p}=0} E(A/\mathfrak{p}) \to \ldots . \to \oplus_{\text{ht}\mathfrak{p}=d-1} E(A/\mathfrak{p}) \to E(A/m) \to 0.$$

PROOF. The first statement follows directly from 10.1.2 (iii) and the above theorem. Using this and the definition of the μ^i after 3.3.1, one obtains the result about the minimal injective resolution of A.

We now shall give a characterization of Gorenstein rings in terms of irreducible ideals. Recall that a proper ideal of a ring A is called irreducible if it cannot be expressed as the intersection of two strictly larger ideals. First, we need

10.1.7 LEMMA. Let (A,m,k) be a noetherian local ring and let $\mathfrak{a}$ be an m-primary ideal of A. Then the following statements are equivalent:

(i) $\mathfrak{a}$ is irreducible;

(ii) The k-vectorspace $\mathrm{Hom}_A(k,A/\mathfrak{a}) \simeq (\mathfrak{a}:m)/\mathfrak{a}$ is one-dimensional;

(iii) The set of all ideals of A which properly contain $\mathfrak{a}$ admits $(\mathfrak{a}:m)$ as smallest element.

PROOF. Obviously $\mathrm{Hom}_A(k,A/\mathfrak{a}) \simeq (\mathfrak{a}:m)/\mathfrak{a}$ is a nonnull k-vectorspace which for convenience we abbreviate to V. Notice that $(\mathfrak{a}:m)$ is a proper ideal if and only if $\mathfrak{a} \neq m$.

(i) ⇒ (ii): If $\dim_k V > 1$, then $(0) \subset V$ would be the intersection of two nonnull subspaces so that there would exist ideals $\mathfrak{a}'$ and $\mathfrak{a}''$ of A with $\mathfrak{a} \subsetneq \mathfrak{a}' \subset (\mathfrak{a}:m)$, $\mathfrak{a} \subsetneq \mathfrak{a}'' \subset (\mathfrak{a}:m)$ and $\mathfrak{a} = \mathfrak{a}' \cap \mathfrak{a}''$. This contradicts the fact that $\mathfrak{a}$ is irreducible.

(ii) ⇒ (iii): Let $\mathfrak{a}'$ be an ideal of A for which $\mathfrak{a} \subsetneq \mathfrak{a}' \subset m$. We must show that $\mathfrak{a}' \supset (\mathfrak{a}:m)$. Since the A-module $A/\mathfrak{a}$ is of finite length, by taking a composition series, we can find an ideal $\mathfrak{a}''$ such that $\mathfrak{a} \subsetneq \mathfrak{a}'' \subset \mathfrak{a}'$ and such that $m\mathfrak{a}'' \subset \mathfrak{a}$. Hence $\mathfrak{a} \subsetneq \mathfrak{a}'' \subset (\mathfrak{a}:m)$. But by hypothesis the A-module $(\mathfrak{a}:m)/\mathfrak{a}$ has length one so that $\mathfrak{a}'' = (\mathfrak{a}:m)$. Therefore $\mathfrak{a}' \supset (\mathfrak{a}:m)$, as required.

(iii) ⇒ (i): Since $\mathfrak{a} \subsetneq (\mathfrak{a}:\mathfrak{m})$ and every ideal which contains $\mathfrak{a}$ properly has to contain $(\mathfrak{a}:\mathfrak{m})$, the implication is immediate.

The next result comes from Bass' influential paper [Bas].

10.1.8 THEOREM. Let $(A,\mathfrak{m},k)$ be a noetherian local ring of dimension d. Then the following statements are equivalent:

(i) A is Gorenstein;

(ii) A is Cohen-Macaulay and $\mu^d(A) = 1$;

(iii) A is Cohen-Macaulay and some parameter ideal of A is irreducible;

(iv) All parameter ideals of A are irreducible.

NOTE. The Cohen-Macaulayness of A in (ii) can be deleted, as proved by P. Roberts [Ro 83]. We shall not pursue this sharpening.

PROOF. (i) ⇒ (ii): This is immediate from 10.1.3 and 10.1.6.
(ii) ⇒ (iv): Let $x_1,\ldots,x_d$ be a system of parameters for A. Since A is Cohen-Macaulay, the x's form a regular sequence in A. If $\bar{A}$ denotes the ring $A/(x_1,\ldots,x_d)$, it follows from 3.3.5 that $\mu^0_{\bar{A}}(\bar{A}) = \mu^d_A(A) = 1$. But $\dim_k \mathrm{Hom}_{\bar{A}}(k,\bar{A}) = \mu^0_{\bar{A}}(\bar{A})$ so $k \simeq \mathrm{Hom}_{\bar{A}}(k,\bar{A}) \simeq \mathrm{Hom}_A(k,\bar{A})$, hence the ideal generated by the x's is irreducible by 10.1.7.
(iii) ⇒ (i): Suppose that A is Cohen-Macaulay and let there be given an irreducible ideal, generated by a s.o.p. $x_1,\ldots,x_d$. By 10.1.2 (ii), A is Gorenstein if and only if $\bar{A} = A/(x_1,\ldots,x_d)$ is Gorenstein. We shall prove that $\bar{A}$ is Gorenstein. Since (0) is irreducible in $\bar{A}$, $E_{\bar{A}}(\bar{A})$ is an indecomposable injective module by 3.2.2, hence $E_{\bar{A}}(\bar{A}) \simeq E_{\bar{A}}(k)$ by 3.2.5. Now

$\bar{A}$ is of finite length and $E_{\bar{A}}(k)$ is its Matlis dual, so has the same length by 3.4.5. Thus the inclusion $\bar{A} \to E_{\bar{A}}(\bar{A})$ is in fact an isomorphism, so $\bar{A}$ is self-injective and hence Gorenstein.

(iv) ⇒ (iii): We only need to prove that A is Cohen-Macaulay. Clearly, by dividing out a maximal regular sequence, we may assume that dp A = 0. We must prove that dim A = 0. Suppose not, then there exists a system of parameters $x_1,\dots,x_d$, where $d \geq 1$. For each $t \geq 1$, let $\mathfrak{a}_t$ be the ideal generated by $x_1^t,\dots,x_d^t$; then we obtain a strictly descending chain $\mathfrak{a}_1 \supsetneq \cdots \supsetneq \mathfrak{a}_t \supsetneq \cdots$ of irreducible $\mathfrak{m}$-primary ideals of A such that $\cap_{t\geq 1} \mathfrak{a}_t = 0$. By Lemma 10.1.7 we find that $(\mathfrak{a}_{t+1}:\mathfrak{m}) \subset \mathfrak{a}_t$ for all $t \geq 1$. Hence $(0:\mathfrak{m}) = (\cap_{t\geq 1} \mathfrak{a}_{t+1}):\mathfrak{m} = \cap_{t\geq 1}(\mathfrak{a}_{t+1}:\mathfrak{m}) \subset \cap_{t\geq 1} \mathfrak{a}_t = 0$. It follows that $\mathfrak{m} \notin \mathrm{Ass}\, A$, contradicting the assumption dp A = 0, and we are done.

We next characterize Gorenstein rings in terms of certain subcategories of their module category. For convenience, we introduce some notation. Let A be a local ring. We define $\mathcal{F}$ (resp. $\mathcal{J}$) to be the full subcategory of the category of A-modules, having as objects the A-modules of finite flat dimension (resp. finite injective dimension). Similarly, we define $\mathcal{F}_{<\infty}$ (resp. $\mathcal{J}_{<\infty}$) to be the full subcategory of $\mathcal{F}$ (resp. $\mathcal{J}$), having as objects the A-modules $M \in \mathcal{F}$ (resp. $M \in \mathcal{J}$) with T-codp $M < \infty$ (or, equivalently, with E-dp $M < \infty$, see 6.1.8). In these terms, the characterization reads:

10.1.9 THEOREM. Let $(A,\mathfrak{m},k)$ be a noetherian local ring. Then the following statements are equivalent:

(i) A is Gorenstein;

(ii) $\mathrm{id}^k M < \infty$ for some $M \in \mathcal{F}_{<\infty}$;

(iii) $\mathrm{fd}^k M < \infty$ for some $M \in \mathcal{J}_{<\infty}$;

(iv) $\mathcal{J}_{<\infty} \cap \mathcal{F}_{<\infty} \neq \emptyset$;

(v) $\mathcal{J}_{<\infty} = \mathcal{F}_{<\infty}$;

(vi) $\mathcal{J} = \mathcal{F}$.

PROOF. We obviously have the implications (vi) ⇒ (v) ⇒ (iv) ⇒ (iii). In view of 7.1.1, the Matlis duality functor takes care of the equivalence between (ii) and (iii). The implication (ii) ⇒ (i) immediately follows from 7.1.3, where we now take N = A: the assumption (ii) yields that $\mathrm{id}^k A < \infty$ and hence id A < ∞ which means that A is Gorenstein. The only thing left to prove is (i) ⇒ (vi). Let A be Gorenstein and assume that $M \in \mathcal{F}$. It follows from 7.1.2 with N = A, that also id M < ∞. Hence $\mathcal{F} \subset \mathcal{J}$. Conversely, if $M \in \mathcal{J}$, then $M^\vee \in \mathcal{F}$ by 7.1.1. Since we just have proved that $\mathcal{F} \subset \mathcal{J}$, it follows that $M^\vee \in \mathcal{J}$. But by 7.1.1 (ii) this means that $M \in \mathcal{F}$. Hence also $\mathcal{J} \subset \mathcal{F}$. This finishes the proof of the theorem.

10.1.10 COROLLARY. Let (A,m,k) be a d-dimensional local Cohen-Macaulay ring. Then A is Gorenstein if and only if $H^d_m(A)$ is injective. In this case $H^d_m(A) \simeq E(k)$.

PROOF. Let A be Gorenstein. Notice that for $\mathfrak{p} \neq m$, $m^t \not\subset \mathfrak{p}$ and hence $\mathrm{Hom}_A(A/m^t, E(A/\mathfrak{p})) = 0$, see 3.2.9. It follows that $L_m(E(A/\mathfrak{p})) = 0$ for $\mathfrak{p} \neq m$. Furthermore, $L_m(E(k)) = E(k)$ by 3.2.9. Therefore, using the minimal injective resolution $E^\bullet$ of A, whose specific shape is described in 10.1.6, we see that $L_m(E^\bullet)$ is a complex concentrated in degree d at which spot E(k) features. It follows that $H^d_m(A)$, which is $H^d(L_m(E^\bullet))$ by definition, is equal to E(k).

Conversely, let A be Cohen-Macaulay and let $H^d_m(A)$ be injective. Then certainly $H^d_m(A) \in \mathcal{J}$. Since $H^d_m(A) \in \mathcal{F}$ by 9.1.4 and obviously $\text{E-dp } H^d_m(A) = 0 < \infty$, we find that $H^d_m(A) \in \mathcal{J}_{<\infty} \cap \mathcal{F}_{<\infty}$. So A is Gorenstein by 10.1.9.

This corollary contains the last ingredient we need to prove the Local Duality Theorem for Gorenstein rings. This will be carried out in 10.2.1.

We close the section with a few comments on the characterization of Gorenstein rings in Theorem 10.1.9. Let (A,m,k) be a d-dimensional noetherian local ring. We know that fd M $\leq$ pd M for every A-module M, while if M is finitely generated, $\mathrm{fd}^k M = \mathrm{fd}\, M = \mathrm{pd}\, M = \mathrm{pd}^k M$ as recalled in Chapters 7 and 9. In general, there is however the following result, [Fo 77a, Cor. 3.4]: if fd M $< \infty$, then pd M $\leq$ d. This implies that the category $\mathcal{F}$ in 10.1.9 is the same as the category $\mathcal{P}$ of all modules of finite projective dimension. As a consequence of 7.1.6 and 10.1.9 the reader may prove

10.1.11 EXERCISE. Let M be a finitely generated module over a d-dimensional noetherian local ring. If pd M and id M are both finite, then $\mu^i(M) = \beta_{d-i}(M)$ for all i.

10.2 LOCAL DUALITY

This section deals with some aspects of local duality theory. We should emphasize that we only treat part of the theory. Extended and stronger versions as well as other applications can be found in e.g. [Gro 67], [Ha], [HK] and [Sc 82a].

We shall begin with the Local Duality Theorem for Gorenstein rings. As a consequence, we characterize the Krull dimension of a finitely generated module over an arbitrary noetherian local ring in terms of local cohomology. This parallels an analogous result for depth in 5.3.15. Secondly, we introduce and discuss dualizing complexes. This results in a duality theorem for rings which are a homomorphic image of a Gorenstein local ring. From this it follows that certain ideals in such rings are not nilpotent, a result which we shall use in section 11.2.

10.2.1 THEOREM. (Local Duality for Gorenstein rings) Let $(A,\mathfrak{m},k)$ be a Gorenstein local ring of dimension d. Then, with the usual notations for Matlis dual and $\mathfrak{m}$-adic completion,

(i) $H^{d-i}_{\mathfrak{m}}(M)^{\vee} \simeq \mathrm{Ext}^i_A(M,\hat{A})$ for all A-modules M and all i;

(ii) $H^{d-i}_{\mathfrak{m}}(M) \simeq \mathrm{Ext}^i_A(M,A)^{\vee}$ for all finitely generated A-modules M and all i.

PROOF. Observe from 10.1.10 and the Local Duality Theorem for Cohen-Macaulay rings 9.1.4 that $H^{d-i}_{\mathfrak{m}}(M) \simeq \mathrm{Tor}^A_i(M,E(k))$ for all i and all A-modules M. The result now is a direct consequence of the Ext-Tor-duality theorem 3.4.14 (in one of its statements M is required to be finitely generated) and the facts that $E(k)^{\vee} \simeq \hat{A}$ and $A^{\vee} \simeq E(k)$.

10.2.2 COROLLARY. Let $(A,\mathfrak{m},k)$ be a noetherian local ring and let M be an A-module of dimension d. Then

(i) $H^i_{\mathfrak{m}}(M) = 0$ for $i > d$;

(ii) If M is nonnull and finitely generated, $H^d_{\mathfrak{m}}(M) \neq 0$.

PROOF. The dimension of a submodule of M is $\leq$ d. Furthermore, M is a direct limit of its finitely generated submodules. Since local cohomology commutes with direct limits, 4.1.8, it follows easily that we can reduce (i) to the case where M is finitely generated. In this case we now prove (i) and (ii) simultanously.

First of all, since $\hat{A}\otimes_A H^i_m(M) \simeq H^i_{\hat{m}}(\hat{M})$ for all i by 4.1.7 and since $A \to \hat{A}$ is faithfully flat, we may assume that A is complete in the m-adic topology. By the Cohen Structure Theorem 9.3.1 and the comments following 9.3.3, (A,m) is a homomorphic image of a complete regular local ring $(R,\mathfrak{M})$. Since M can be viewed as a finitely generated d-dimensional R-module and since $H^i_{\mathfrak{M}}(M) \simeq H^i_m(M)$ by 4.1.4, we see that we have reduced the problem to the case where the ring is regular. So we may assume that A is regular (and hence Gorenstein) of dimension say D. Since the Matlis duality functor is faithfully exact, 10.2.1 yields that we must prove that $\sup\{i | \operatorname{Ext}^{D-i}_A(M,A) \neq 0\} = d$. But $\sup\{i | \operatorname{Ext}^{D-i}_A(M,A) \neq 0\} =$ $D - \inf\{i | \operatorname{Ext}^i_A(M,A) \neq 0\} = D - \operatorname{gr} M$. Hence we must prove that $D - \operatorname{gr} M = d$. This follows immediately from 9.1.8 and the fact that A is regular and hence Cohen-Macaulay.

This important result and its proof are due to Grothendieck. It belongs to a family of theorems which assert that "a respectable local cohomology vanishes beyond the dimension of the variety but not at the dimension". An alternative proof of (ii) avoids local duality but uses the other Cohen Structure Theorem 9.3.3 rather than 9.3.1. [HK, Satz 4.12]. There also exists an elementary proof [MS, Th. 2.2]. In combination with 5.3.15, Corollary 10.2.2 states that the local cohomology of a finitely generated module can only live between its depth and its dimension, and certainly does at either extreme; in between, however, anything may happen [Mac], [Sh 75], subject to 4.1.11. To round off 10.1.10 we offer

10.2.3 EXERCISE. A d-dimensional noetherian local ring (A,m,k) is

a) Cohen-Macaulay if and only if $H^i_m(A) = 0$ for $i \neq d$;

b) Gorenstein if and only of $H^i_m(A) = 0$ for $i \neq d$ and $H^d_m(A) \simeq E(k)$.

We now come to the dualizing complexes. There are various ways of introducing these. We have chosen one which for our purposes is the most direct.

10.2.4 DEFINITION. Let A be a noetherian local ring. A complex $D_A^\bullet$ of injective A-modules is called a dualizing complex for A if $D^i = \oplus E(A/\mathfrak{p})$, where $d = \dim A$ and the summation ranges over all primes $\mathfrak{p}$ with $\dim A/\mathfrak{p} = d-i$, and all its homologies $H^i(D_A^\bullet)$ are finitely generated.

The simplest example of a dualizing complex occurs for Gorenstein rings: the minimal injective resolution of A is a dualizing complex by 10.1.6. Not every ring admits a dualizing complex. However, we shall see that a large class of rings does. To explain the name, we mention that if A admits a dualizing complex and if $C^\bullet$ is a finite complex of A-modules for which all homologies are finitely generated, then there is a natural homomorphism of complexes $C^\bullet \to \mathrm{Hom}_A(\mathrm{Hom}_A(C^\bullet, D_A^\bullet), D_A^\bullet))$ which induces an isomorphism on the homology groups. In fact, a complex having this property is often taken as the definition of a dualizing complex. In a certain derived category sense, there are uniqueness statements for these complexes (up to shifts). See e.g. [Ha, Ch. 5, §3]. A complex $D_A^\bullet$ as in 10.2.4 in this context is sometimes called a standard dualizing complex.

The following proposition enables us to produce lots of rings having a dualizing complex. Recall from 3.3.5 the notation for shifting a complex. If $C^\bullet$ is a complex of A-modules, we define the complex $C^\bullet[n]$, $n \in \mathbb{Z}$ to be: (i) $C^\bullet[n]^i = C^{n+i}$ for all i, (ii) the boundary map $C^\bullet[n]^i \to C^\bullet[n]^{i+1}$ of $C^\bullet[n]$ is just the boundary map $C^{n+i} \to C^{n+i+1}$ of $C^\bullet$.

10.2.5 PROPOSITION. Let A be a homomorphic image of a noetherian local ring R, let d = dim A and D = dim R. If R admits a dualizing complex $D_R^\bullet$, then the complex $D_A^\bullet = \operatorname{Hom}_R(A, D_R^\bullet[D-d])$ is a dualizing complex for A.

PROOF. We have $A = R/\mathfrak{a}$ for a certain ideal $\mathfrak{a}$ of R. Let $\mathfrak{p}$ be a prime in k. If $\mathfrak{a} \not\subset \mathfrak{p}$, then $\operatorname{Hom}_R(A, E_R(R/\mathfrak{p})) = 0$ by 3.2.9. Conversely, if $\mathfrak{a} \subset \mathfrak{p}$, then $\operatorname{Hom}_R(A, E_R(R/\mathfrak{p})) = E_A(A/\bar{\mathfrak{p}})$, where $\bar{\mathfrak{p}}$ is the image of $\mathfrak{p}$ in A (again, 3.2.9). One sees that the D_A^i have the desired form, so the only thing left to prove is that the homology of the complex $D_A^\bullet$ is finitely generated.

But this is true for any complex $I_M^\bullet = \operatorname{Hom}_R(M, I^\bullet)$ with M a finitely generated R-module and $I^\bullet$ a complex of injectives with $I^i = 0$ for i sufficiently negative and finitely generated homology. A proof may run as follows. Let $0 \to K \to L \to M \to 0$ be exact with L a finitely generated free R-module. From the exact sequence of complexes $0 \to I_M^\bullet \to I_L^\bullet \to I_K^\bullet \to 0$ we obtain exactness of $H^{i-1}(I_K^\bullet) \to H^i(I_M^\bullet) \to H^i(I_L^\bullet)$ for each i in the long homology sequence. Since the result is obviously true for i sufficiently negative, we may assume by induction that $H^{i-1}(I_K^\bullet)$ is finitely generated. Also $H^i(I_L^\bullet)$, being a finite direct sum of copies of $H^i(I^\bullet)$, is finitely generated, whence $H^i(I_M^\bullet)$ is.

10.2.6 COROLLARY. Let A be a homomorphic image of a local Gorenstein ring R and let $D_R^\bullet$ be a minimal injective resolution of the R-module R. Then $D_R^\bullet$ is a dualizing complex for R and hence $D_A^\bullet$, as defined in the above proposition, is a dualizing complex for A.

PROOF. Is immediate from 10.1.6 and 10.2.5.

Notice that by the Cohen Structure Theorem 9.3.1, this result in particular applies to complete local rings. It has been conjectured by Sharp that only homomorphic images of Gorenstein rings possess a dualizing complex, and this has been verified in certain cases [AG]. Schenzel has shown that for certain purposes another complex can fill the role of a dualizing complex [Sc 82a]. We now prove a variant of the Local Duality Theorem:

10.2.7 THEOREM. Let $(A,\mathfrak{m},k)$ be a d-dimensional local noetherian ring which is a homomorphic image of a Gorenstein local ring R. Let $D_R^\bullet$ be a minimal injective revolution of R and let $D_A^\bullet$ be the dualizing complex for A as defined in 10.2.5 and let $-^\vee$ denote the Matlis duality functor $\mathrm{Hom}_A(-,E_A(k))$. Then $H_{\mathfrak{m}}^{d-i}(M) \simeq (H^i\mathrm{Hom}_A(M,D_A^\bullet))^\vee$ for all i and all finitely generated A-modules M. Moreover, putting dim R = D, we obtain isomorphisms $H_{\mathfrak{m}}^{D-i}(M) \simeq \mathrm{Ext}_R^i(M,R)^\vee$ for all i.

PROOF. Notations as above. Let $-^\sim$ denote the functor $\mathrm{Hom}_R(-,E_R(R/\mathfrak{M}))$, where $\mathfrak{M}$ is the maximal ideal of R. Since $A^\sim = E_A(k)$ by 3.2.9, we see that $N^\sim \simeq \mathrm{Hom}_A(N,A^\sim) = N^\vee$ for any A-module N. Furthermore, by 4.1.4, $H_{\mathfrak{m}}^i(M) \simeq H_{\mathfrak{M}}^i(M)$ for all i. Using these considerations and 10.2.1 (ii), we find $H_{\mathfrak{m}}^{D-i}(M) \simeq H_{\mathfrak{M}}^{D-i}(M) \simeq \mathrm{Ext}_R^i(M,R)^\sim \simeq \mathrm{Ext}_R^i(M,R)^\vee$. On the other hand, by construction of $D_A^\bullet$, we have $\mathrm{Hom}_A(M,D_A^\bullet) \simeq \mathrm{Hom}_A(M,\mathrm{Hom}_R(A,D_R^\bullet[D-d])) \simeq \mathrm{Hom}_R(M\otimes_A A,D_R^\bullet[D-d]) \simeq \mathrm{Hom}_R(M,D_R^\bullet[D-d])$. Since $D_R^\bullet$ is an R-injective resolution of R, we find that $H^i\mathrm{Hom}_A(M,D_A^\bullet) \simeq \mathrm{Ext}_R^{D-d+i}(M,R)$ and hence $H_{\mathfrak{m}}^{d-i}(M) \simeq (H^i\mathrm{Hom}_A(M,D_A^\bullet))^\vee \simeq \mathrm{Ext}_R^{D-d+i}(M,R)^\vee$. This finishes the proof of the theorem.

10.2.8 COROLLARY. Let $(A,\mathfrak{m},k)$ be a d-dimensional noetherian local ring, complete in its $\mathfrak{m}$-adic topology, and let $D^\bullet = D_A^\bullet$ be as above. Let $L^\bullet$ be a finite ascending complex of finitely generated free A-modules. If $H^i(L^\bullet)$ has finite length for every i, then the complex $L^\bullet\otimes_A H_{\mathfrak{m}}^j(A)$ has finite length homology for every j. More precisely, $\ell(H^i(L^\bullet\otimes_A H_{\mathfrak{m}}^j(A))) =$

$\ell(H^i(\mathrm{Hom}_A(L^\bullet, H^{d-j}(D^\bullet))))$ for every i and j.

PROOF. Since A is complete, the Matlis dual is a faithfully exact functor which affords a duality between finitely generated and artinian modules, 3.4.2 and 3.4.12. Thus the complex $L^\bullet \otimes_A H^j_{\mathfrak{m}}(A)$ has finite length homology precisely when its Matlis dual has, and these lengths are equal by 3.4.5. But $(L^\bullet \otimes_A H^j_{\mathfrak{m}}(A))^\vee \simeq \mathrm{Hom}_A(L^\bullet, H^j_{\mathfrak{m}}(A)^\vee)$, and in view of 4.1.11 the module $H^j_{\mathfrak{m}}(A)^\vee$ is finitely generated. Since the complex $L^\bullet_{\mathfrak{p}}$ is split exact for each prime $\mathfrak{p} \neq \mathfrak{m}$, this homology is at most supported in $\mathfrak{m}$, and therefore has finite length. Furthermore, by 10.2.7, $H^j_{\mathfrak{m}}(A)^\vee \simeq H^{d-j}(\mathrm{Hom}_A(A, D^\bullet))^{\vee\vee} \simeq H^{d-j}(D^\bullet)$ and we are through.

We next prove that for homomorphic images of Gorenstein rings certain ideals are not nilpotent. Recall from 7.3.1 that for a noetherian local ring $(A,\mathfrak{m})$, the ideals $\mathfrak{c}^i(A)$ are defined as $\mathrm{Ann}\, H^i_{\mathfrak{m}}(A)$. The result is due to P. Roberts [Ro 76, Prop. 1], see also [Sc 79, Prop. 1].

10.2.9 PROPOSITION. Let $(A,\mathfrak{m})$ be a noetherian local ring of dimension d, which is a homomorphic image of a Gorenstein ring. Then the ideals $\mathfrak{c}^i(A)$, $0 \leq i \leq d-1$, are not nilpotent and neither is their product $\mathfrak{c}$.

PROOF. Putting M = A in 10.2.7, we find that $H^{d-i}_{\mathfrak{m}}(A) \simeq H^i(D^\bullet)^\vee$ for all i where we have again written $D^\bullet$ for $D^\bullet_A$. In view of 3.4.2, it is enough to find a prime ideal $\mathfrak{p}$ such that $\mathrm{Ann}\, H^i(D^\bullet) \not\subset \mathfrak{p}$ $(1 \leq i \leq d)$, since this result will imply that all $\mathrm{Ann}\, H^i(D^\bullet)$ and their product (whence the $\mathfrak{c}^i$ and $\mathfrak{c}$) are not nilpotent. Well, let $\mathfrak{p}$ be a prime such that $\dim A/\mathfrak{p} = d$. Then $(D^\bullet{}_{\mathfrak{p}})$ degenerates to the complex $0 \to (D^0{}_{\mathfrak{p}}) \to 0 \to \ldots \to 0$ so for all $i \geq 1$ we have $H^i((D^\bullet{}_{\mathfrak{p}})) = 0$. Since localization is exact, it commutes with taking homology. Hence $(H^i(D^\bullet))_{\mathfrak{p}} = 0$ for all $i \geq 1$. But by definition, the $H^i(D^\bullet)$ are finitely generated so for all i $(1 \leq i \leq d)$ we find that $\mathrm{Ann}\, H^i(D^\bullet) \not\subset \mathfrak{p}$ and we are done.

This proposition will be used in the next chapter to prove the Big Cohen-Macaulay Module Conjecture in equal characteristic $p > 0$.

10.3 THE DIRECT SUMMAND AND MONOMIAL CONJECTURES IN EQUAL CHARACTERISTIC

We shall begin this section with a result which is based on 10.2.2 and which states that the Monomial Conjecture is true for high powers of systems of parameters. As a consequence, we derive the Monomial Conjecture in equal characteristic $p > 0$. Furthermore, if A is a Gorenstein local ring and M is an A-module, we shall give a criterion for the purity of an injection $A \to M$. This has as a corollary that the Monomial Conjecture implies the Direct Summand Conjecture. In consequence both conjectures are true in equal characteristic.

Recall the notation of 5.1.11.

10.3.1 PROPOSITION. Let A be a d-dimensional noetherian local ring and let $x = x_1, \ldots, x_d$ be a s.o.p. for A. Then $\mathfrak{a}_{x(s)} \neq A$ for all large integers s, i.e. the monomial conjecture is true for $x(s) = x_1^s, \ldots, x_d^s$ for all large s.

REMARK. In 11.2.4 we shall give an alternative proof of this result. Here, we only use the highest nonvanishing local cohomology group. In 11.2.4, we shall use 10.2.9, i.e. we shall use all the lower local cohomology groups. This has the advantage that we get a uniform bound on s, valid for every s.o.p..

PROOF. Let $\mathfrak{a}$ be the ideal generated by the x's. Since $\mathfrak{a}$ is m-primary, $H^i_{\mathfrak{m}}(A) \simeq H^i_{\mathfrak{a}}(A)$ for all i and hence $H^d_{\mathfrak{a}}(A) \neq 0$ by 10.2.2. On the other hand, $H^d_{\mathfrak{m}}(A)$ is the direct limit of the maps $.x^t\colon A/\mathfrak{a}_s \to A/\mathfrak{a}_{s+t}$ as we saw in section 4.3. We find that for s large enough all maps $A/\mathfrak{a}_s \to H^d_{\mathfrak{a}}(A)$ have a nonzero image and therefore all maps $.x^t\colon A/\mathfrak{a}_s \to A/\mathfrak{a}_{s+t}$ are nonnull. In particular $.x^{st}\colon A/\mathfrak{a}_s \to A/\mathfrak{a}_{s(t+1)}$ are nonnull and the result follows.

10.3.2 COROLLARY. Let A be a noetherian local ring of equal characteristic $p > 0$. Then the Monomial Conjecture is true for A, i.e. $\mathfrak{q}_x \neq A$ for every s.o.p. x of A.

PROOF. Let $x = x_1,\ldots,x_d$ be a s.o.p. for A and assume that $\mathfrak{q}_x = A$, i.e. there is a positive integer t such that $x_1^t\ldots x_d^t \in (x_1^{t+1},\ldots,x_d^{t+1})$. Applying the Frobenius endomorphism $x \mapsto x^p$ to this relation, we find that $x_1^{tp^e}\ldots x_d^{tp^e} \in (x_1^{(t+1)p^e}\ldots x_d^{(t+1)p^e})$ for all exponents e. Whence $\mathfrak{q}_{x(p^e)} = A$ for all e. This contradicts 10.3.1.

We now use the description of Gorenstein rings in 10.1.8 to prove, in the notation of 5.1.11,

10.3.3 PROPOSITION. Let (A,m,k) be a local Gorenstein ring. Let M be an A-module and $f\colon A \to M$ be an injective A-module homomorphism. Let $x = x_1,\ldots,x_d$ be a s.o.p. in A and let $\mathfrak{a}$ denote the ideal generated by the x's. Write v for an element of A such that its image in $(\mathfrak{a}:\mathfrak{m})/\mathfrak{a}$ generates this one-dimensional k-vector space. Then f is pure if and only if $f(v) \notin Q_x(M)$.

REMARK. Notice that v modulo $\mathfrak{a}$ is unique up to multiplication with units of A.

PROOF. Assume that f is pure and that $f(v) \in Q_x(M)$, say $x^t f(v) \in \mathfrak{a}_{t+1}M$. Consider the commutative diagram

$$\begin{array}{ccc} A/\mathfrak{a} & \xrightarrow{\ .x^t\ } & A/\mathfrak{a}_{t+1} \\ \Big\downarrow {\scriptstyle f\otimes_A A/\mathfrak{a}} & & \Big\downarrow {\scriptstyle f\otimes_A A/\mathfrak{a}_{t+1}} \\ M/\mathfrak{a}M & \xrightarrow[\ .x^t\ ]{} & M/\mathfrak{a}_{t+1}M \end{array}$$

By the purity of f, the vertical maps are injective. Hence $x^t f(v) \in \mathfrak{a}_{t+1}M$ implies $x^t v \in \mathfrak{a}_{t+1}$. Since A is x-pre-regular, this means that $v \in \mathfrak{a}$, contradiction.

Conversely, assume that $f(v) \notin Q_x(M)$. Observe that for $t \geq 1$, $x_1^t,\ldots,x_d^t$ is a s.o.p. in A so that the k-vector space $(\mathfrak{a}_t:m)/\mathfrak{a}_t$ is one-dimensional. Since $x^{t-1}v \notin \mathfrak{a}_t$ (use pre-regularity) and obviously $mx^{t-1}v \subset \mathfrak{a}_t$, the image of the element $x^{t-1}v$ in $A/\mathfrak{a}_t$ generates the unique minimal ideal of $A/\mathfrak{a}_t$.

To reach our conclusion, we need to show that tensoring f with any A-module N, and it is enough to test for finitely generated N, preserves injectivity. So let N be a finitely generated A-module and consider the commutative diagram in the notation of section 2.2

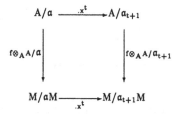

$$\begin{array}{ccc} N & \xrightarrow{f\otimes N} & M\otimes_A N \\ \Big\downarrow{\scriptstyle \tau_N} & & \Big\downarrow{\scriptstyle \tau_{M\otimes_A N}} \\ \hat{N} & \xrightarrow[(f\otimes N)\hat{}]{} & (M\otimes_A N)\hat{} \end{array}$$

Since N is finitely generated, N is Hausdorff in its $\mathfrak{m}$-adic topology and by 2.2.5 τ_N is monic. Therefore we are done if we can prove that $(f\otimes_A N)\hat{}$ is monic. But $(f\otimes_A N)\hat{} = \varprojlim_t f_t\otimes_A N$: $A/\mathfrak{a}_t\otimes_A N \to M/\mathfrak{a}_t M\otimes_A N$, where $f_t = f\otimes_A A/\mathfrak{a}_t$ and since both $\varprojlim$ and $\otimes$ are additive functors, we need only prove that all maps f_t: $A/\mathfrak{a}_t \to M/\mathfrak{a}_t M$, $t \geq 1$, are split injective. First, notice that f_t is injective, for if $A \cap f^{-1}(\mathfrak{a}_t M)$ would be larger than $\mathfrak{a}_t$, its image in $A/\mathfrak{a}_t$ would contain the unique minimal ideal $(x^{t-1}v)/\mathfrak{a}_t$. But then $x^{t-1}f(v)$ would be in $\mathfrak{a}_t M$, contradicting our assumption on f(v). Hence the $A/\mathfrak{a}_t$-linear map f_t is injective and therefore splits, because $A/\mathfrak{a}_t$ is an artinian local Gorenstein ring (observe that $x_1^t,\ldots,x_d^t$ form a regular sequence in the Cohen-Macaulay ring A and use 10.1.2 (ii)) whence self-injective. This finishes the proof.

10.3.4 COROLLARY. Let B be a noetherian local ring and assume that $Q_x(B) \neq B$ for all systems of parameters x of B (i.e. the Monomial Conjecture is true for B). Then every inclusion of local rings $A \subset B$ with A regular and B a module finite A-algebra, splits (i.e. the Direct Summand Conjecture 9.3.4 is true for B).

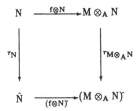

PROOF. In view of 2.1.8 it is enough to show that the injection f: $A \to B$ is pure. The maximal ideal $\mathfrak{m} \subset A$ can be generated by a regular sequence $x = x_1,\ldots,x_d$, see 9.2.3. Since $\mathfrak{a} = \mathfrak{m}$, the element v of 10.3.3 can be taken to be the unit element. Hence by 10.3.3 we are done if we can prove that $1 - f(1) \notin Q_x(B)$. But x is also a system of parameters for B, 8.2.10, which clinches the argument.

As a last result of this section we state

10.3.5 PROPOSITION. Both the Monomial Conjecture and the Direct Summand Conjecture are true for noetherian local rings of equal characteristic.

PROOF. By 9.3.7, it is enough to prove the Direct Summand Conjecture in equal characteristic. The equal characteristic zero case is done in 9.3.5. The equal characteristic $p > 0$ case follows immediately from 10.3.2 and 10.3.4.

In mixed charateristic, these conjectures remain undecided, despite much effort [Ho 83].

11 FROBENIUS AND BIG COHEN-MACAULAY MODULES IN CHARACTERISTIC p

In this chapter we shall construct a pre-regular module M for any given system of parameters $x = x_1,\ldots,x_d$ in a d-dimensional noetherian local ring $(A,\mathfrak{m},k)$ of positive characteristic. By 9.1.1 we then know that the $\mathfrak{m}$-adic completion $\hat{M}$ of M is regular for every s.o.p., in other words, is a balanced Big Cohen-Macaulay module over A.

We outline the procedure. In the first section we "construct" a module M which is x-pre-regular, except that perhaps $M = \mathfrak{a}M$ for $\mathfrak{a} = (x_1,\ldots,x_d)$. In order to show that indeed $M \neq \mathfrak{a}M$, we need two ingredients.

In 11.2 we compare certain annihilator ideals with the ideal $\mathfrak{c}$ introduced in section 7.3 and whose nonnilpotency in a complete ring was established in 10.2.9.

In the next section we introduce the Frobenius functor, a powerful tool in positive characteristic. We develop some of its properties, in particular on finite free complexes, since they will be needed in section 13.1. In the last two paragraphs of 11.3 we consider a few simple functorial properties of the Frobenius.

The latter, together with the nonnilpotency results of 11.2, allow one to finally prove that $M \neq \mathfrak{a}M$ for the module constructed in 11.1. This is explained in 11.4, and by completion we obtain in 11.5 the result we are after.

11.1 MODIFICATIONS

If M is an A-module, recall the definition of the submodule $B_x(M)$ and the notation $\mathfrak{b}_x = B_x(A)$, 5.1.11. Then $(x_1,\ldots,x_d)M \subset B_x(M) \subset M$ and these inclusions describe natural transformations between functors in the module category. According to 5.1.12 it is enough to construct a module M with $(x_1,\ldots,x_d)M = B_x(M) \neq M$, for such a module is x-pre-regular.

The basic idea is quite simple; it is a variant of Hochster's original modification procedure [Ho 75a, Ch. 4]. To him must be credited the marvellous insight that such a "brute force" approach might actually work. Our discussion is taken from [Ba, Ch. II, §5] which in turn refined [BS, §2], where we spoke of monomial modifications. Much in the treatment of our modifications in this section is purely formal so we take A to be any commutative ring and $x_1,\ldots,x_n$ a fixed n-tuple of elements. We write $\mathfrak{a}$ for the ideal $(x_1,\ldots,x_n)$ and suppose we have a module M with $\mathfrak{a}M \subsetneq B(M)$, suppressing the subscript x until we shall need to consider different tuples x. There is an obvious way to get rid of any given "bad" element $m_0 \in B(M)$ which we now describe. Here, for purpose of exposition, we call the elements of B(M) "bad" and we want to render them "harmless" by mapping them into $\mathfrak{a}M'$ for some other module M' (notice that the elements of $\mathfrak{a}M$ are both bad and harmless). Specifically, we wish to construct a morphism of modules $f\colon M \to M'$ with $f(m_0) \in \mathfrak{a}M'$, such that whenever $\phi\colon M \to W$ is a homomorphism with $\phi(m_0) \in \mathfrak{a}W$, there exists a homomorphism $\phi'\colon M' \to W$ with $\phi' \circ f = \phi$, thus:

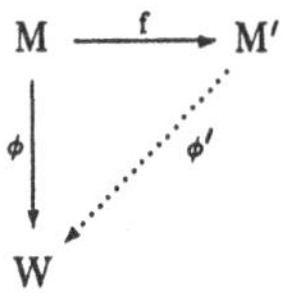

A universal way of achieving this is to put $M' = (M\oplus A^n)/Av$ with $v = (-m_0, x_1, \ldots, x_n) \in M\oplus A^n$, and to define $f(m) = (m, 0, \ldots, 0) \bmod Av$. For if $\phi(m_0) = \Sigma_{i=1}^n x_i w_i$ with $w_i \in W$, the homomorphism Φ: $M\oplus A^n \to W$ defined by $\Phi(m, a_1, \ldots, a_n) = \phi(m) + \Sigma_{i=1}^n a_i w_i$ carries the element v to 0, hence factors through ϕ': $M' \to W$ such that $\phi' \circ f = \phi$.

We say that f: $M \to M'$ is a modification of the bad element $m_0 \in B(M)$ and may now characterize the submodule $B(M) \subset M$ of the bad elements as follows: an element $m \in M$ is bad if and only if there exists a modification f: $M \to M'$ of a bad element m_0 such that $f(m) \in \mathfrak{a}M'$. The proof of this equivalence, which is in fact valid for any functorially defined submodule C(M) with $\mathfrak{a}M \subset C(M) \subset M$, is left to the reader.

Our aim is to start from the module A where we have the chain of ideals $\mathfrak{a} \subset \mathfrak{b} \subset A$; if $\mathfrak{a} = \mathfrak{b}$ we need not modify and the ring A is Cohen-Macaulay as we have observed in section 9.1. If not, we successively modify all the bad elements which are not in $\mathfrak{a}$. In the resulting module, we again modify the bad elements and carrying on in this way, we eventually hope to create a module $M \neq \mathfrak{a}M$ with all its bad elements in $\mathfrak{a}M$, i.e. with $\mathfrak{a}M = B(M) \neq M$. To make this precise, we shall now assume our ring A is noetherian and consider a sequence of modifications

$$(A,1) = (M_0,u_0) \overset{f_0}{\to} (M_1,u_1) \overset{f_1}{\to} \ldots \overset{f_{r-1}}{\to} (M_r,u_r) \overset{f_r}{\to} (M_{r+1},u_{r+1}) \overset{f_{r+1}}{\to} \ldots$$

where f_r is a modification of a bad element $m_r \in B(M_r)\backslash\mathfrak{a}M_r$ and $f_r(u_r) = u_{r+1}$ for all r, so that the u_r are the successive images of u_0 which we have chosen to be $1 \in A$.

11.1.1 PROPOSITION. For an ideal $\mathfrak{a} = (x_1,\ldots,x_n)$ in a noetherian ring A, the following three properties are equivalent:

(i) There exists an A-module W with $\mathfrak{a}W = B(W) \neq W$;

(ii) In every sequence of modifications as above $u_r \notin \mathfrak{a}M_r$ for all r;

(iii) For some such sequence the direct limit $M = \varinjlim_r M_r$ has property i).

PROOF. Since (iii) patently implies (i), suppose we have a module W as in (i) and a sequence of modifications. Take a $w \in W\backslash\mathfrak{a}W$ and define $\phi_0: M_0 \to W$ by $\phi_0(u_0) = w$. Since $\phi_0(B(M_0)) \subset B(W) = \mathfrak{a}W$, we can extend ϕ_0 to $\phi_1: M_1 \to W$ with $\phi_1 \circ f_0 = \phi_0$, whichever bad element in $\mathfrak{b}\backslash\mathfrak{a}$ we have modified. By induction we find $\phi_r: M_r \to W$ with $\phi_r(u_r) = w$. Now if $u_r \in \mathfrak{a}M_r$, then $w \in \mathfrak{a}W$, which is not the case. So (i) implies (ii).

To prove that (ii) implies (iii), all we need to do is to choose a sequence of modifications in such a way that its limit M satisfies $\mathfrak{a}M = B(M)$. For we already know that $\lim u_r = u \notin \mathfrak{a}M$ because condition (ii) holds at every finite stage r. To accomplish this, choose bad elements $g_0^0,\ldots,g_{h_0-1}^0$ such that $\mathfrak{b} = (\mathfrak{a},g_0^0,\ldots,g_{h_0-1}^0)$. First modify g_0^0 in $M_0 = A$, and then successively modify the images of g_i^0 in M_i, $i = 0,\ldots,h_0-1$. In the noetherian module M_{h_0} the submodule $B(M_{h_0})$ is finitely generated and we can choose $g_0^1,\ldots,g_{h_1-1}^1$ to be bad elements in this module such that $B(M_{h_0}) = \mathfrak{a}M_{h_0} + Ag_0^1 + \ldots + Ag_{h_1-1}^1$. In turn we modify $g_0^1,\ldots,g_{h_1-1}^1$ until we arrive at the module $M_{h_0+h_1}$ and we begin all over again, and so on. A little reflection shows that, for each module M_r occurring in the sequence of modifications, all the bad elements of $B(M_r)\backslash\mathfrak{a}M_r$ inevitably get modified somewhere along the way, so becomes harmless in $M = \varinjlim_r M_r$; hence $\mathfrak{a}M = B(M)$, which finishes the proof of the proposition.

In order for this construction to yield a pre-regular module, we therefore have to verify condition (ii) assuming that $x_1,\ldots,x_d$ is a s.o.p.

in a noetherian local ring $(A,\mathfrak{m},k)$. Then these elements are analytically independent, in the sense that $1 \notin \mathfrak{b}$, 8.3.4. By our characterization of bad elements, we conclude that $u_1 \notin \mathfrak{a}M_1$, whichever bad element of $\mathfrak{b}\backslash\mathfrak{a}$ we have modified to obtain $f_0: M_0 \to M_1$. Of course, we would like to know that $u_r \notin \mathfrak{a}M_r$ for every r, and it is enough to know that $u_{r-1} \notin B(M_{r-1})$.

11.1.2 PROPOSITION. Let $x = x_1,\ldots,x_d$ be a s.o.p. in a local noetherian ring and $\mathfrak{a}$ and $\mathfrak{b}$ have their usual meaning. Suppose $I \subset A$ is an ideal such that $I^r \not\subset \mathfrak{a}$ but $I\mathfrak{b} \subset \mathfrak{a}$. Then $u_r \notin \mathfrak{a}M_r$ in every sequence of modifications of bad elements.

PROOF. If $\phi_{j-1}: M_{j-1} \to A$ is an A-homomorphism and $m_{j-1} \in M_{j-1}$ is a bad element, then $a\phi_{j-1}$ carries m_{j-1} to an element of the ideal $a\mathfrak{b} \subset \mathfrak{a}$ for any choice of $a \in I$. If $f_{j-1}: M_{j-1} \to M_j$ is a modification of m_{j-1}, its universal property gives us a homomorphism $\phi_j: M_j \to A$ with $\phi_j \circ f_{j-1} = a\phi_{j-1}$. For arbitrary elements $a_1,\ldots,a_{j-1}$ in I we may begin with $\mathrm{id} = \phi_0: M_0 \to A$ and proceed by induction to obtain $\phi_{r-1}: M_{r-1} \to A$ such that $\phi_{r-1}(u_{r-1}) = a_1 \ldots a_{r-1}$. But we may choose $a_1,\ldots,a_{r-1},a_r$ in I such that their product is in $I^r\backslash\mathfrak{a}$. If $u_{r-1} \in B(M_{r-1})$, then $a_1 \ldots a_{r-1} \in \mathfrak{b}$, hence $a_1 \ldots a_{r-1}a_r \in \mathfrak{a}$, contradicting the choice of the a_j's.

In section 11.2 we shall show that, at least if the ring A has positive dimension and is complete in its $\mathfrak{m}$-adic topology, there exists a nonnilpotent ideal I which satisfies $I\mathfrak{b}_x \subset (x_1,\ldots,x_d) = \mathfrak{a}_x$ for all s.o.p. This means that for an arbitary positive integer r we can find a s.o.p. such that in every sequence of modifications of bad elements we know that $u_r \notin \mathfrak{a}M_r$. For since $(x_1^t,\ldots,x_d^t) \subset \mathfrak{m}^t$ and $\cap_{t=1}^{\infty} \mathfrak{m}^t = (0)$ in view of 2.1.4, we can choose t so large that $(0) \neq \mathfrak{m}^r \not\subset (x_1^t,\ldots,x_d^t)$. Then $x_1^t,\ldots,x_d^t$ is such a s.o.p. Of course, this is not good enough: we need to show that given an arbitrary s.o.p., the element $u_r \notin \mathfrak{a}M_r$ for *all* $r \geq 1$ in every sequence of modifications of bad elements, while it is enough to prove the existence of a single s.o.p. with this property in view of 9.1.1 and 11.1.1. This we shall achieve in positive characteristic in section 11.4, using properties of the Frobenius functor which will be discussed in section 11.3.

In low dimensions the programme can be carried through without the help of 11.1.2 [Ba, Ch. II, Cor. 5.3]:

11.1.3 EXERCISE. Any noetherian local ring of dimension less than three has an x-pre-regular module for every s.o.p. x [Ba, Ch. III, Cor. 3.6].

11.2 RELATIONS BETWEEN ANNIHILATORS

Our objective in this section is to construct an ideal $\mathfrak{I}$ with the properties called for in Proposition 11.1.2. Let $(A,\mathfrak{m},k)$ throughout be a d-dimensional noetherian local ring and let for every s.o.p. $x = x_1,\dots,x_d$ the ideals $\mathfrak{a}_x$, $\mathfrak{b}_x$ and $\mathfrak{q}_x$ have their usual meaning. Define $\mathfrak{I}$ to be the ideal $\cap_x(\mathfrak{a}_x:\mathfrak{b}_x)$ where the intersection is taken over all s.o.p.'s. Then $\mathfrak{I}\mathfrak{b}_x \subset \mathfrak{a}_x$ for every s.o.p., but if $\mathfrak{I}$ is nonnilpotent we can, given $r \geq 1$, choose an x such that $\mathfrak{I}^r \not\subset \mathfrak{a}_x$. We shall show that this is the case if A is complete, varying an approach due to Schenzel [Sc 79], [Sc 82a, §2.4]. This in turn represented an elaboration on Hochster's theme of "amiability", [Ho 75a, Lemma 4.4].

Let us introduce Schenzel's ideal

$$\mathfrak{r} = \cap_x \operatorname{Ann}[((x_1,\dots,x_{i-1}) : x_i)/(x_1,\dots,x_{i-1})]$$

where the intersection is taken over all s.o.p.'s $x = x_1,\dots,x_d$ and $1 \leq i \leq d$.

Notice that $\mathfrak{r} = A$ precisely when A is a Cohen-Macaulay ring, and the worse the ring, the smaller one may expect this ideal to be. The point of the following considerations is to show that, provided the ring is complete, this ideal cannot become too small. Our first remark is that we may as well restrict the intersection to $i = d$. For suppose a is in this seemingly larger ideal, and let $yx_i \in (x_1,\ldots,x_{i-1})$ for some s.o.p. $x_1,\ldots,x_d$, $1 \leq i \leq d$. Then $yx_i \in (x_1,\ldots,x_{i-1},x_{i+1}^t,\ldots,x_d^t)$ for all $t \geq 1$. Since $x_1,\ldots,x_{i-1}$, $x_{i+1}^t,\ldots,x_d^t,x_i$ is a s.o.p., we know that $ay \in (x_1,\ldots,x_{i-1},x_{i+1}^t,\ldots,x_d^t)$ and consequently in the intersection over all $t \geq 1$. This intersection is just $(x_1,\ldots,x_{i-1})$ as one sees by observing that the images of $x_{i+1},\ldots,x_d$ are all in the maximal ideal of the noetherian local ring $A/(x_1,\ldots,x_{i-1})$. Thus $a \in \mathfrak{r}$.

Recall the ideals $\mathfrak{c}_i = \operatorname{Ann} H^i_{\mathfrak{m}}(A)$ of 7.3.1 and their product $\mathfrak{c} = \mathfrak{c}_0\mathfrak{c}_1\cdots\mathfrak{c}_{d-1}$.

11.2.1 THEOREM. In a d-dimensional noetherian local ring A with $d > 0$

$$\begin{array}{ccccc} \mathfrak{c} & \subset & \bigcap_x \operatorname{Ann} H_1(x,A) & \subset & \mathfrak{r} \\ \cup & & & & \cup \\ \mathfrak{r}^d & \subset & \bigcap_x(\mathfrak{a}_x:\mathfrak{q}_x) & \subset & \mathfrak{I} \end{array}$$

where the intersections are taken over all s.o.p.'s x. Thus all these ideals have the same radical.

PROOF. The horizontal inclusions are the ones of importance to us, as they establish that $\mathfrak{c}^d \subset \mathfrak{I}$. In the complete case, $\mathfrak{c}$ is nonnilpotent and hence so is $\mathfrak{I}$, the result we are after. We shall therefore prove the horizontal inclusions in full, confining ourselves to a brief indication of proof or a reference for the remaining ones.

$\mathfrak{r}^d \subset \mathfrak{c}$: This follows from [Sc 82a, Satz 2.4.5].

$\mathfrak{c} \subset \bigcap_x \operatorname{Ann} H_1(x,A)$: Apply 7.3.3 to $K_\bullet(x,A)$. Since the Koszul complex is a finite complex of finitely generated free modules with finite length homology, the ideal $\mathfrak{c}_0\mathfrak{c}_1\cdot\ldots\cdot\mathfrak{c}_i$ annihilates $H_{d-i}(x,A)$ for $0 \le i \le d-1$ and all x. Our statement is the case $i = d-1$.

$\bigcap_x \operatorname{Ann} H_1(x,A) \subset \mathfrak{r}$: The short exact sequence of 4.2.4 (i) for p = 1 yields a surjection $H_1(x_1,\ldots,x_d,A) \to H_1(K_\bullet(x_d,A) \otimes_A H_0(x_1,\ldots,x_{d-1},A))$. The right hand module is just $[(x_1,\ldots,x_{d-1}) : x_d]/(x_1,\ldots,x_{d-1})$. The result follows, if one bears in mind the description of $\mathfrak{r}$ given above.

$\mathfrak{r}^d \subset \bigcap_x(\mathfrak{a}_x:\mathfrak{q}_x)$: Given a s.o.p. $x = x_1,\ldots,x_d$ and a d-tuple $a_1,\ldots,a_d \in \mathfrak{r}$, it is enough to show that if $q \in \mathfrak{q}_x$ and $x_1^t\ldots x_i^t a_{i+1}\ldots a_d q \in (x_1^{t+1},\ldots,x_i^{t+1},x_{i+1},\ldots,x_d)$, then $x_1^t\ldots x_{i-1}^t a_i a_{i+1}\ldots a_d q \in (x_1^{t+1},\ldots,x_{i-1}^{t+1},x_i,\ldots,x_d)$. Namely, using descending induction, from $q \in \mathfrak{q}_x$ (i.e. $x_1^t\ldots x_d^t q \in (x_1^{t+1},\ldots,x_d^{t+1})$ for some $t \ge 1$) we can then derive that $a_1\cdot\ldots\cdot a_d q \in (x_1,\ldots,x_d)$, the result we are after. We have $x_i^t(x_1^t\ldots x_{i-1}^t a_{i+1}\ldots a_d q - x_i y) \in (x_1^{t+1},\ldots,x_{i-1}^{t+1},x_{i+1},\ldots,x_d)$ for a certain $y \in A$. Now $x_1^{t+1},\ldots,x_{i-1}^{t+1},x_{i+1},\ldots,x_d,x_i^t$ is a s.o.p., hence $a_i(x_1^t\ldots x_{i-1}^t a_{i+1}\ldots a_d q - x_i y) \in (x_1^{t+1},\ldots,x_{i-1}^{t+1},x_{i+1},\ldots,x_d)$ because $a_i \in \mathfrak{r}$. Therefore $x_1^t\ldots x_{i-1}^t a_i a_{i+1}\ldots a_d q \in (x_1^{t+1},\ldots,x_{i-1}^{t+1},x_i,\ldots,x_d)$ and the induction step has been taken.

$\bigcap_x(\mathfrak{a}_x:\mathfrak{q}_x) \subset \mathfrak{I}$: This reflects the inclusion $\mathfrak{b}_x \subset \mathfrak{q}_x$.

$\mathfrak{I} \subset \mathfrak{r}$: Observe that for any s.o.p. $x = x_1,\ldots,x_d$ the ideal $(x_1,\ldots,x_{d-1}) : x_d$ is contained in $\mathfrak{b}_x$ in view of 5.1.10, so if $a \in \mathfrak{I}$, then for all $t \ge 1$ $a[(x_1,\ldots,x_{d-1}) : x_d] \subset a[(x_1,\ldots,x_{d-1}) : x_d^t] \subset (x_1,\ldots,x_{d-1},x_d^t)$ and the result follows. This finishes the proof of the theorem.

Several remarks are in order. As we have seen, all the ideals in 11.2.1 are the unit ideal precisely when A is a Cohen-Macaulay ring. In fact if we define V as the set of those prime ideals which contain $\mathfrak{c}$ (or any of the other ideals occurring in 11.2.1), then Schenzel has proven that V consists of those prime ideals $\mathfrak{p}$ such that $\operatorname{dp} A_\mathfrak{p} + \dim A/\mathfrak{p} < d$ [Sc 82a, Satz 2.4.6] provided A has a dualizing complex. Thus if A is equidimensional i.e. $\dim A/\mathfrak{p} = \dim A$ for every minimal prime $\mathfrak{p}$, then V is the "set of non Cohen-Macaulay points". Furthermore all these results can be extended to finitely generated A-modules, cf. [Sc 82a, §2.4], where other interesting observations can be found. We shall merely record the properties of $\mathfrak{I}$ as needed in 11.1.2, using the nonnilpotency of $\mathfrak{c}$, 10.2.9, in a complete ring. In this case none of the ideals in 11.2.1 is nilpotent.

11.2.2 COROLLARY. If $(A,\mathfrak{m})$ is a noetherian local ring, complete in its $\mathfrak{m}$-adic topology, then for any $r \geq 1$ there exists a s.o.p. x such that $\mathfrak{I}^r \not\subset \mathfrak{a}_x$ and $\mathfrak{I}\mathfrak{b}_x \subset \mathfrak{a}_x$.

Since a s.o.p. remains one after completion, we may as well carry through our modifications for a complete noetherian local ring. Two further consequences are worth recording, following [Ba, Ch. III, §2.6] where we use notation of 5.1.14. Such a result was also proved by Dutta [Du 89, Th. 2.3].

11.2.3 COROLLARY. Situation as above. There exists a nonnull ideal $\mathfrak{I}$ such that the Monomial Conjecture is true for all s.o.p.'s x with $\mathfrak{I} \not\subset \mathfrak{a}_x$ i.e. if y is a s.o.p. with $\mathfrak{a}_y \subset \mathfrak{a}_x$, $\underline{y} = T\underline{x}$ and $\det T = \Delta$, then Δ: $A/\mathfrak{a}_x \to A/\mathfrak{a}_y$ is nonnull.

PROOF. Write $\mathfrak{I} = \cap_x(\mathfrak{a}_x:\mathfrak{q}_x)$, then $\mathfrak{I}\mathfrak{q}_x \subset \mathfrak{a}_x$ for all x. Then $\mathfrak{q}_x$ cannot be A if $\mathfrak{I} \not\subset \mathfrak{a}_x$, so the Monomial Conjecture is true for such an x, see 8.3.5. In the complete case, $\mathfrak{I} \neq (0)$ satisfies.

11.2.4 COROLLARY. In a noetherian local ring there exists an integer N such that for every s.o.p. $x = x_1,\ldots,x_d$ the powers $x_1^s,\ldots,x_d^s$ satisfy the monomial conjecture whenever $s \geq N$.

PROOF. Since a possible violation $x_1^t \ldots x_d^t = \Sigma_{i=1}^d a_i x_i^{t+1}$ remains one under m-adic completion, we need only prove the statement in the complete case. Then $I \neq (0)$. Now $\mathfrak{a}_{x(s)} \subset m^s$ for any $n \geq 1$, while $\cap_{s=1}^{\infty} m^s = (0)$, so just choose N large enough that $I \not\subset m^N$.

11.3 THE FROBENIUS FUNCTOR

In this section and the next, we work with a ring A of positive prime characteristic p, so that we have the Frobenius endomorphism $x \mapsto x^p$ available. First we show how to extend this to a Frobenius functor which carries left A-modules to left A-modules. This functor has several nice properties, a few of which we develop in this section, even beyond what we shall need in sections 11.4 and 13.1. In the next section then, we show how to use this functor to succesfully carry through the modifications of section 11.1, utilizing Corollary 11.2.2. It should be pointed out that the Frobenius functor also played an important role in Peskine-Szpiro's proof [PS, Ch. I, §1] of the Intersection Theorem 8.4.4, and that a detailed discussion of its properties may be found in [HSb, §1].

Let (A,1) be an abelian group isomorphic to A but whose left and right A-module structures are distinct: its left structure is given by the identity on A, its right structure by the Frobenius endomorphism. We write the elements of (A,1) as pairs (a,1), $a \in A$, where the 1 is just a dummy keeping track of the structure: $x.(a,1) = (xa,1)$ and $(a,1).x = (x^p a,1)$ for $x \in A$. If M is a left A-module, we define $F(M) = (A,1) \otimes_A M$, giving it the structure of a left A-module by the action on (A,1). Clearly then $F(A) = A$ as a left A-module, and we can write $F^e(M) = (A,e)\otimes_A M$ in which we calculate according to the rules $x.((a,e)\otimes m) = (xa,e)\otimes m$, $x^{p^e}.((a,e)\otimes m) = (x^{p^e}a,e)\otimes m = (a,e)\otimes xm$ for any positive integer e and $m \in M$.

Now we examine what the Frobenius functor does to certain types of maps. If $.x: M \to M$ is scalar multiplication by $x \in A$ on a module M, then $F(x): F(M) \to F(M)$ is multiplication by x^p. In particular, if a homomorphism $\phi: A^m \to A^n$ between finitely generated free modules is described by an $n\times m$-matrix (x_{ij}), then $F(\phi)$ is described by the matrix (x_{ij}^p). This leads to an interpretation of F(M) for an arbitrary finitely generated module M. For let $M = A^n/(A\alpha_1+\ldots+A\alpha_m)$, where $\alpha_j \in A^n$ is a column vector $(x_{1j},\ldots,x_{nj})$, $j = 1,\ldots,m$. Then $F(M) = A^n/(A\alpha_1^p+\ldots+ A\alpha_m^p)$, where α_j^p is the column vector $(x_{1j}^p,\ldots,x_{nj}^p)$. This is seen by letting M be the cokernel of the map $A^m \to A^n$ described by the matrix (x_{ij}). Right exactness of the Frobenius functor, as a tensor product, and the above observation prove our contention. In particular, if $\mathfrak{a} \subset A$ is an ideal generated by elements $x_1,\ldots,x_m$, then $F(A/\mathfrak{a}) = A/(x_1^p,\ldots,x_m^p)$, a statement which is also easily proved for an infinite number of generators.

We next show that the Frobenius functor behaves well under taking fractions.

11.3.1 PROPOSITION. Let $S \subset A$ be a multiplicative system. Then $S^{-1}F(M) \simeq F(S^{-1}M)$ for every A-module M.

PROOF. We need to show that $S^{-1}A\otimes_A(A,1)\otimes_A-$ and $(S^{-1}A,1)\otimes_{S^{-1}A}S^{-1}A\otimes_A-$ are isomorphic functors. Now $(S^{-1}A,1)\otimes_{S^{-1}A}S^{-1}A$ is just $(S^{-1}A,1)$ while $x/s\otimes(a,1) \mapsto (xa/s,1)$ for $x, a \in A$, $s \in S$, determines an isomorphism $S^{-1}A\otimes_A(A,1) \to (S^{-1}A,1)$.

11.3.2 COROLLARY. Supp $F(M) \subset$ Supp M for every A-module M. Equality holds for finitely generated modules.

PROOF. Since $F(M)_{\mathfrak{p}} = F(M_{\mathfrak{p}})$ for every prime ideal $\mathfrak{p}$, $\mathfrak{p} \in$ Supp $F(M)$ implies $M_{\mathfrak{p}} \neq (0)$. If M is finitely generated and $\mathfrak{p} \in$ Supp M, there is a surjection $M_{\mathfrak{p}} \to A_{\mathfrak{p}}/\mathfrak{p}A_{\mathfrak{p}}$. But $F(A_{\mathfrak{p}}/\mathfrak{p}A_{\mathfrak{p}}) \neq (0)$ by previous remarks, and $F(M_{\mathfrak{p}})$ maps onto this, so $\mathfrak{p} \in$ Supp $F(M)$.

We shall now study a case in which the Frobenius functor is exact. First, however, we note that this is not true in general:

11.3.3 EXAMPLE. Let k be a field of characteristic p and $A = k[X]/(X^p)$. Observe that $(x) \simeq A/(x^{p-1})$ and $k = A/(x)$ where we have written x for the image of X in A. Applying F to the short exact sequence of A-modules $0 \to (x) \to A \to k \to 0$ yields that $F((x)) \to F(A) \to F(k) \to 0$ is exact. Now $F(k) = A/(x^p) = A$, hence $F(A) \to F(k)$ is an isomorphism. On the other hand, $F((x)) \simeq A/(x^{(p-1)p}) = A$. So the functor F is not exact. Furthermore, $\mathrm{pd}_A k = \infty$ since A is not regular, but $F(k)$ is free.

In contrast with this, we have

11.3.4 PROPOSITION. Let $0 \to L_s \overset{d_s}{\to} L_{s-1} \overset{d_{s-1}}{\to} \dots \overset{d_1}{\to} L_0$ be a complex $L_\bullet$ of finitely generated free modules over a noetherian ring A. Then $L_\bullet$ is exact if and only if the complex $F(L_\bullet)$ is.

PROOF. This is an easy application of the Buchsbaum-Eisenbud criterion (6.2.3 and 6.2.6) for exactness, to which we refer for notation. Recall that the chain modules in both complexes are the same, and that if

d_i is described by a rectangular matrix X_i, then the matrix Y_i of $F(d_i)$ consists of the same entries raised to the p-th power. If now Δ is a k×k-minor of X_i, then Δ^p is the corresponding minor of Y_i. Thus $\mathfrak{R}(I_k(d_i)) = \mathfrak{R}(I_k(F(d_i)))$ and in consequence $dp(I(d_i),A) = dp(I(F(d_i)),A)$ for $i = 1,\ldots,s$ and also $rk\ d_i = rk\ F(d_i)$. The result now follows from the characterization of exact complexes in the criterion.

The above argument moreover shows that in the case of a local ring $(A,\mathfrak{m})$ the complex $L_\bullet$ is a minimal resolution of $M = \text{coker}\ d_1$, i.e. all the entries in each matrix X_i are in $\mathfrak{m}$, if and only if $F(L_\bullet)$ is a minimal resolution of $F(M)$.

11.3.5 THEOREM [PS 73, Ch. I, Th. 1.7]. Let M be a finitely generated module over a noetherian ring A with $pd_A M < \infty$. Then $pd_{A_\mathfrak{p}} F(M_\mathfrak{p}) = pd_{A_\mathfrak{p}} M_\mathfrak{p}$ for all prime ideals $\mathfrak{p} \subset A$.

PROOF. Immediate in view of 11.3.2 and the above. Example 11.3.3 shows that it is incorrect to assume that $pd_A F(M) < \infty$ instead of $pd_A M < \infty$.

This theorem and 11.3.4 imply that F is exact on the subcategory consisting of such modules. Since F, being a tensor product, commutes with direct limits, we conclude

11.3.6 COROLLARY. For a regular ring, F is an exact functor.

We next consider the effect of the Frobenius on finite free complexes with finite length homology. If $L_\bullet = 0 \to L_s \to \ldots \to L_0 \to 0$ is such a complex, one is interested in its Euler-Poincaré characteristic $\chi(L_\bullet) =$

$\Sigma_{i=0}^{s}(-1)^i \ell(H_i(L_\bullet))$. Once we have chosen bases for the free modules and written the maps in terms of rectangular matrices, the maps in $F(L_\bullet)$ are succinctly described by raising each entry in these matrices to its p-th power. Not so clear however is what happens to χ. A plausible conjecture was raised by Szpiro [Sz, Conj. C.2]:

11.3.7 CONJECTURE. Let $L_\bullet$ be a finite free complex over a d-dimensional local ring. If it has finite length homology, then $\chi(F(L_\bullet)) = p^d\chi(L_\bullet)$.

This was recently confirmed by Gillet and Soulé in case the ring is a complete intersection [GS, Th. B] but is not true in general, though the order of magnitude is correct [Sei, Prop. 1]. We shall not discuss their work, but turn to a weaker, asymptotic result on the Euler-Poincaré characteristic as we iterate Frobenius. To this end we need a useful lemma, due to P. Roberts [Ro 87a, Lemma 5]. For reasons of exposition we formulate it for an ascending complex.

11.3.8 LEMMA. Let $(A,\mathfrak{m},k)$ be a noetherian local ring and $L^\bullet = 0 \to L^0 \to \ldots \to L^s \to 0$ be a complex of finitely generated free A-modules with $\dim H^i(L^\bullet) = 0$ for $i = 0,\ldots,s$. Suppose M is a finitely generated A-module of dimension $\leq r$. Then there exists a positive integer C, depending on M, such that for all $e \geq 0$ and all i, $\ell(H_i(\mathrm{Hom}_A(F^e(L^\bullet),M))) \leq p^{er}C$.

PROOF. First observe that the H_i's are finitely generated modules, at most supported in $\mathfrak{m}$, so have finite length. There is a finite filtration of M such that all subquotients are of type $N = A/\mathfrak{p}$ with $\mathfrak{p}$ a prime in A and with $\dim N \leq r$, and it is clearly enough to establish the inequality for such modules N. We induce on r; for $r = 0$ we are dealing with $N = k$ and $H_i(\mathrm{Hom}_A(F^e(L^\bullet),k))$ is a vector space of dimension $\mathrm{rk}\, L^i = C$.

For $N = A/\mathfrak{p}$ of positive dimension r, there exists a $y \in \mathfrak{m}\backslash\mathfrak{p}$ such that the localized complex $L^\bullet_y$ splits, since all the homology of $L^\bullet$ is annihilated by a power of $\mathfrak{m}$. This means that there exists an endomorphism σ of $L^\bullet_y$ of degree -1 such that $\mathrm{id} = \sigma d + d\sigma$ where d is the boundary map of $L^\bullet_y$. But then, for some positive integer n, the map $y^n\sigma$ lives on $L^\bullet$, so putting $x = y^n$, we find that $x\colon L^\bullet \to L^\bullet$ is homotopic to 0. So is then for each e the map $.x^{p^e}\colon F^e(L^\bullet) \to F^e(L^\bullet)$. The L^i being free, the short exact sequence

$$0 \to N \overset{x^{p^e}}{\to} N \to N/x^{p^e}N \to 0$$

induces exactness of

$$H_i(\mathrm{Hom}_A(F^e(L^\bullet),N)) \overset{x^{p^e}}{\to} H_i(\mathrm{Hom}_A(F^e(L^\bullet),N)) \to H_i(\mathrm{Hom}_A(F^e(L^\bullet),N/x^{p^e}N)).$$

Since the first map is 0, this implies that the length of the second term is bounded above by that of the third. In the filtration

$N \supset xN \supset x^2N \supset \ldots \supset x^{p^e}N$ all the successive factor modules are isomorphic to N/xN, since x acts as a non zerodivisor on the integral domain N. Now $\dim N/xN \le r-1$, so the induction hypothesis tells us that $\ell(H_i(\mathrm{Hom}_A(F^e(L^\bullet),N/xN))) \le p^{e(r-1)}C$ where the constant C does not depend on e. The lemma follows.

For a complex of finitely generated free modules $L_\bullet = 0 \to L_s \to \ldots \to L_0 \to 0$ with finite length homology we put $\chi_\infty(L_\bullet) = \lim_{e\to\infty} p^{-ed}.\chi(F^e(L_\bullet))$, where d is the dimension of our local ring. This type of asymptotic invariant was first considered by Dutta [Du 83]. Its present use for the Euler-Poincaré characteristic is due to P. Roberts [Ro 87a]. That the limit exists is a consequence of the result of Seibert already mentioned [Se, Prop. 1]. Szpiro remarked on the existence of this limit provided one has a certain local Riemann-Roch theorem available, [Sz, §4, Cor.]. The latter was established by Roberts in [Ro 87a, Lemma 2], who based himself on the local Riemann-Roch theorem in [Fu, Ex. 18.3.12]. See also [Ro 89]. A reader not wishing to take these developments for granted may define $\chi_\infty(L_\bullet)$ to be the lim inf in the next theorem, which is good enough for its application in 13.1.2.

Two further remarks are worth making. Once we have our Big Cohen-Macaulay module in 11.5.1, we may replace the condition "$H_0(L_\bullet) \neq 0$" below by "$L_\bullet$ is nonexact" by virtue of 6.1.4. Secondly, the result is remarkable in view of Roberts' construction in [Ro 89, §4] of a three-dimensional integrally closed Cohen-Macaulay domain with a complex $L_\bullet$ as below such that $\chi(L_\bullet) = -1$. By his asymptotic result to follow, $\chi_\infty(L_\bullet) > 0$, so that Szpiro's Conjecture 11.3.7 fails for this ring.

11.3.9 THEOREM. Let $L_\bullet = 0 \to L_d \to \dots \to L_0 \to 0$ be a complex of finitely generated free modules over a d-dimensional local ring (A,m,k). If $\ell(H_t(L_\bullet)) < \infty$ for $t = 0,\dots,d$ and $H_0(L_\bullet) \neq 0$, then $\chi_\infty(L_\bullet) > 0$.

PROOF. We may as well assume A is complete, since completion is exact on finitely generated modules and the length of the homology is preserved. Positivity of χ_∞ follows right away from the claim:

$$\lim_{e\to\infty} p^{-ed}.\ell(H_t(F^e(L_\bullet)) > 0 \text{ for } t = 0 \text{ and is } 0 \text{ for } t = 1,\dots,d.$$

We first treat the positivity. Frobenius being right exact, we have $H_0(F^e(L_\bullet)) = F^e(H_0(L_\bullet))$ and the latter has finite positive length by 11.3.2. A surjection $H_0(L_\bullet) \to k$ is preserved by F, so $\ell(F^e(H_0(L_\bullet))) \geq \ell(F^e(k)) \geq \ell(A/m^{p^e})$. Although we have not treated this subject in our book, it is well known that for large n the length of A/m^n is a Hilbert-Samuel polynomial in n of degree d and that its leading term is of the shape $a.n^d$ with $a > 0$ [AM, Th. 11.4]. This proves the claim for $t = 0$.

To continue for positive t, we shift $L_\bullet$ to an ascending complex

$L^\bullet = 0 \to L^0 \to \ldots \to L^d \to 0$ with $L^t = L_{d-t}$ and appropriate transcriptions of the boundary maps. Let $x = x_1,\ldots,x_d$ be a s.o.p. in A and $\mathfrak{a}$ the ideal generated by this, and write $K^\bullet$ for the complex $K^\bullet_\infty(x,A)$ introduced in 4.3.2. It is enough to show that for each t there is a positive integer C, not depending on e, such that $\ell(H^t(F^e(L^\bullet))) \le p^{et}C$. As in 7.3.8 we may consider the double complex $F^e(L^\bullet)\otimes_A K^\bullet$ to estimate this length. By that proposition, we need only show that $\ell(H^i(F^e(L^\bullet)\otimes_A H^j(A))) \le p^{et}C$ whenever $i+j = t$. The complete ring A has a dualizing complex, $D^\bullet$, so by 10.2.8 it suffices to prove that $\ell(H_i(\mathrm{Hom}_A(F^e(L^\bullet),H^{d-j}(D^\bullet)))) \le p^{et}C$. Now $D^{d-j} = \oplus\, E(A/\mathfrak{p})$ where the sum is taken over all primes $\mathfrak{p}$ with $\dim A/\mathfrak{p} = j$. In view of 3.2.7, the chain module D^{d-j} is only supported at primes containing some such $\mathfrak{p}$. Therefore the finitely generated homology module $H^{d-j}(D^\bullet)$ has dimension $\le j$. Since $j \le t \le d$, the inequality is now a consequence of Lemma 11.3.8, and the claim follows.

To finish this section we take another look at Frobenius:

11.3.10 REMARK. For a reduced ring A, on which the Frobenius endomorphism is injective, one can describe F more naively as follows. Write $A^{(p)} \subset A$ for the subring consisting of all p-th powers; it is isomorphic to A. To each A-module M attach an $A^{(p)}$-module $\tilde{M}$ which is isomorphic to M as an abelian group by the mapping $m \mapsto \tilde{m}$. The $A^{(p)}$-structure of $\tilde{M}$ is defined by putting $a^p.\tilde{m} = \widetilde{am}$. It is now easily seen that the A-module $A\otimes_{A^{(p)}}\tilde{M}$ is isomorphic to F(M). In [Ku 71, Th. 2.1] Kunz proved that a reduced ring A is flat over $A^{(p)}$ if and only if A is regular. In fact Corollary 11.3.6 is the first half of his result, albeit proved by different means. See also [Her, Kor. 3.3].

11.3.11 We finally discuss the connection between the Frobenius functor and p-linear homomorphisms. An additive homomorphism $\lambda: M \to N$ between two A-modules is called p-linear if $\lambda(xm) = x^p\lambda(m)$ for all $x \in A$, $m \in M$; an example is the Frobenius endomorphism of A. Another example is the map $\mu: M \to F(M)$ defined by sending m to $(1,1)\otimes m$, $m \in M$. This is universal in the following sense: for λ p-linear as above, there exists a unique A-linear homomorphism $\theta_\lambda: F(M) \to N$ with $\theta_\lambda \circ \mu = \lambda$; it is defined by $\theta_\lambda((a,1)\otimes m) = a\lambda(m)$, $a \in A$, $m \in M$.

On the other hand, if $f: M \to N$ is A-linear, then

$$\begin{array}{ccc} M & \xrightarrow{f} & N \\ \downarrow{\scriptstyle \mu_M} & & \downarrow{\scriptstyle \mu_N} \\ F(M) & \xrightarrow[F(f)]{} & F(N) \end{array}$$

is a commutative square, in which the horizontal maps are linear and the vertical ones p-linear. Similar observations hold for the iterated functors F^e.

11.4 PRE-REGULAR MODULES IN CHARACTERISTIC p

Using the functorial properties of the Frobenius functor just discussed, we can succesfully carry through the modifications of 11.1. Specifically, let (A,m) be a noetherian local ring of positive characteristic p and $x = x_1,..,x_d$ be a fixed s.o.p. Let us make a few preliminary remarks on how Frobenius behaves with respect to modifications.

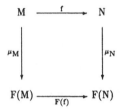

Let m_0 be a bad element in the A-module M, with respect to x. This means that $x_1^t \dots x_d^t m_0 \in (x_1^{t+1},\dots,x_d^{t+1})\alpha^{dt-t-1}M$ for some $t \geq 1$, where $\alpha = (x_1,\dots,x_d)$. Then $(x_1^t \dots x_d^t)^p \mu(m_0) \in (x_1^{(t+1)p},\dots,x_d^{(t+1)p})\alpha^{(dt-t-1)p}F(M)$, so that $\mu(m_0)$ is a bad element in F(M) w.r.t. the s.o.p. $x^p = x_1^p,\dots,x_d^p$.

If f: M → M′ is a modification of m_0 w.r.t. x, we may apply F to the commutative diagram

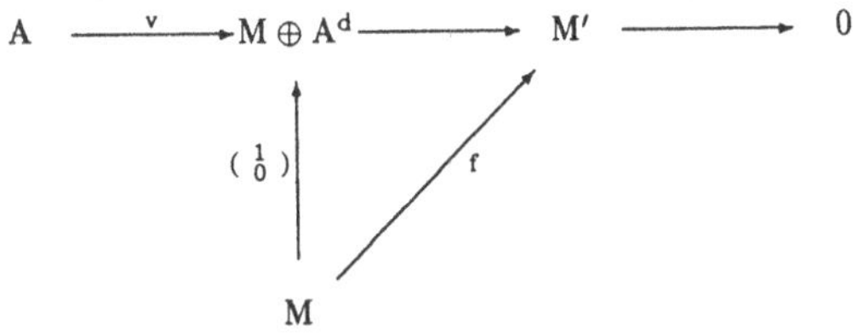

in which the row is exact and $v = (-m_0, x_1,\dots,x_d)$. Right exactness of F and the diagram ending 11.3 show that F(f): F(M) → F(M′) is the modification of the bad element $\mu(m_0)$ in F(M) w.r.t. the s.o.p. x^p.

We are now ready to clinch the argument.

11.4.1 THEOREM. Let $x = x_1,\dots,x_d$ be an arbitrary s.o.p. in a noetherian local ring (A,m) of prime characteristic p. Then there exists an x-pre-regular A-module.

PROOF. Let (M_r, u_r) be a sequence of modifications of bad elements as in Proposition 11.1.1. Suppose $u_r \in \alpha M_r$ for some $r \geq 1$, then we shall derive a contradiction. Observe that $\mu^e(u_r) \in (x_1^{p^e},\dots,x_d^{p^e})F^e(M)$ for all $e \geq 0$. Now $F^e(M_j, \mu^e(u_j))$ is a sequence of modifications of bad elements w.r.t. the s.o.p. $x^{p^e} = x_1^{p^e},\dots,x_d^{p^e}$. Recall that we may as well suppose A is complete in its m-adic topology. Then I, the ideal of 11.2.1, is nonnilpotent by 11.2.2. Thus we can choose e large enough so that $I^r \not\subset \alpha^{p^e}$; a fortiori $I^r \not\subset (x_1^{p^e},\ldots,x_d^{p^e})$. By definition we have $\mathfrak{b}_{x^{p^e}} \subset (x_1^{p^e},\ldots,x_d^{p^e})$. But then 11.1.2. tells us that $\mu^e(u_r) \notin (x_1^{p^e},\ldots,x_d^{p^e})F^e(M)$, which is our contradiction. Therefore $u_r \notin \mathfrak{a}M_r$ for all r and $M = \varinjlim_r M_r$ is pre-regular w.r.t. x by Proposition 11.1.1.

11.5 BALANCED BIG COHEN-MACAULAY MODULES IN CHARACTERISTIC p

Completing the module of 11.4.1 in its $\mathfrak{m}$-adic topology, we obtain the result we are after, in view of 9.1.1 and 5.2.3.

11.5.1 THEOREM. Every local noetherian ring of prime characteristic p has balanced Big Cohen-Macaulay modules. There are such modules M for which $(x_1,\ldots,x_i)M$ is closed in the $\mathfrak{m}$-adic topology of M for all s.o.p. $x_1,\ldots,x_d$, $0 \le i \le d$, and where M is complete in its $\mathfrak{m}$-adic topology.

Hochster was the first to construct Big Cohen-Macaulay modules in [Ho 75a, Th. 5.1], after having realized their importance for a host of homological conjectures, and he conjectured their existence also in mixed characteristic [Ho 75a, Ch. 2, Conj. (7)]. In [Ho 75b, Cor. 5.8] he gave a construction of balanced Big Cohen-Macaulay modules, different from ours. Needless to say, our treatment should he considered a refinement of his pioneering work.

A final remark is that the above theorem remains valid for a ring A such that A_{red} is of characteristic p. For a s.o.p. remains one after factoring out the nilradical.

12 BIG COHEN-MACAULAY MODULES IN EQUAL CHARACTERISTIC 0

by

L.P.D. van den Dries

12.0 INTRODUCTION

Fix the positive integer n for the remainder of this Introduction. We are going to describe a strategy for the proof of the following fundamental result due to M. Hochster.

THEOREM I. Let A be a noetherian ring containing a field and let $x_1,\ldots,x_n$ in A generate an ideal of height n. Then there exists an $(x_1,\ldots,x_n)$-regular A-module.

In the preceding chapter we proved this theorem in the case A contains a field of characteristic $p > 0$. (In fact, the result was only stated there for local A with $x_1,\ldots,x_d$ a system of parameters, but the more general formulation used here follows easily by localization.)

In this chapter we introduce an important general method to reduce algebraic problems "in equal characteristic 0" to similar problems "in equal characteristic $p > 0$". In order to apply this method in proving Theorem I we have to reformulate the problem in terms of polynomials.

To do this we use the (sometimes subscripted) expression ξ, or $\xi(X,Y)$, to denote a finite system of polynomial equations over $\mathbb{Z}$:

$$\xi(X,Y) : F_1(X,Y) = \dots = F_h(X,Y) = 0$$

where $X = (X_1,\dots,X_n)$, $Y = (Y_1,\dots,Y_q)$, $F_i(X,Y) \in \mathbb{Z}[X,Y]$ for $i = 1,\dots,h$. Note that n is the integer fixed above, and that q and h depend on the system $\xi = \xi(X,Y)$. Given a ring A we call an (n+q)-tuple (x,y) in A a solution of $\xi(X,Y)$ in A if

$$F_1(x,y) = \dots = F_h(x,y) = 0.$$

(Here we interpret the integer coefficients of the F's by their images in A.)

The next proposition reduces the problem of proving Theorem I to one on systems of polynomials.

PROPOSITION II. There is a family $(\xi_\lambda)_{\lambda \in \Lambda}$ of finite systems of polynomial equations over $\mathbb{Z}$ (of the form specified above) such that for each ring A and each $x = (x_1,\dots,x_n) \in A^n$ the following are equivalent:

(i) There exists an x-regular A-module;

(ii) No system $\xi_\lambda = \xi_\lambda(X,Y)$ has a solution (x,y) in A.

A construction of these systems ξ_λ, $\lambda \in \Lambda$, was actually given by Hochster in [Ho 75a, Prop. 4.2]. In §1 we prove the proposition as a special case of a more general result. Our argument is completely different from Hochster's.

Suppose now that A is a noetherian ring containing a field of characteristic 0, that $x = (x_1,\dots,x_n) \in A^n$ satisfies $ht(x_1,\dots,x_n) = n$, and that, contrary to Theorem I, there is no x-regular A-module. Then, by Proposition II, one of the systems $\xi_\lambda(X,Y)$ has a solution (x,y) in A.

We will get a contradiction from this by proving that then the same system ξ_λ must have a solution (x',y') in some noetherian ring A' containing a field of characteristic $p > 0$, with $ht(x') = n$. Indeed, by the preceding chapter, there exists in this case an x'-regular A'-module, and we have a contradiction with Proposition II.

The remaining step in this argument (i.e. "by proving that then ...") is provided by Theorem III below.

THEOREM III. Let $\xi = \xi(X,Y)$ be a finite system of polynomial equations over $\mathbb{Z}$ and suppose that ξ has a solution (x,y) in a noetherian ring A containing a field of characteristic 0, such that $ht(x_1,\dots,x_n) = n$. Then there exists a domain A′ containing a finite field and finitely generated over that field such that ξ has a solution (x′,y′) in A′ with $ht(x_1',\dots,x_n') = n$.

The proof of Theorem III splits into two steps:

LEMMA IV. Assume that the hypotheses of Theorem III hold and suppose moreover that A is finitely generated over a field of characteristic 0. Then there is a domain A′ containing a finite field and finitely generated over that field such that ξ has a solution (x′,y′) in A′ with $ht(x_1',\dots,x_n') = n$.

Lemma IV is obviously a special case of Theorem III. That we can reduce to this special case follows from the next lemma.

LEMMA V. Under the same hypotheses as in Theorem III, there is a domain A′ containing a field of characteristic 0 and finitely generated over that field, such that ξ has a solution (x′,y′) in A′ with $ht(x_1',\dots,x_n') = n$.

From the discussion above, the implications (II & III) ⇒ I and (IV & V) ⇒ III should be clear. As said before we will prove Proposition II in section 1. After this we are reduced to proving Lemmas IV and V. In sections 2 and 3 we prove IV. Lemma V is the harder one and requires (or so it seems) a deep approximation theorem due to M. Artin. We treat this very useful tool in section 6, with the necessary preliminaries on Weierstrass Theorems and henselization in sections 4 resp. 5. In the short section 7 we can finally prove Lemma V.

Sections 2, 4, 5 and 6 can be read independently from the remainder of this book. These sections introduce methods and tools which, apart from their intrinsic interest, seem to become more and more important in several branches of algebra.

12.1 HOCHSTER ALGEBRAS AND EQUATIONAL CONSTRAINTS

12.1.1 We fix again the positive integer n and write $X = (X_1,\ldots,X_n)$. So $\mathbf{Z}[X] = \mathbf{Z}[X_1,\ldots,X_n]$.

To prove Proposition II of the Introduction we have to find the equational constraints on pairs (A,x), A a ring, $x \in A^n$, which are necessary and sufficient for existence of an x-regular A-module. Formally we identify such a pair (A,x) with the $\mathbf{Z}[X]$-algebra A whose structural morphism $\mathbf{Z}[X] \to A$ is given by $X_i \mapsto x_i$ for $i = 1,\ldots,n$. Under this identification a morphism $(A,x) \to (B,y)$ of $\mathbf{Z}[X]$-algebras is a ring homomorphism $A \to B$ mapping x_i to y_i, $i = 1,\ldots,n$.

12.1.2 DEFINITION. A Hochster algebra (of type n) is a $\mathbf{Z}[X]$-algebra

(A,x) such that there exists an x-regular A-module.

All we need to know about Hochster algebras is that they form a class of $\mathbb{Z}[X]$-algebras with the following two properties.

12.1.3 LEMMA.

(a) If $(A,x) \to (B,y)$ is a $\mathbb{Z}[X]$-algebra morphism and (B,y) is a Hochster algebra, then (A,x) is a Hochster algebra.

(b) If each finitely generated subalgebra (A',x) of a $\mathbb{Z}[X]$-algebra (A,x) is a Hochster algebra, then (A,x) is a Hochster algebra.

PROOF.

(a) If M is a B-module, then its submodule $(y_1,\ldots,y_k)M$ is, as a set, equal to the A-module $(x_1,\ldots,x_k)M$, $1 \le k \le n$. Hence M is regular if and only if M is x-regular as an A-module.

(b) Let $(A_\gamma,x)_{\gamma\in\Gamma}$ be the family of finitely generated subalgebras of (A,x), and suppose each (A_γ,x) is a Hochster algebra. We shall define an ideal $\mathfrak{a}$ of the product algebra $(\Pi A_\gamma,x)$, prove that the factor algebra $((\Pi A_\gamma)/\mathfrak{a},x/\mathfrak{a})$ is a Hochster algebra, and finally indicate a $\mathbb{Z}[X]$-algebra morphism $(A,x) \to((\Pi A_\gamma)/\mathfrak{a},x/\mathfrak{a})$. By (a) this suffices to conclude that (A,x) is a Hochster algebra.

Call a subset Δ of Γ large, if for some finite subset Y of A the set Δ contains all $\gamma \in \Gamma$ with $Y \subset A_\gamma$. It is obvious that Γ itself is large, that Δ' is large if $\Delta' \supset \Delta$ and Δ is large, and that the large sets are nonempty. Most importantly, the large sets are closed under finite intersections (in other words, they form a filter).

Put $\mathfrak{a}_\Delta = \{a = (a_\gamma)_{\gamma\in\Gamma} \in \Pi A_\gamma \mid a_\gamma = 0 \text{ for all } \gamma \in \Delta\}$ for large Δ. Then each $\mathfrak{a}_\Delta$ is an ideal of ΠA_γ, hence their union $\mathfrak{a}$ is an ideal of ΠA_γ.

By assumption there is for each $\gamma \in \Gamma$ an x-regular A_γ-module M_γ. For large Δ, Let N_Δ be the submodule $\{m = (m_\gamma)_{\gamma\in\Gamma} \in \Pi M_\gamma \mid m_\gamma = 0 \text{ for all } \gamma \in \Delta\}$ of the ΠA_γ-module ΠM_γ and let N be the union of all N_Δ; so N is a submodule of ΠM_γ with $\alpha(\Pi M_\gamma) \subset N$. Therefore $(\Pi M_\gamma)/N$ is a $(\Pi A_\gamma)/\alpha$-module and it is straightforward to check that it is an x/α-regular module, where $x/\alpha = (x_1/\alpha,\ldots,x_n/\alpha)$. Hence $((\Pi A_\gamma)/\alpha, x/\alpha)$ is a Hochster algebra.

Finally, we define ϕ: $A \to (\Pi A_\gamma)/\alpha$ as follows: For any $a \in A$, let Δ be a large set such that $a \in A_\gamma$ for all $\gamma \in \Delta$. Put $\phi(a) = (a_\gamma)/\alpha$, where $a_\gamma = a$ for $\gamma \in \Delta$ and a_γ is arbitrary otherwise. It is easily verified that ϕ is well defined and that it is a $\mathbf{Z}[X]$-algebra morphism.

12.1.4 More generally, consider instead of $\mathbf{Z}[X]$ a noetherian ring Z and, in stead of the class of Hochster algebras, a class $\mathscr{C}$ of Z-algebras with the following properties:

(a) if $A \to B$ is a Z-algebra morphism and $B \in \mathscr{C}$, then $A \in \mathscr{C}$;

(b) if each finitely generated subalgebra of the Z-algebra A belongs to $\mathscr{C}$, then $A \in \mathscr{C}$.

(So for $Z = \mathbf{Z}[X]$ and $\mathscr{C}$ = class of Hochster algebra, conditions (a) and (b) hold by the previous lemma.)

Under these assumptions we can prove:

12.1.5 PROPOSITION. There exists a set Ξ of systems of polynomial equations over Z:

$$\xi(Y) \ : \ F_1(Y) = \ldots = F_h(Y) = 0$$

($F_i(Y) \in Z[Y]$, $Y = (Y_1,\ldots,Y_q)$, h and q depending on the system $\xi(Y)$) such that for each Z-algebra A: $A \in \mathscr{C} \Leftrightarrow$ no system $\xi(Y)$ of Ξ has a solution in A.

PROOF. Define Ξ as the set of all systems

$\xi(Y)$: $F_1(Y) = \ldots = F_h(Y) = 0$ ($F_i(Y) \in Z[Y]$, $Y = (Y_1,\ldots,Y_q)$) which have no solution in any Z-algebra belonging to $\mathscr{C}$. We only have to show that conversely a Z-algebra A in which no system $\xi(Y)$ of Ξ is soluble, belongs to $\mathscr{C}$.

In view of condition (b) we may reduce to the case that A is finitely generated as a Z-algebra, say $A = \bar{Z}[y]$, $y = (y_1,\ldots,y_q)$, $\bar{Z}$ = image of structural morphism $Z \to A$. Let $Y = (Y_1,\ldots,Y_q)$ and let $F_1(Y),\ldots,F_h(Y)$ in $Z[Y]$ generate the kernel of the Z-algebra morphism $Z[Y] \to A$ with $Y_i \mapsto y_i$.

Then $F_1(y) = \ldots = F_h(y) = 0$, so the system of equations $F_1(Y) = \ldots = F_h(Y) = 0$ does not belong to Ξ. By definition of Ξ there exists a Z-algebra B in $\mathscr{C}$ in which the system has a solution $y' = (y'_1,\ldots,y'_q)$, which means that the Z-algebra morphism $Z[Y] \to B$ with $Y_i \mapsto y'_i$, $i = 1,\ldots,q$, has $F_1(Y),\ldots,F_q(Y)$ in its kernel. So it induces a Z-algebra morphism $A \simeq Z[Y]/(F_1(Y),\ldots,F_h(Y)) \to B$. Hence A belongs to $\mathscr{C}$ by condition (a).

Applying 12.1.5 to the class of Hochster algebras yields Proposition II of the Introduction.

Our treatment here is in the spirit of model theory and does not give a method to construct the actual equational "constraints". Such a method is implicit in [Ho 75a, p. 22] by Hochster who first realized that the existence of "universal equational constraints" can be used for reduction to prime characteristic.

In Exercise 6 below we indicate another important class of $\mathbf{Z}[X]$-algebras satisfying conditions (a) and (b).

12.1.6 EXERCISES.

1. (Converse to 12.1.5) Let Z be any ring and $\mathscr{C}$ any class of Z-algebras for which Proposition 12.1.5 holds. Then conditions (a) and (b) of

12.1.4 are satisfied by $\mathscr{C}$.

2. Suppose that Z is a noetherian ring and $\mathscr{C}$ a class of noetherian Z-algebras such that (a) and (b) of 12.1.4 hold, with A a noetherian Z-algebra in both conditions. Check that 12.1.5 is still valid for noetherian Z-algebras A.

Let in the next two exercises Z and $\mathscr{C}$ satisfy 12.1.4.

3. Show that $\mathscr{C}$ is closed under taking direct limits. (Hint: use 12.1.5.)

4. Suppose that $A \in \mathscr{C} \Rightarrow A_{red} \in \mathscr{C}$. Show that Ξ as in 12.1.5 may be taken to consist only of systems $F_1(Y) = \ldots = F_h(Y) = 0$, such that $(F_1(Y),\ldots,F_h(Y))$ is a radical ideal of Z[Y].

5. (Case n = 1) Let A be a ring, $x \in A$. Write $\mathrm{Ann}^{\infty}(x)$ for $\bigcup_{m=1}^{\infty} \mathrm{Ann}(x^m)$, the union of the ascending chain of ideals $\mathrm{Ann}(x^m)$. Show the following to be equivalent:

(i) (A,x) is a Hochster algebra;

(ii) $(x) + \mathrm{Ann}^{\infty}(x) \neq A$;

(iii) No equation $x^{m+1}y = x^m$, $m > 0$, has a solution y in A.

(Hints: to prove (iii) $\Rightarrow$ (ii), suppose that (ii) does not hold, i.e. $xy_1 + y_2 = 1$, $x^m y_2 = 0$ for certain y_1, $y_2 \in A$, $m > 0$, and derive a solution for $x^{m+1}y = x^m$; to prove (ii) $\Rightarrow$ (i), show that (ii) implies that $A/\mathrm{Ann}^{\infty}(x)$ is x-regular; for (i) $\Rightarrow$ (iii), suppose that $x^{m+1}y = x^m$, $m > 0$, and that x is not a zerodivisor on the A-module M; conclude that $xM = M$.) Note that for A a domain (ii) means: x is neither a unit, nor zero, while for A a local ring (ii) is the same as: x is neither a unit, nor nilpotent.

The following class of $\mathbb{Z}[X]$-algebras was introduced by Hochster in [Ho 75b, §2], in connection with the mixed characteristic case of his conjecture.

DEFINITION. Call $x = (x_1,\dots,x_n) \in A^n$, A a ring, a system of quasi-parameters if for each finitely generated $\mathbb{Z}[X]$-subalgebra (A',x) of (A,x) there exists a $\mathbb{Z}[X]$-algebra morphism (A',x) → (B,y) with B local noetherian and y a s.o.p. for B. (Completing and dividing out a prime of maximal coheight one can always take for B a complete local, noetherian domain.) Let us call the $\mathbb{Z}[X]$-algebra (A,x) a quasi-parameter algebra if x is a system of quasi-parameters.

12.1.6 EXERCISES.

6. Show that the class of quasi-parameter algebras satisfies the conditions (a) and (b) of 12.1.4.

7. Show: if (A,x) is a quasi-parameter algebra, then $(A,x)_{red}$ is also a quasi-parameter algebra.

Comment. Exercise 6 combined with 12.1.5, shows that there exists a "complete" set of equational constraints defining systems of quasi-parameters. It would be important to "know" these constraints explicitly; do they express something natural, say on the ideals which one can form from the x's using various operations? Some connections with Hochster algebras are contained in the following exercises.

12.1.6 EXERCISES.

8. Show that Hochster's conjecture, see section 11.5, is equivalent with: all quasi-parameter algebras are Hochster algebras.

9. Show that for n = 1 Hochster algebras and quasi-parameter algebras are the same.

12.2 CONSTRUCTIBLE PROPERTIES AND TRANSFER TO FINITE FIELDS

12.2.1 In this section we introduce a method - the method of constructible properties - which enables one to reduce certain problems over fields of characteristic 0 to similar problems over finite fields. This method will then be used in section 3 to prove Lemma IV of the Introduction. As motivation we first consider an example where a reduction to finite fields is apparent.

12.2.2 Consider a system of linear equations

$$\begin{array}{ccccc} c_{11}U_1 & + \cdots + & c_{1N}U_N & = & c_1 \\ \vdots & & \vdots & & \vdots \\ c_{M1}U_1 & + \cdots + & c_{MN}U_N & = & c_M \end{array}$$

where the U's are the unknowns and the c's are integers. We shall show how to compute a prime number p such that the system is soluble in $\mathbb{Q}$ if and only if it is soluble in $\mathbb{F}_p$.

We know that the system is soluble if and only if

$$\text{rank}\begin{pmatrix} c_{11} & \cdots & c_{1N} \\ \vdots & & \vdots \\ c_{M1} & \cdots & c_{MN} \end{pmatrix} = \text{rank}\begin{pmatrix} c_{11} & \cdots & c_{1N} & c_1 \\ \vdots & & \vdots & \vdots \\ c_{M1} & \cdots & c_{MN} & c_m \end{pmatrix} \qquad (*$$

(This is true for c's in K and solubility in K, where K is any field.) The rank of a matrix is the least $k \geq 0$ such that all $(k+1)\times(k+1)$ minors vanish. Take for p a prime number not dividing any of the nonzero minors of the two matrices in (*). Then for both matrices in (*) we have:

$$\text{rank over } \mathbb{Q} = \text{rank modulo } p.$$

By (*) this implies that the system is soluble in $\mathbb{Q}$ if and only if it is soluble in $\mathbb{F}_p$.

12.2.3 The crucial point in this argument is that we could express solubility by certain polynomial equations and inequations over **Z** in the c's (certain minors are zero and others are nonzero) and that these (in)equations are independent of the field K in which the coefficients c_{ij}, c_j lie.

A similar result holds for arbitrary finite systems of polynomial equations, provided one considers solubility in the algebraic closure of the field of coefficients. (This will be proved in 12.2.10.) The simplest nonlinear example is the equation $c_0U^2 + c_1U + c_2 = 0$. Its solubility in the algebraic closure of the field K, where $c_i \in K$, can be expressed by: $(c_0 \neq 0)$ or $(c_1 \neq 0)$ or $(c_2 = 0)$.

More generally we want to consider properties of systems of polynomials over fields which can be expressed by equations and inequations over **Z** in their coefficients. Polynomials are determined by their vector of coefficients, so we may consider instead properties of vectors $c \in K^M$, K a field, M > 0. The idea is that a "property" is something which holds for certain fields and M-tuples over that field, and does not hold for other fields and M-tuples, This suggests:

12.2.4 DEFINITION. Let $C = (C_1,\ldots,C_M)$ be a vector of distinct variables, M > 0. A C-property is a map which assigns to each pair (K,c), K a field and $c \in K^M$, the value "true" or the value "false".

C-properties will be denoted by $\phi(C)$, $\psi(C)$, etc. If $\phi(C)$ assigns to the pair (K,c) the value true, we also say that "K satisfies $\phi(c)$" or "$\phi(c)$ holds in K", and we write $K \models \phi(c)$.

We shall deal with C-properties in an informal way:

12.2.5 EXAMPLES. Let M > 0, $C = (C_1,\ldots,C_M)$ as above.

(1) Let $\phi(C)$ be : "$X^M + C_1X^{M-1} + \dots + C_M$ has M distinct zeroes in the algebraic closure". A formally precise definition of $\phi(C)$ is as follows. For any field K and any $c = (c_1,\dots,c_M) \in K^M$ we put: $K \models \phi(c)$ if and only if $X^M + c_1X^{M-1} + \dots + c_M$ has M distinct zeroes in the algebraic closure of K.

(2) Let $f_1(C,T)$, $f_2(C,T) \in \mathbb{Z}[C,T]$. Then we define a C-property $\psi(C)$ by: "$f_1(C,T)$ and $f_2(C,T)$ have a common zero in some extension field". So $K \models \psi(c)$ iff there is an extension field L of K and $t \in L$ such that $f_1(c,t) = f_2(c,t) = 0$.

12.2.6 We introduce the following operations on C-properties. Given C-properties $\phi_1(C)$ and $\phi_2(C)$ we define the C-properties $\phi_1(C) \vee \phi_2(C)$, $\phi_1(C) \wedge \phi_2(C)$, $\neg\phi_1(C)$ by putting, for any field K and $c \in K^M$:

$K \models \phi_1(c) \vee \phi_2(c)$ iff $K \models \phi_1(c)$ or $K \models \phi_2(c)$,

$K \models \phi_1(c) \wedge \phi_2(c)$ iff $K \models \phi_1(c)$ and $K \models \phi_2(c)$,

$K \models \neg\phi_1(c)$ iff not $K \models \phi_1(c)$.

The binary operations $\vee$ and $\wedge$ on C-properties are clearly commutative and associative, and distributive over each other. Also $\neg(\phi_1(C) \vee \phi_2(C)) = (\neg\phi_1(C)) \wedge (\neg\phi_2(C))$, and similarly with "$\vee$" and "$\wedge$" interchanged. It will not surprise the reader that we call $\phi_1(C) \vee \dots \vee \phi_k(C)$ the disjunction of the C-properties $\phi_1(C),\dots,\phi_k(C)$, similarly $\phi_1(C) \wedge \dots \wedge \phi_k(C)$ their conjunction, and $\neg\phi(C)$ the negation of $\phi(C)$.

12.2.7 From 12.2.2 and the discussion in 12.2.3 it will be clear that our main interest is in C-properties expressible by equations and inequations over $\mathbb{Z}$ in the variables C. For instance the C-property $\phi(C)$ of 12.2.5 Example (1) is of this type: Let D(C) be the discriminant of

$X^M + C_1X^{M-1} + \ldots + C_M$ considered as a polynomial in X. Then $\phi(C)$ coincides with the C-property "$D(C) \neq 0$" (which is true for a pair (K,c) iff $D(c) \neq 0$). (For a description of "discriminant" see [Lan, Ch. V, §9].) The technical term for a property expressible by equations and inequations over $\mathbb{Z}$ is: constructible property. A precise definition follows.

12.2.8 DEFINITION. Let $M > 0$, let $C = (C_1,\ldots,C_M)$ be a vector of M distinct variables.

(i) A basic C-property is a C-property "$\alpha(C) = 0$", where $\alpha(C) \in \mathbb{Z}[C]$. (It is true for a pair (K,c) iff $\alpha(c) = 0$.)

(ii) Let $\alpha_1,\ldots,\alpha_p,\beta_1,\ldots,\beta_q \in \mathbb{Z}[C]$, $p + q > 0$, and let α and β denote the sequences $(\alpha_1,\ldots,\alpha_p)$ and $(\beta_1,\ldots,\beta_q)$ respectively. Then the C-property "$\alpha_1(C) = \ldots = \alpha_p(C) = 0$ and $\beta_1(C) \neq 0$ and ... and $\beta_q(C) \neq 0$" will be denoted by $(\alpha(C);\beta(C))$ and is called a C-system.

(iii) A constructible C-property is by definition a disjunction of finitely many C-systems. (By convention, the disjunction of an empty set of C-properties is the C-property "$1 = 0$" which is never satisfied.)

The following exercises state simple facts about constructible properties. The last two are intended for readers familiar with constructible sets. In the remainder of this section C will denote a vector $(C_1,\ldots,C_M)$ of M distinct variables, $M > 0$.

12.2.9 EXERCISES. Prove:

1. The negation of a C-system is a constructible C-property.

2. The constructible C-properties form a boolean algebra with respect to $\vee$, $\wedge$, and $\neg$. Its greatest element is the basic C-property "$0 = 0$" which is always satisfied, and its least element is the basic C-property "$1 = 0$", which never holds. (Hint: use Exercise 1.)

3. Let $\phi(C)$ be a constructible C-property, K and L fields, $K \subset L$ and $c \in K^M$. Then $K \models \phi(c)$ iff $L \models \phi(c)$.

4. The boolean algebra of constructible C-properties is isomorphic with the boolean algebra of constructible subsets of Spec($\mathbf{Z}[C]$), [Ma 80, Ch. 2, §6], under the isomorphism which associates to the basic C-property "$\alpha(C) = 0$" ($\alpha(C) \in \mathbf{Z}[C]$) the Zariski-closed set $\{\mathfrak{p} \in \mathrm{Spec}(\mathbf{Z}[C]) \mid \alpha(C) \in \mathfrak{p}\}$.

5. Let $\phi(C)$ be a constructible C-property and $\hat{\phi}(C)$ the constructible subset of Spec($\mathbf{Z}[C]$) corresponding to $\phi(C)$ under the isomorphism in 4. Let K be a field and $c \in K^M$. Then $K \models \phi(c)$ iff the ring homomorphism $\mathbf{Z}[C] \to K$ given by $C_i \mapsto c_i$ has a kernel belonging to $\hat{\phi}(C)$.

The following fact explains why constructible properties are so useful for reducing characteristic zero problems to problems over finite fields: If $\phi(C)$ is a constructible C-property and $K \models \phi(c)$ for some field K of characteristic 0 and some $c \in K^M$, then there is a finite field K′ and a $c' \in (K')^M$ such that $K' \models \phi(c')$. This and similar results are derived from the following crucial theorem.

12.2.10 THEOREM. Let $U = (U_1,\dots,U_N)$ be a sequence of new variables, $N > 0$, and let $f_1(C,U),\dots,f_k(C,U)$, $g_1(C,U),\dots,g_j(c,U) \in \mathbf{Z}[C,U]$, $k+j > 0$. Then the C-property $\Phi(C|f_1,\dots,f_k;g_1,\dots,g_j)$ defined by: "the system of equations and inequations

$$f_1(C,U) = \dots = f_k(C,U) = 0,\ g_1(C,U) \neq 0,\ \dots\ ,\ g_j(C,U) \neq 0$$

has a solution $u \in L^N$ in an extension field L" is a constructible C-property.

REMARK. In fact, we shall explicitly construct a constructible C-property $\psi(C)$ which coincides with $\Phi(C|f_1,\dots,g_j)$. Our argument is easily adapted to showing that $\psi(C)$ also coincides with the C-property

$\Phi'(C|f_1,\ldots,f_k;g_1,\ldots,g_j)$ defined by: "the system

$$f_1(C,U) = \ldots = f_k(C,U) = 0,\ g_1(C,U) \neq 0,\ \ldots\ ,g_j(C,U) \neq 0$$

is soluble for U within the algebraic closure". (One simply replaces everywhere in the proof Φ by Φ' and the phrase "in an extension field L" by: "in the algebraic closure")

PROOF. Consider first the case N = 1. $\Phi(C|f_1,\ldots,f_k;g_1,\ldots,g_j)$ clearly coincides with $\Phi(C|f_1,\ldots,f_k,0;\Pi g_i)$, so we may assume $k \geq 1$ and $j = 1$. (If no polynomial g_i occurs in Φ, take $g_1 = 1$.) Without loss of generality we may also assume that $\deg_{U_1} f_1 \geq \deg_{U_1} f_2 \geq \ldots \geq \deg_{U_1} f_k$, where $\deg_{U_1} 0 = 0$ by convention. Write

$$f_1(C,U_1) = \alpha_d(C)U_1^d + \alpha_{d-1}(C)U_1^{d-1} + \ldots + \alpha_0(C),$$

$$g_1(C,U_1) = \beta_e(C)U_1^e + \beta_{e-1}(C)U_1^{e-1} + \ldots + \beta_0(C),$$

where $d = \deg_{U_1} f_1$, $e = \deg_{U_1} g_1$.

We use induction on the integer D defined by: $D = \Sigma_{i=1}^k \deg_{U_1} f_i + e$.

Trivial case: At most one term of the sequence $f_1,\ldots,f_k,g_1$ has positive degree in U_1. This term (if any) is either f_1 or g_1. If it is f_1 or if there is none, then the polynomials $f_2,\ldots,f_k,g_1$ belong to $\mathbb{Z}[C]$ and $\Phi(C|f_1,\ldots,f_k;g_1)$ equals the C-property

$$(\alpha_0(C) = 0 \vee \alpha_1(C) \neq 0 \vee \ldots \vee \alpha_d(C) \neq 0) \wedge f_2(C) = 0 \wedge \ldots \wedge$$
$$\wedge\ f_k(C) = 0 \wedge g_1(C) \neq 0,$$

which is clearly constructible. If it is g_1, then $f_1,\ldots,f_k$ belong to $\mathbb{Z}[C]$, and $\Phi(C|f_1,\ldots,f_k;g_1)$ equals the constructible C-property

$$f_1(C) = \ldots = f_k(C) = 0 \wedge (\beta_0(C) \neq 0 \vee \ldots \vee \beta_e(C) \neq 0).$$

Main case: At least two terms of the sequence $f_1,\ldots,f_k,g_1$ have positive degree in U_1. The idea is to lower D by applying the Euclidean algorithm to these two polynomials.

Subcase 1: $d > e > 0$.

Write $\bar{g}_1 = g_1 - \beta_e(C)U_1^e = \beta_{e-1}(C)U_1^{e-1} + \ldots + \beta_0(C)$. By the

Euclidean algorithm there are $q, r \in \mathbb{Z}[C,U_1]$ such that $\beta_e^{d-e+1}.f_1 = q.g_1+r$ and $\deg_{U_1} r < e$. Then we have: $\Phi(C|f_1,\ldots,f_k;g_1) = \Phi(C|f_1,\ldots,f_k,\beta_e;g_1) \vee$ $\Phi(C|f_1,\ldots,f_k;\beta_e.g_1) = \Phi(C|f_1,\ldots,f_k,\beta_e;\bar{g}_1) \vee \Phi(C|f_1,\ldots,f_k;\beta_e.r) \vee$ $\Phi(C|q,f_2,\ldots,f_k,r;\beta_e.g_1)$. Now $\deg_{U_1} q + \deg_{U_1} r < \deg_{U_1} f_1$, hence by the induction hypothesis $\Phi(C|f_1,\ldots,f_k;g_1)$ is the disjunction of three constructible C-properties, so is constructible.

Subcase 2: $e \geq d > 0$. Define:

$$\bar{f}_1 = f_1 - \alpha_d(C)U_1^d = \alpha_{d-1}(C)U_1^{d-1} + \ldots + \alpha_0(C),$$

$$\hat{g}_1 = \alpha_d(C)g_1 - \beta_e(C)U_1^{e-d}f_1,$$

so $\deg_{U_1}\bar{f}_1 < d$, and $\deg_{U_1}\hat{g}_1 < e$. Then we have:

$$\Phi(C|f_1,\ldots,f_k;g_1) = \Phi(C|f_1,\ldots,f_k,\alpha_d;g_1) \vee \Phi(C|f_1,\ldots,f_k;\alpha_d.g_1) = \Phi(C|\bar{f}_1,f_2,\ldots,f_k,\alpha_d;g_1) \vee \Phi(C|f_1,\ldots,f_k;\hat{g}_1).$$

Again the induction hypothesis implies that $\Phi(C|f_1,\ldots,f_k;g_1)$ is a constructible C-property. These two subcases cover $e > 0$.

Subcase 3: $e = 0$, $d \geq h > 0$, where $h = \deg_{U_1} f_2$.

Write $f_2 = \gamma_h(C)U_1^h + \gamma_{h-1}(C)U_1^{h-1} + \ldots + \gamma_0(C)$,

and

$$\hat{f}_1 = \gamma_h(C)f_1 - \alpha_d(C)U_1^{d-h}f_2 \qquad (\text{so } \deg_{U_1}\hat{f}_1 < d),$$

$$\bar{f}_2 = f_2 - \gamma_h(C)U_1^h \qquad (\text{so } \deg_{U_1}\bar{f}_2 < h).$$

Then we have:

$$\Phi(C|f_1,\ldots,f_k;g_1) = \Phi(C|f_1,f_2,\ldots,f_k,\gamma_h;g_1) \vee \Phi(C|f_1,\ldots,f_k;\gamma_h.g_1) = \Phi(C|f_1,\bar{f}_2,\ldots,f_k,\gamma_h;g_1) \vee \Phi(C|\hat{f}_1,\ldots,f_k;\gamma_h.g_1).$$

As before, the induction hypothesis implies that $\Phi(C|f_1,\ldots,f_k;g_1)$ is a constructible C-property. This finishes the proof for $N = 1$.

Suppose now that $N = N'+1$, $N' > 0$, and that the statement of the theorem is true for N' instead of N (induction hypothesis). Then we consider the (C,U_1)-property $\Phi((C,U_1)|f_1,\ldots,f_k;g_1,\ldots,g_j)$. By the induction hypothesis this (C,U_1)-property is constructible, hence equals a disjunction of (C,U_1)-systems:

$(\alpha_1(C,U_1);\beta_1(C,U_1)) \vee \ldots \vee (\alpha_m(C,U_1));\beta_m(C,U_1))$, where for each $i = 1,\ldots,m$ the (C,U_1)-system $(\alpha_i(C,U_1);\beta_i(C,U_1))$ equals $\alpha_{i_1}(C,U_1) = \ldots = \alpha_{i_{k(i)}}(C,U_1) = 0 \wedge \beta_{i_1}(C,U_1) \neq 0 \wedge \ldots \wedge \beta_{i_{j(i)}}(C,U_1) \neq 0$ (the α_{i_j}'s and β_{i_j}'s belonging to $\mathbb{Z}[C,U_1]$). Define for $i = 1,\ldots,m$ the C-property $\Phi_i(C)$ as $\Phi(C|\alpha_{i_1},\ldots,\alpha_{i_{k(i)}};\beta_{i_j},\ldots,\beta_{i_{j(i)}})$, which is constructible by what we proved for $N = 1$.

We claim that $\Phi(C|f_1,\ldots,f_k;g_1,\ldots,g_j) = \Phi_1(C) \vee \ldots \vee \Phi_m(C)$. For suppose K is a field, $c \in K^M$ and $K \models \Phi(c|f_1,\ldots,f_k;g_1,\ldots,g_j)$. This means that there is an extension field L of K and $u = (u_1,\ldots,u_N) \in L^N$ such that $f_1(c,u) = \ldots = f_k(c,u) = 0$, $g_1(c,u) \neq 0, \ldots, g_j(c,u) \neq 0$. Then $(u_2,\ldots,u_N) \in L^{N-1}$ is a solution of $f_1(c,u_1,U_2,\ldots,U_N) = \ldots = f_k(c,u_1,U_2,\ldots,U_N) = 0$, $g_1(c,u_1,U_2,\ldots,U_N) \neq 0, \ldots, g_j(c,u_1,U_2,\ldots,U_N) \neq 0$, so $L \models \Phi((c,u_1)|f_1,\ldots,f_k;g_1,\ldots,g_j)$. This implies $L \models (\alpha_i(c,u_1);\beta_i(c,u_1))$ for some $i \in \{1,\ldots,m\}$. By definition of $\Phi_i(C)$, this means that $K \models \Phi_i(c)$, so $K \models \Phi_1(c) \vee \ldots \vee \Phi_m(c)$.

The converse: given a field K and $c \in K^M$ such that $K \models \Phi_1(c) \vee \ldots \vee \Phi_m(c)$, then $K \models \Phi(c|f_1,\ldots,f_k;g_1,\ldots,g_j)$, is routine and left to the reader.

This establishes our claim, and finishes the proof of the theorem.

12.2.11 REMARKS. Note that the C-property $\psi(C)$ of 12.2.5, Example (2), is constructible by Theorem 12.2.10.

For readers familiar with constructible sets we remark that Theorem 12.2.10 corresponds, by Exercise 12.2.9, 5, to the case of Chevalley's constructibility theorem which says that the image of a constructible set under the canonical map $\mathrm{Spec}(\mathbb{Z}[C,U]) \to \mathrm{Spec}(\mathbb{Z}[C])$ is constructible [Ma 80, Ch. 2, Th. 6].

But our proof gives extra information as we noted in the remark following the statement of the theorem. A consequence is a version of Hilbert's Nullstellensatz.

12.2.12 COROLLARY. Let K be a field, $f_1,\dots,f_k,g_1,\dots,g_j \in K[U]$, where $U = (U_1,\dots,U_N)$. If the system of equations and inequations $f_1(U) = \dots = f_k(U) = 0$, $g_1(U) \neq 0, \dots, g_j(U) \neq 0$ has a solution in an extension field of K, then it has a solution in the algebraic closure of K.

PROOF. Replace the different nonzero coefficients $c_1,\dots,c_M$ in the k+j polynomials by distinct variables $C_1,\dots,C_M$ and apply Theorem 12.2.10 and the remark following its statement.

The next result is the basis of our method of reduction to finite fields.

12.2.13 PROPOSITION. Let $\phi(C)$ be a constructible C-property. If there is a field K of characteristic 0 and $c \in K^M$ such that $K \models \phi(c)$, then there is a finite field K' and $c' \in (K')^M$ such that $K' \models \phi(c')$.

PROOF. It suffices to consider the case that $\phi(C)$ is a C-system $(\alpha(C);\beta(C))$, where $\alpha = (\alpha_1,\dots,\alpha_k)$, $\beta = (\beta_1,\dots,\beta_j)$, $\alpha_i,\beta_h \in \mathbb{Z}[C]$. So we assume that $\alpha_1(c) = \dots = \alpha_k(c) = 0$, $\beta_1(c) \neq 0, \dots, \beta_j(c) \neq 0$ in K. Let $d_1,\dots,d_m$ be the distinct nonzero coefficients of the polynomials $\alpha_1,\dots,\alpha_k,\beta_1,\dots,\beta_j$. Recall that these coefficients are integers.

We replace in $\alpha_1,\dots,\alpha_k,\beta_1,\dots,\beta_j$ each coefficient d_i by a new variable D_i so that we obtain polynomials $A_1,\dots,A_k,B_1,\dots,B_j$ in $\mathbb{Z}[D_1,\dots,D_m,C_1,\dots,C_M]$ with $A_i(d,C) = \alpha_i(C), \dots, B_h(d,C) = \beta_h(C)$, where

$d = (d_1,\ldots,d_m) \in \mathbb{Z}^m$, $i = 1,\ldots,k$, $h = 1,\ldots,j$.

Using the notation of 12.2.10 we can express $\alpha_1(c) = \ldots = \alpha_k(c) = 0$, $\beta_1(c) \neq 0$, ... , $\beta_j(c) \neq 0$ by $\mathbb{Q} \models \Phi(d|A_1,\ldots,A_k;B_1,\ldots,B_j)$.

But $\Phi(D|A_1,\ldots,A_k;B_1,\ldots,B_j)$ is, by 12.2.10, a disjunction of D-systems, so for one of these disjuncts, say $(\gamma(D);\delta(D))$, with $\gamma = (\gamma_1,\ldots,\gamma_r)$, $\delta = (\delta_1,\ldots,\delta_s)$, we have

$$\gamma_1(d) = \ldots = \gamma_r(d) = 0,\ \delta_1(d) \neq 0,\ \ldots\ ,\delta_s(d) \neq 0.$$

Let p be a prime number not dividing any of the $\delta_1(d),\ldots,\delta_s(d)$. Let $d_1',\ldots,d_m'$ be the images of the integers $d_1,\ldots,d_m$ in $\mathbb{Z}/(p) = \mathbb{F}_p$, and put $d' = (d_1',\ldots,d_m')$. Then clearly $\mathbb{F}_p \models (\gamma(d');\delta(d'))$, and because $(\gamma(D);\delta(D))$ is a disjunct of $\Phi(D|A_1,\ldots,A_k;B_1,\ldots,B_j)$, we obtain $\mathbb{F}_p \models \Phi(d'|A_1,\ldots,A_k;B_1,\ldots,B_j)$.

By the remark following 12.2.10 this implies that in the algebraic closure of $\mathbb{F}_p$ there are $c_1',\ldots,c_M'$ such that, with $c' = (c_1',\ldots,c_M')$: $A_1(d',c') = \ldots = A_k(d',c') = 0$, $B_1(d',c') \neq 0$, ... , $B_j(d',c') \neq 0$, i.e. $\alpha_1(c') = \ldots = \alpha_k(c') = 0$, $\beta_1(c') \neq 0$, ... , $\beta_j(c') \neq 0$, so $K' \models \phi(c')$ where $K' = \mathbb{F}_p(c')$. The proof is finished.

12.2.14 REMARK. The proof of 12.2.13 shows that we can choose any sufficiently large prime number p as the characteristic of K'.

12.2.15 EXERCISES. Prove:

1. If K is a field and L an extension field which is finitely generated as a K-algebra, then $[L : K] < \infty$. (Hint: use 12.2.12.)

2. A field which is finitely generated as a ring (i.e. as a $\mathbb{Z}$-algebra) is a finite field. (Hint: use 12.2.13 to show that the characteristic is not 0; then apply 1.)

3. A nonempty constructible subset of $\mathrm{Spec}(\mathbb{Z}[C])$ contains a prime ideal $\mathfrak{p}$ such $\mathbb{Z}[C]/\mathfrak{p}$ is a finite field. (Hint: use 12.2.12, 12.2.13.)

REMARK. Theorem 12.2.10 closely resembles a well-known theorem of Tarski. See for instance [Ja, Ch. 5, §6]. After completion of the manuscript we noticed that a related concept of constructible property has been developed by Grothendieck in a scheme theoretic context, cf. [Gro 61, Ch. 0, §9].

12.3 PROOF OF LEMMA IV.

12.3.1 The problem arises how to apply 12.2.13 to prove Lemma IV of the Introduction. Let us sketch an approach.

A finitely generated algebra A over a field K can be written as a quotient $K[U]/(f_1(U),\dots,f_k(U))$, $U = (U_1,\dots,U_N)$. Given $x = (x_1,\dots,x_n) \in A^n$, $y = (y_1,\dots,y_q) \in A^q$, we introduce polynomials $\chi_1(U),\dots,\chi_n(U)$, $\eta_1(U),\dots,\eta_q(U)$ in $K[U]$ such that $x_i = \chi_i + (f_1,\dots,f_k)$, $y_j = \eta_j + (f_1,\dots,f_k)$, $i = 1,\dots,n$, $j = 1,\dots,q$.

Let $c \in K^M$ be the vector of distinct nonzero coefficients of the $k+n+q$ polynomials $f_1,\dots,f_k$, $\chi_1,\dots,\chi_n$, $\eta_1,\dots,\eta_q$.

One can show:

(1) $ht_A(x_1,\dots,x_n) = n \Leftrightarrow K \models ht(c)$, where $ht(C)$ is a certain constructible C-property independent of K;

(2) $F_1(x,y) = \dots = F_h(x,y) = 0 \Leftrightarrow K \models \phi(c)$, where $\phi(C)$ is also constructible and independent of K.

Under the hypotheses of Lemma IV, i.e. K has characteristic 0, $ht_A(x_1,\dots,x_n) = n$, and $F_1(x,y) = \dots = F_h(x,y) = 0$, we can now apply 12.2.13

and obtain a finite field K′ and $c' \in (K')^M$ such that, if we put $A' = K'[U]/(f'_1,\dots,f'_k)$, then $ht_{A'}(x'_1,\dots,x'_n) = n$ and $F_1(x',y') = \dots = F_h(x',y') = 0$, where the polynomials F'_λ, χ'_i, η'_j are obtained from the polynomials f_λ, χ_i, η_j by replacing c by c′ and where $x'_i = \chi'_i + (f'_1,\dots,f'_k)$, $y'_j = \eta'_j + (f'_1,\dots,f'_k)$.

So the conclusion of Lemma IV is satisfied, except that A′ may not be a domain, but this objection is removed by dividing out a suitable minimal prime. The single remaining problem is to prove (1) and (2) above. Alas, this being somewhat outside the scope of the book, we just make a few remarks before presenting a complete proof by a somewhat different method. In the "model theoretic" study of fields one considers the boolean algebra (under $\vee,\wedge$) of so-called elementary C-properties of which the constructible C-properties form a proper subalgebra. An often useful method to prove that a C-property is elementary is to represent it as a disjunction of a set of elementary C-properties, and simultaneously as a conjunction of a set of elementary C-properties. (Both sets are allowed to be infinite.) One can then apply the following converse of 12.2.9 Exercise 3 to find out whether it is constructible: an elementary C-property $\phi(C)$ is constructible iff for any fields K, L with $K \subset L$ and any $c \in K^M$ one has: $K \models \phi(c) \Leftrightarrow L \models \phi(c)$.

For further details and references on this subject, see [Dr], from which it is easy to extract a proof of (1) and (2) above.

Here we have to take another route, avoiding (1) and (2). The following lemma enables us to convert a height condition into "constructible" form, at the cost of introducing extra variables and polynomials. Recall that $\mathfrak{R}(\mathfrak{a})$ denotes the radical of an ideal $\mathfrak{a}$.

12.3.2 LEMMA. Let K be a field, D a domain $\supset$ K and finitely generated as a K-algebra, with dim D = n. Then an ideal $\mathfrak{a}$ of D has height n if and only if there are $u_1,\dots,u_n \in D$, algebraically independent over K, and an irreducible polynomial $p \in K[U_1,\dots,U_n,V]$ and a $v \in D$ such that:

(i) D is integral over its subring $K[u_1,\dots,u_n]$;

(ii) $p(0,\dots,0,0) = p(u_1,\dots,u_n,v) = 0$, and $\mathfrak{R}(\alpha) = \mathfrak{R}(u_1,\dots,u_n,v)$ in D.

PROOF. Suppose first that ht $\alpha = n$. By Noether normalization, [Ma 80, (14.G)], there are K-algebraically independent $y_1,\dots,y_n$ in D such that D is integral over $K[y_1,\dots,y_n] = K[y]$. Then ht $\alpha = \mathrm{ht}_{K[y]}(\alpha \cap K[y])$ by [Ma 80, (5.E) & (13.C)], so there are $u_1,\dots,u_n \in \alpha \cap K[y]$ such that K[y] is integral over $K[u_1,\dots,u_n] = K[u]$, by [Ma 80, (14.F)]. Then D is integral over K(u), so its fraction field Q(D) is algebraic over K(u). Because dim D = $\mathrm{tr.deg}_K Q(D) = n$, this implies that $u_1,\dots,u_n$ are algebraically independent over K.

Now $(u_1,\dots,u_n)D \subset \alpha$, and both ideals have height n in D. (For the first ideal this follows from [Ma 80, (5.E) (i) & (v), (13.B) (3)].) So there exists a v which is in all minimal primes of α, but outside each minimal prime of $(u_1,\dots,u_n)D$ which is not also minimal over α. For this v we have: $\mathfrak{R}(\alpha) = \mathfrak{R}(u_1,\dots,u_n,v)$, because α and $(u_1,\dots,u_n,v)$ have the same minimal primes (all of which are also maximal).

Because D is integral over K[u] we can take $p \in K[U_1,\dots,U_n,V]$ monic in V such that $p(u,v) = 0$. Taking p of minimal degree in V, it follows from Gauss' Lemma that p is irreducible in $K[U_1,\dots,U_n,V]$. Also $U_1,\dots,U_n,V$ and p generate a proper ideal in $K[U_1,\dots,U_n,V]$ because their images $u_1,\dots,u_n,v$ and 0 in $K[u_1,\dots,u_n,v] \subset D$ do so in D. Subtracting terms in p involving $U_1,\dots,U_n,V$ we see that this can only happen if the constant term $p(0,\dots,0,0)$ of p vanishes. This finishes the proof of one half of the lemma.

As for the other direction: consider the kernel of the natural map $K[U_1,\dots,U_n,V] \to K[u_1,\dots,u_n,v] \subset D$. This kernel contains the irreducible polynomial p, and is a prime ideal of height 1, because dim $K[u_1,\dots,u_n,v]$ +

$ht(\text{kernel}) = \text{tr.deg}_K K(u_1,\ldots,u_n,v) + ht(\text{kernel}) = n + ht(\text{kernel}) = \dim K[U_1,\ldots,U_n,V] = n+1$, by [Ma 80, (14.H)].

Therefore the kernel is in fact generated by p. Now, because $p(0,\ldots,0,0) = 0$, we have $p \in (U_1,\ldots,U_n,V)K[U_1,\ldots,U_n,V]$, so the image of the proper ideal $(U_1,\ldots,U_n,V)K[U_1,\ldots,U_n,V]$ in $K[u_1,\ldots,u_n,v]$ is a proper ideal, i.e. $1 \notin (u_1,\ldots,u_n,v)K[u_1,\ldots,u_n,v]$. Because D is integral over $K[u_1,\ldots,u_n,v]$, this implies $1 \notin (u_1,\ldots,u_n,v)D$, by [Ma 80, (5.E)].

Hence we have $n = \dim D \geq ht_D(u_1,\ldots,u_n,v) = ht_D\alpha \geq ht_D(u_1,\ldots,u_n) = ht(u_1,\ldots,u_n)K[u_1,\ldots,u_n] = n$, so $ht_D\alpha = n$, which ends the proof.

We need two easy lemmas before we can apply 12.3.2.

12.3.3 LEMMA. Let $U = (U_1,\ldots,U_N)$, $N > 0$, and $f = f(C,U) \in \mathbb{Z}[C,U]$. Define the C-property Abs.Irr.$(C|f)$ by putting, for any field K and $c \in K^M$: $K \models \text{Abs.Irr.}(c|f) \Leftrightarrow f(c,U)$ is irreducible in $\bar{K}[U]$. ($\bar{K}$ = algebraic closure of K, "Abs.Irr." is shorthand for "absolutely irreducible".) Then Abs.Irr$(C|f)$ is a constructible C-property.

PROOF. Suppose that we specify which of the coefficients of $f(c,U)$ are zero and which are nonzero. (This amounts to requiring that a certain conjunction of equations and inequations in C holds in (K,c).) This specification determines the total degree of $f(c,U)$ in U. Then $f(c,U)$ is irreducible in $\bar{K}[U]$ if and only if this degree is positive and $f(c,U)$ is not the product of two polynomials in $\bar{K}[U]$ of lower degree; this last condition can be expressed by the nonsolubility in $\bar{K}$ of a certain system of equations.

We leave it to the reader to make the above precise (which is trivial, but messy), and to conclude the required result from 12.2.10.

12.3.4 REMARK. Note that we have to use here absolute irreducibility (i.e. irreducibility over the algebraic closure), in order not to violate 12.2.9 Exercise 3: an irreducible polynomial like $X^2 + 1 \in \mathbb{Q}[X]$ becomes reducible over an extension field.

Note also that 12.3.3 implies, by 12.2.14, the wellknown fact that an absolutely irreducible polynomial in $\mathbb{Z}[U] \subset \mathbb{Q}[U]$ remains absolutely irreducible modulo sufficiently large primes.

12.3.5 LEMMA. Let $f_1,\dots,f_k,\ g_1,\dots,g_j \in \mathbb{Z}[C,U]$, $U = (U_1,\dots,U_N)$, $N > 0$, and define the C-property $[C|\mathfrak{R}(f_1,\dots,f_k) = \mathfrak{R}(g_1,\dots,g_j)]$ by putting, for any field K and $c \in K^M$: $K \models [c|\mathfrak{R}(f_1,\dots,f_k) = \mathfrak{R}(g_1,\dots,g_j)] \Leftrightarrow$ $\mathfrak{R}(f_1(c,U),\dots,f_k(c,U)) = \mathfrak{R}(g_1(c,U),\dots,g_j(c,U))$ in $K[U]$. Then $[C|\mathfrak{R}(f_1,\dots,f_k) \cdot \mathfrak{R}(g_1,\dots,g_j)]$ is constructible.

PROOF. Clearly the C-property $[C|\mathfrak{R}(f_1,\dots,f_k) = \mathfrak{R}(g_1,\dots,g_j)]$ is the conjunction of the k+j C-properties $[C|f_1 \in \mathfrak{R}(g_1,\dots,g_j)]$, ... , $[C|f_k \in \mathfrak{R}(g_1,\ \dots\ ,\ g_j)]$, $[C|g_1 \in \mathfrak{R}(f_1,\ \dots\ ,\ f_k)]$, ... , $[C|g_j \in \mathfrak{R}(f_1,\dots,f_k)]$ (which are defined in the obvious way).

Each of these is constructible. Consider, say, the first one, and observe that by Hilbert's Nulstellensatz we have the equivalence (for any field K and $c \in K^M$): $f_1(c,U) \in \mathfrak{R}(g_1(c,U),\dots,g_j(c,U))$ in $K[U] \Leftrightarrow$ there is no extension field of K in which the system $g_1(c,U) = \dots = g_j(c,U) = 0$, $f_1(c,U) \neq 0$ has a solution for the U's. This last statement reads, in the notation of Theorem 12.2.10: $K \models \neg\Phi(c|g_1,\dots,g_j;f_1)$.

So we have shown that the C-property $[C|f_1 \in \mathfrak{R}(g_1,\dots,g_j)]$ equals the constructible C-property $\neg\Phi(C|g_1,\dots,g_j;f_1)$.

12.3.6 We finally need one new operation on (constructible) properties. Let $(C,U) = (C_1,\ldots,C_M,U_1,\ldots,U_N)$ and let $\phi = \phi(C,U)$ be a (C,U)-property. We obtain from ϕ a C-property $\phi_U(C)$ by defining, for any field K and $c \in K^M$: $K \models \phi_U(c) \Leftrightarrow K(U_1,\ldots,U_N) \models \phi(c,U_1,\ldots,U_N)$, where $U_1,\ldots,U_N$ are algebraically independent over K. We ask the reader to verify the following trivialities.

12.3.7 EXERCISE. Let (C,U) be as above.

(i) Show that for (C,U)-properties ϕ and ψ $(\phi \vee \psi)_U(C) = \phi_U(C) \vee \psi_U(C)$, $(\neg\phi)_U(C) = \neg(\phi_U(C))$.

(ii) Let $\phi(C,U)$ be the basic (C,U)-property "$\alpha(C,U) = 0$", where $\alpha(C,U) = \sum_{i_1+\ldots+i_N \le d} \alpha_{i_1\ldots i_N}(C)U_1^{i_1}\ldots U_N^{i_N} \in \mathbb{Z}[C,U]$. Show that $\phi_U(C)$ is the conjunction of the finitely many basic C-properties "$\alpha_{i_1\ldots i_N}(C) = 0$", $i_1+\ldots+i_N \le d$.

(iii) Conclude from (i) and (ii) that if $\phi(C,U)$ is a constructible (C,U)-property, then $\phi_U(C)$ is a constructible C-property.

12.3.8 We now have all the ingredients to prove Lemma IV. The proof which follows is rather lengthy. However, once some reductions are made, the remaining steps follow the pattern of 12.3.1. For a different approach, see [Ho 78, pp. 891-892].

12.3.9 PROOF OF LEMMA IV.

Recall that A is a finitely generated algebra over a field K of characteristic 0, that $\xi = \xi(X,Y)$ is a system of equations over $\mathbb{Z}$: $\xi = \xi(X,Y)$: $F_1(X,Y) = \ldots = F_h(X,Y) = 0$ where $F_i(X,Y) \in \mathbb{Z}[X,Y]$, $X = (X_1,\ldots,X_n)$, $Y = (Y_1,\ldots,Y_q)$, $n > 0$, and that ξ has a solution $(x,y) = (x_1,\ldots,x_n,y_1,\ldots,y_q) \in A^{n+q}$ with $ht(x_1,\ldots,x_n) = n$.

We have to show that a similar algebra exists over a finite field. First we prove that we may reduce to the case that A is a domain of dimension n and K is algebraically closed.

We make A a domain by dividing out a suitable minimal prime. Suppose then that dim A > n. Take a minimal prime ideal $\mathfrak{p}$ of $(x_1,\ldots,x_n)$ of height n. Then dim $A/\mathfrak{p} > 0$, so there exists $u \in A$ whose image in $A/\mathfrak{p}$ is transcendental over K. Let $S = K[u]\setminus\{0\}$. Then S is a multiplicative subset of A and disjoint from $\mathfrak{p}$. Hence $S^{-1}A$ is a domain containing K(u), and is a finitely generated algebra over this field. Moreover, $\dim S^{-1}A = \dim A - 1$ (because $Q(S^{-1}A) = Q(A)$ and $\mathrm{tr.deg}_{K(u)}Q(A) = \mathrm{tr.deg}_K Q(A) - 1$), and one checks easily that $\mathrm{ht}_{S^{-1}A}(x_1,\ldots,x_n) = \mathrm{ht}_A(x_1,\ldots,x_n) = n$.

Applying this localization trick finitely many times we reduce to the case that for the "new" domain A we have dim A = n. However, the field K may not be algebraically closed. This is remedied as follows. Let $\bar{K}$ be the algebraic closure of K, and consider the finitely generated $\bar{K}$-algebra $\bar{A} = A\otimes_K\bar{K}$ obtained by base change. If we consider $\bar{A} = A\otimes_K\bar{K}$ as an extension of $A = A\otimes_K K$, then $\bar{A}$ is integral over A and flat as an A-module, because $\bar{K}$ is integral over K and flat as a K-module. Therefore, by [Ma 80, (13.B) (3) & (13.C)], we have $\mathrm{ht}_{\bar{A}}(x_1,\ldots,x_n)\bar{A} = \dim \bar{A} = n$.

Now $\bar{A}$ may not be a domain any more, but this is easily restored by dividing out a suitable minimal prime. So from now on we may, and shall, assume that A is a domain of dimension n and that K is algebraically closed.

By 12.3.2 there are K-algebraically independent $u_1,\ldots,u_n$ in A, an irreducible polynomial $p = p(U_1,\ldots,U_n,V_1) \in K[U_1,\ldots,U_n,V_1]$ and $v_1 \in A$ such that:

(i) A is integral over $K[u_1,\ldots,u_n]$;

(ii) $p(0,\ldots,0,0) = p(u_1,\ldots,u_n,v_1) = 0$ and $\mathfrak{R}(x_1,\ldots,x_n) = \mathfrak{R}(u_1,\ldots,u_n,v_1)$ in A.

Let us write $A = K[u_1,\ldots,u_n,v_1,\ldots,v_m]$ and use the following shorthand: $u = (u_1,\ldots,u_n)$, $v = (v_1,\ldots,v_m)$, $U = (U_1,\ldots,U_n)$, $V = (V_1,\ldots,V_m)$.

We next introduce polynomials $f_1,\dots,f_k$, $\chi_1,\dots,\chi_n$, $\eta_1,\dots,\eta_q$ in K[U,V] with the following properties:

(a) $f_1,\dots,f_k$ generate the kernel of the K-algebra morphism $K[U,V] \to A$ given by $U_i \mapsto u_i$, $V_j \mapsto v_j$, where we choose the f's such that for each $j = 1,\dots,m$ at least one of $f_1,\dots,f_k$ is in $K[U,V_j]$ and is monic in V_j (to express that v_j is integral over K[u]).

(b) $\chi_1(u,v) = x_1, \dots, \chi_n(u,v) = x_n,$

$\eta_1(u,v) = y_1, \dots, \eta_q(u,v) = y_q.$

Before proceeding let us note certain properties of these polynomials. Firstly, it follows immediately from (b) that the polynomials $F_i(\chi_1,\dots,\chi_n,\eta_1,\dots,\eta_q) \in K[U,V]$, $i = 1,\dots,h$, vanish at (u,v). Secondly, from $\mathfrak{R}(x_1,\dots,x_n) = \mathfrak{R}(u_1,\dots,u_n,v_1)$ in $K[u,v] = A$, and (a), (b), we get: $\mathfrak{R}(\chi_1,\dots,\chi_n,f_1,\dots,f_k) = \mathfrak{R}(U_1,\dots,U_n,V_1,f_1,\dots,f_k)$ in K[U,V].

We are ready to introduce our C-property: let $c_1,\dots,c_M \in K$ be the distinct nonzero coefficients of the polynomials $f_1,\dots, f_k$, $\chi_1,\dots,\chi_n$, $\eta_1,\dots,\eta_q$ and p, which are all in K[U,V]. Let $c = (c_1,\dots,c_M)$, $C = (C_1,\dots,C_M)$. Replace each coefficient c_i in these polynomials by C_i, so that we get polynomials $f_1^C,\dots,f_k^C$, $\chi_1^C,\dots,\chi_n^C$, $\eta_1^C,\dots,\eta_q^C$, p^C in $\mathbb{Z}[C,U,V]$.

We let $\theta(C)$ be the C-property obtained by defining for any field F and $d \in F^M$: $F \models \theta(d)$ if and only if the following four conditions hold:

(1) $p^C(d,U,V)$ is absolutely irreducible and $p^C(d,0,\dots,0,0) = 0$.

(2) $\mathfrak{R}(\chi_1^C(d,U,V),\dots,\chi_n^C(d,U,V),f_1^C(d,U,V),\dots,f_k^C(d,U,V)) =$
$= \mathfrak{R}(U_1,\dots,U_n,V_1,f_1^C(d,U,V),\dots,f_k^C(d,U,V))$ in F[U,V].

(3) For each $j = 1,\dots,m$ at least one of $f_1^C(d,U,V),\dots,f_k^C(d,U,V)$ is in $F[U,V_j]$ and is monic in V_j.

(4) In some extension field of $F(U_1,\dots,U_n)$ there exists a solution for the unknowns $V_1,\dots,V_m$ of the following system of equations over $F(U_1,\dots,U_n)$:

$$\begin{array}{l}
F_1(\chi_1^c(d,U,V), \ \ldots\ , \chi_n^c(d,U,V), \eta_1^c(d,U,V), \ \ldots\ , \eta_q^c(d,U,V)) = 0 \\
\vdots \\
F_h(\chi_1^c(d,U,V), \ \ldots\ , \chi_n^c(d,U,V), \eta_1^c(d,U,V), \ \ldots\ , \eta_q^c(d,U,V)) = 0 \\
f_1^c(d,U,V) = 0 \\
\vdots \\
f_k^c(d,U,V) = 0 \\
p^c(d,U,V) = 0.
\end{array}$$

Because $f_1^c(c,U,V) = f_1$, etc., we have $K \models \theta(c)$, by the properties of $f_1,\ldots,f_k$, $\chi_1,\ldots,\chi_n$, $\eta_1,\ldots,\eta_q$, p.

But $\theta(C)$ is constructible: clauses (1) and (2) are handled by 12.3.3 and 12.3.4 respectively, (3) is obvious, and for (4) one uses 12.3.7 (iii) and 12.2.10.

By 12.2.13 we may conclude that there is a finite field K′ and $c' \in (K')^M$ such that $K' \models \theta(c')$. In particular, let $v' = (v_1',\ldots,v_m')$ be a solution of the system in (4) above, with K′,c′ instead of F,d. Write $A' = K'[U,v']$, $x_i' = \chi_i^c(c',U,v')$, $y_j' = \eta_j^c(c',U,v')$, $i = 1,\ldots,n$, $j = 1,\ldots,q$, and $p' = p^c(c',U,V_1) \in K'[U,V_1]$. Then $K' \models \theta(c')$ implies: A′ is a domain (by (4) and the definition of v′), $p'(U,V_1) \in K'[U,V_1]$ is irreducible and $p'(0,\ldots,0,0) = 0$ (by (1)), A′ is integral over its subring K′[U] (by (3)), $F_i(x_1',\ldots,x_n',y_1',\ldots,y_q') = 0$ for $i = 1,\ldots,h$ and $p'(U,v_1') = 0$ (by (4)), and $\mathfrak{R}(x_1',\ldots,x_n') = \mathfrak{R}(U_1,\ldots,U_n,v_1')$ in A′ (by (2) and (4)).

Again we can use 12.3.2 to conclude that $ht(x_1',\ldots,x_n') = n$ in A′, and we have finished the proof of Lemma IV.

12.4 THE WEIERSTRASS THEOREMS

12.4.1 The Weierstrass Theorems often enable us to prove results for (rings of) power series by induction on the number of variables. In section 12.6 they will be an essential tool in deriving an approximation theorem of M. Artin.

In this section a pair (R,m) will always denote a ring R together with an ideal $m \subset \text{rad } R$. Note that an element r is invertible in R iff its image $\bar{r}$ in R/m is invertible.

12.4.2 WEIERSTRASS DIVISION THEOREM FOR FORMAL POWER SERIES.
Let (R,m) be such that R is separated and complete in the m-adic topology. Suppose $f = \Sigma_{i=0}^{\infty} f_i X^i \in R[[X]]$ is regular in X of order $s \in \mathbb{N}$, i.e. $f_0,\ldots,f_{s-1} \in m$, and f_s is invertible. Then for each $g \in R[[X]]$ there is a unique pair (q,r) with $g = qf + r$ and $q \in R[[X]]$, $r \in R[X]$, $\deg_X r < s$.

PROOF. Existence: Let $u = f_s + f_{s+1}X + \ldots + f_{s+n}X^n + \ldots$. Then u is invertible in $R[[X]]$ and $u^{-1}f = u^{-1}(\Sigma_{i<s} f_i X^i + X^s u) = (u^{-1}.\Sigma_{i<s} f_i X^i) + X^s$. So replacing f by $u^{-1}f$, we may assume $f = X^s - F$ with $F \in m[[X]]$.

CLAIM: For each $k = 0,1,\ldots,$ and each $g \in m^k[[X]]$ there are $q \in m^k[[X]]$, $r \in m^k[X]$ with $\deg_X r < s$ and $T(g) \in m^{k+1}[[X]]$ such that $g = qf + r + T(g)$. (The reader will see from the claim's proof that the notation $T(g)$ is justified.)

Accepting the claim for a moment, we can write for any $g \in R[[X]]$:

$$\begin{array}{rcccccccl} g & = & q_0 f & + & r_0 & + & T(g) & & (q_i \in m^i[[X]], \\ T(g) & = & q_1 f & + & r_1 & + & T^2(g) & & r_i \in m^i[X],\ \deg_X r_i < s, \\ \vdots & & \vdots & & \vdots & & \vdots & & T^i(g) \in m^i[[X]]) \\ \hline g & = & qf & + & r, & & & + & \text{where } q = \Sigma\, q_i,\ r = \Sigma\, r_i. \end{array}$$

So we need only prove the claim. Given $g = \Sigma\, g_i X^i \in m^k[[X]]$, we put: $q = \Sigma_{i\geq s}\, g_i X^{i-s}$, $r = \Sigma_{i<s}\, g_i X^i$, $T(g) = F.(\Sigma_{i\geq s}\, g_i X^{i-s})$, and it is clear that $g = q.(X^s - F) + r + T(g)$.

Uniqueness: Suppose the pairs (q_1, r_1) and (q_2, r_2) satisfy the required conditions. Then putting $q = q_1 - q_2$ and $r = r_2 - r_1$ we have $qf = r$ with $q \in R[[X]]$, $r \in R[X]$, $\deg_X r < s$. It clearly suffices to derive from this that $q = 0$. Suppose $q = \Sigma\, q_i X^i \in m^k[[X]]$. For any $i \in \mathbb{N}$ the coefficient of X^{s+i} in qf is 0, so

$$0 = q_i f_s + \Sigma_{j<i}\, q_j f_{s+i-j} + \Sigma_{i<j\leq i+s}\, q_j f_{s+i-j},$$

hence (assuming, as in the existence part, $f = X^s - F$, $F \in m[[X]]$), $q_i \in m^{k+1}$. We have shown: $q \in m^k[[X]] \Rightarrow q \in m^{k+1}[[X]]$. Hence $q = 0$.

12.4.3 WEIERSTRASS PREPARATION THEOREM FOR FORMAL POWER SERIES.

Let (R,m) be such that R is separated and complete in the m-adic topology, and let $f \in R[[X]]$ be regular in X of order $s \in \mathbb{N}$. Then there is a unique pair (u,W) with $u \in R[[X]]$ invertible and $W \in R[X]$ monic of degree s in X and $f = uW$.

PROOF. Let $\overline{R} = R/m$ and, for each $r \in R$, let $\overline{r}$ denote its image in $\overline{R}$. Write $f = \Sigma\, f_i X^i$. By 12.4.2 we have:

$X^s = qf + r$, $q = \Sigma\, q_i X^i \in R[[X]]$, $r = r_0 + \ldots + r_{s-1}X^{s-1} \in R[X]$. So

$X^s = (\Sigma\, \overline{q}_i X^i)(\overline{f}_s X^s + \overline{f}_{s+1} X^{s+1} + \ldots) + (\overline{r}_0 + \ldots + \overline{r}_{s-1} X^{s-1})$ in $\overline{R}[[X]]$, which implies $\overline{r}_0 = \ldots = \overline{r}_{s-1} = 0$.

Also $\overline{q}_0 . \overline{f}_s = 1$, so q_0 is invertible in R, and hence q is invertible in $R[[X]]$. Putting $u = q^{-1}$ and $W = X^s - r$, we obtain $f = uW$, and the pair (u,W) satisfies the requirements. If these are also satisfied by (u',W'), one obtains from $f = u'(X^s - (X^s - W'))$ that $X^s = (u')^{-1}f + (X^s - W')$, which implies by the uniqueness in 12.4.2: $q = (u')^{-1}$, $r = X^s - W'$, i.e. $u = u'$ and $W = W'$.

12.4.4 EXERCISE. Suppose the assumptions of 12.4.3 hold. Prove:

1. Given (u,W) as in 12.4.3, W is of the form $W = X^s + a_{s-1}X^{s-1} + \ldots + a_0$ where $a_0,\ldots,a_{s-1} \in m$.

2. $R[[X]]/(f)$ is a free R-module on the basis $1, X, \ldots, X^{s-1}$.

12.4.5 LEMMA. Let A be a ring, $n > 0$, and $0 \neq f \in A[[X_1,\ldots,X_n]]$. Then there is an integer $d \geq 1$ with $f(T^{e(n-1)},T^{e(n-2)},\ldots,T^{e(0)}) \neq 0$ in $A[[T]]$ where $e(k) = d^k$.

PROOF. Write $X(i)$ for the monomial $X_1^{i_1}\ldots X_n^{i_n}$ and order all monomials in $A[[X_1,\ldots,X_n]]$ lexicographically: $X(i) < X(j)$ if $i_1 = j_1, \ldots, i_{k-1} = j_{k-1}$ but $i_k < j_k$ for some $k \in \{1,\ldots,n\}$. Let $aX(i)$ be the lowest nonnull monomial occurring in f. Choose an integer d larger than each of the $i_1,\ldots,i_n$. Now apply the substitutions $X_k \mapsto T^{e(n-k)}$, $k = 1,\ldots,n$, to the power series f. Then $aX(i)$ is transformed to $aT^{d(i)}$ with $d(i) = \Sigma_{k=1}^n i_k d^{n-k}$, which is clearly less than the exponent $d(j)$ which occurs for any other nonnull monomial $bX(j)$ in f. Consequently $f(T^{e(n-1)},\ldots,T^{e(0)}) \neq 0$.

In order to apply 12.4.2 and 12.4.3 we need:

12.4.6 LEMMA. Let K be a field, $n > 0$, $0 \neq f \in K[[X_1,\ldots,X_n]]$. Then there is a K-automorphism τ of $K[[X_1,\ldots,X_n]]$ such that $\tau(f)$ is regular in

X_n (of some order $s \in \mathbb{N}$). (Here $g \in K[[X_1,\dots,X_n]]$ is called regular in X_n if it is regular in X_n considered as an element of $R[[X_n]]$, where $(R,m) \overset{\text{def.}}{=} (K[[X_1,\dots,X_{n-1}]],(X_1,\dots,X_{n-1})K[[X_1,\dots,X_{n-1}]])$; note that this means that $g(0,\dots,0,X_n) \neq 0$.)

PROOF. Take an integer d such that $f(T^{e(n-1)},\dots,T^{e(0)}) \neq 0$ with $e(k) = d^k$. Define a K-automorphism τ of $K[[X_1,\dots,X_n]]$ by $\tau(X_k) = X_k + X_n^{e(n-k)}$, $k = 1,\dots,n-1$ and $\tau(X_n) = X_n$. Then $(\tau f)(0,\dots,0,X_n) = f(X_n^{e(n-1)},\dots,X_n^{e(0)}) \neq 0$, which is what we wished to prove.

12.5 HENSELIAN RINGS AND HENSELIZATION

Roughly, a local ring is henselian if it satisfies an algebraic version of the Implicit Function Theorem. As we shall see, every local ring has a "smallest" henselian extension, which we call its henselization.

12.5.1 EXAMPLE. Let $f(T_1,\dots,T_n,X) \in \mathbb{C}\{T_1,\dots,T_n\}[X]$, where $\mathbb{C}\{T_1,\dots,T_n\}$ is the ring of power series in n variables over $\mathbb{C}$ converging in some neighbourhood of $0 \in \mathbb{C}^n$ (which may depend on the series).

Suppose $\alpha \in \mathbb{C}$, $f(0,\dots,0,\alpha) = 0$ and $\frac{\partial f}{\partial X}(0,\dots,0,\alpha) \neq 0$. Then the Implicit Function Theorem for analytic functions says that there is a power series $a(T_1,\dots,T_n) \in \mathbb{C}\{T_1,\dots,T_n\}$ such that $a(0,\dots,0) = \alpha$ and $f(T_1,\dots,T_n, a(T_1,\dots,T_n)) = 0$.

Algebraically, the substitution map $a(T_1,\dots,T_n) \mapsto a(0,\dots,0)$ from $\mathbb{C}\{T_1,\dots,T_n\}$ to $\mathbb{C}$ is just a residue class map. This point of view leads to the following algebraic description of the property of the local ring

$\mathbb{C}\{T_1,\ldots,T_n\}$ we just mentioned.

12.5.2 DEFINITION. A local ring $(R,\mathfrak{m})$ is called henselian if it satisfies Hensel's Lemma 2.3.1 i.e. given any $f(X) \in R[X]$ and any $\alpha \in R$ with $f(\alpha) \in \mathfrak{m}$ and $f'(\alpha)$ invertible, there is an $a \in R$ with $f(a) = 0$ and $a \equiv \alpha$ mod $\mathfrak{m}$.

12.5.3 EXAMPLES.

1. $(\mathbb{C}\{T_1,\ldots,T_n\},(T_1,\ldots,T_n))$ is henselian by Example 12.5.1.

2. If the local ring $(R,\mathfrak{m})$ is separated and complete in the $\mathfrak{m}$-adic topology, then $(R,\mathfrak{m})$ is henselian, Proposition 2.3.1.

12.5.4 LEMMA. Given a local ring $(R,\mathfrak{m})$, the following statements are equivalent:

(i) $(R,\mathfrak{m})$ is henselian;

(ii) Given any $f(X) \in R[X]$, $\alpha \in R$ and $c \in \mathfrak{m}$ with $f(\alpha) = c(f'(\alpha))^2$, there is an $a \in R$ with $f(a) = 0$ and $a \equiv \alpha$ mod $(c.f'(a))$.

PROOF. (i) $\Rightarrow$ (ii): write $f(\alpha+X) = f(\alpha) + f'(\alpha)X + \Sigma_{i\geq 2}\, b_iX^i = c.(f'(\alpha))^2 + f'(\alpha)X + \Sigma_{i\geq 2}\, b_iX^i$, for certain $b_i \in R$. Substitution of $cf'(\alpha)Y$ for X yields $f(\alpha+cf'(\alpha)Y) = c.(f'(\alpha))^2.(1 + Y + \Sigma_{i\geq 2}\, cd_iY^i)$ $(d_i \in R)$. The polynomial $1 + Y + \Sigma_{i\geq 2}\, cd_iY^i$ has a zero y in R by (i) (substitute $Y \mapsto -1$), so $a = \alpha + cf'(\alpha)y$ is a zero of f with $a - \alpha \in c.f'(\alpha)R$.

(ii) $\Rightarrow$ (i): it clearly suffices to apply (ii) to the case that $f'(\alpha)$ is invertible.

12.5.5 EXERCISE. Given a local ring $(R,\mathfrak{m})$, show that each of the following conditions is equivalent to: $(R,\mathfrak{m})$ is henselian.

(1) Each polynomial $1 + Y + a_2Y^2 + \ldots + a_nY^n \in R[Y]$ with $n \geq 2$ and $a_i \in m$ for $i = 2,\ldots,n$, has a zero in R;

(2) Each polynomial $X^n + b_{n-1}X^{n-1} + \ldots + b_1X + b_0 \in R[X]$, with $n \geq 2$, b_1 invertible and $b_0 \in m$, has a zero in m;

(3) Each polynomial $Z^n + Z^{n-1} + a_2Z^{n-2} + \ldots + a_n \in R[Z]$, with $n \geq 2$ and $a_i \in m$ for $i = 2,\ldots,n$, has an invertible zero in R.

(Hints: the proof of 12.5.4 really shows that (1) ⇒ 12.5.4 (ii). That henselian ⇒ (2) follows by substituting $X \mapsto 0$. (2) ⇒ (3): make the transformation $Z \mapsto X - 1$ to obtain a polynomial as in (2). (3) ⇒ (1) follows similarly by an easy transformation.)

12.5.6 REMARK. The property 12.5.4 (ii) of henselian rings is essential in the next section on Artin approximation. Most proofs of Artin approximation use a more difficult version of 12.5.4 (ii) for a system of polynomials in several variables, but we shall bypass this in our treatment.

The following elementary lemma is quite useful.

12.5.7 LEMMA. Let R be a local ring with residue field $k = R/m$, let $f(X) \in R[X]$ be a monic polynomial of degree $n > 0$, and put $R[x] = R[X]/(f)$, $x = X \bmod (f)$. Going modulo m is denoted by a bar. Let $\bar{f} = \prod \bar{f}_i^{e_i}$, $e_i > 0$, be the factorization of $\bar{f}(X) \in k[X]$ into monic irreducible polynomials $\bar{f}_i(X) \in k[X]$, where $f_i(X) \in R[X]$, $\bar{f}_i \neq \bar{f}_j$ for $i \neq j$. Put $m_i = (m, f_i(x))R[x]$.

Then the m_i are exactly the maximal ideals of $R[x]$, $m_i \neq m_j$ for $i \neq j$, and $R[x]/m_i \simeq k[X]/(\bar{f}_i)$.

PROOF. Consider for each i the canonical maps

$$R[x] = R[X]/(f) \to k[X]/(\bar{f}) \to k[X]/(\bar{f}_i).$$

The first map has kernel $mR[x]$, the next map has its kernel generated by

$\bar{f}_i \bmod(\bar{f})$, so the composite map has kernel $m_i = (m, f_i(x))R[x]$, since $f_i(x)$ is a pre-image of $\bar{f}_i \bmod(\bar{f})$ in $R[x]$. The image of the composite map is the field $k[X]/(\bar{f}_i)$, so m_i is indeed a maximal ideal.

If $j \neq i$, then the image of $f_j(x)$ under the composite map is $\bar{f}_j(X \bmod(\bar{f}_i)) = \bar{f}_j(\text{zero of } \bar{f}_i) \neq 0$, so $f_j(x) \notin m_i$. Therefore $m_i \neq m_j$ for $i \neq j$.

To show that the m_i's are all the maximal ideals of $R[x]$ we note that, since $R[x]$ is integral over R, each maximal ideal of $R[x]$ contains $mR[x]$. Now $\Pi\, f_i^{e_i} - f \in mR[X]$ and $f(x) = 0$, so $\Pi\, f_i(x)^{e_i} \in mR[x]$. Therefore each maximal ideal of $R[x]$ contains an element $f_i(x)$, so equals $(m, f_i(x))R[x]$.

12.5.8 Later, in 12.5.21, we prove that an integrally closed local domain R is henselian if and only if each domain $S \supset R$ that is integral over R is local. First we establish one direction of this equivalence: this is (i) $\Rightarrow$ (iv) in the next lemma.

12.5.9 LEMMA. Let R be an integrally closed local domain with fraction field K and residue field $k = R/m$. Then we have (i) $\Rightarrow$ (ii) $\Rightarrow$ (iii) $\Rightarrow$ (iv) for the conditions below.

(i) Each domain $S \supset R$ that is integral over R is local;

(ii) If $f(X) \in R[X]$ is monic, and irreducible in $K[X]$, then $\bar{f} = \phi^e$ for some irreducible monic $\phi \in k[X]$;

(iii) For each monic $f(X) \in R[X]$ and each factorization $\bar{f} = \phi\psi$ where $\phi, \psi \in k[X]$ are monic and relatively prime there is a factorization $f = gh$ such that $g, h \in R[X]$ are monic and $\bar{g} = \phi$, $\bar{h} = \psi$;

(iv) R is henselian.

PROOF. (i) ⇒ (ii): since f is monic, each polynomial $g \in R[X]$ can be written as $g = qf + r$ where $q,r \in R[X]$, deg r < deg f. It follows that $fK[X] \cap R[X] = fR[X]$, so $R[x] = R[X]/(f)$ is a domain. Since R[x] is integral over R it is local by assumption (i). Then 12.5.7 tells us that $\bar{f} = \phi^e$ for some irreducible monic $\phi \in k[X]$.

(ii) ⇒ (iii): factor $f = f_1 \ldots f_m$, where $f_1, \ldots, f_m$ are monic and irreducible in K[X]. Then the coefficients of the f_i's are integral over R since they are elementary symmetric functions of zeros of f_i and hence of f. Hence $f_i \in R[X]$, so by assumption (ii) we have $\bar{f}_i = \phi_i^{e_i}$ with $\phi_i \in k[X]$ monic and irreducible, $e_i > 0$. Now if $\bar{f} = \phi\psi$ with relatively prime ϕ, ψ as in (iii) it follows from $\bar{f} = \phi_1^{e_1}, \ldots, \phi_m^{e_m}$ that ϕ is a product of some factors $\phi_i^{e_i}$ and that ψ is a product of the remaining factors $\phi_j^{e_j}$. Put $g = \prod_{\phi_i | \phi} f_i^{e_i}$ and $h = \prod_{\phi_j | \psi} f_j^{e_j}$.

Then clearly $f = gh$, $\bar{g} = \phi$, $\bar{h} = \psi$.

(iii) ⇒ (iv): by 12.5.5 we only have to show that each monic polynomial $f(X) = X^n + b_{n-1} X^{n-1} + \ldots + b_1 X + b_0 \in R[X]$ with $b_1 \notin m$, $b_0 \in m$ has a zero in m. Now $\bar{f}(X)$ factors as $\bar{f}(X) = X(X^{n-1} + \ldots + \bar{b}_1)$, and the two factors are relatively prime in k[X] since $\bar{b}_1 \neq 0$. Lifting this factorization to one in R[X] we get a linear factor $X - c$, $c \in m$, so $f(c) = 0$.

12.5.10 A crude way to make a local ring henselian is to complete it at its maximal ideal. For many purposes it is necessary to replace completion by a more delicate algebraic construction called "henselization" that better reflects the algebraic properties of the original local ring. We now give the definition of henselization.

12.5.11 DEFINITION.

(i) Let R,S be local rings. Then a *local morphism* ϕ: R → S is a ring homomorphism such that $\phi(m(R)) \subset m(S)$. (Note that then ϕ induces an embedding $R/m(R) \to S/m(S)$ between their residue fields.)

(ii) Let R be a local ring. A *henselization* of a R is a pair (R^h, i) consisting of a local henselian ring R^h and a local morphism $i: R \to R^h$ with the following universal property: if $\phi: R \to S$ is any local morphism into a henselian local ring S then there is a unique local morphism $\phi^h: R^h \to S$ such that the following diagram commutes:

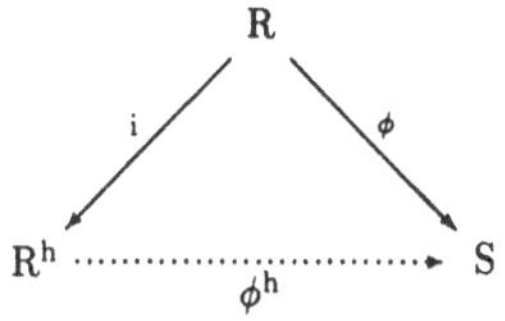

12.5.12 REMARKS.

1. If a henselization of R exists it is clearly unique up to a unique local isomorphism over R. So we may unambiguously refer to *the* henselization of R if R has one.

2. If R is already henselian, then (R,id) is the henselization of R.

3. Roughly speaking, to construct the henselization of R we only have to adjoin to R a zero of each polynomial $f(X) \in R[X]$ whose reduction modulo m has a nonsingular zero in the residue field. To make this idea work we need the following two lemmas.

12.5.13 LEMMA. Let R be a local ring with residue field $k = R/m$, let $f(X) \in R[X]$ and let $\alpha \in k$ be a nonsingular zero of $\bar{f}(X)$, i.e., $\bar{f}(\alpha) = 0$, $\bar{f}'(\alpha) \neq 0$. Then α can be lifted to at most one zero of f in R.

PROOF. If $f(a) = 0$, $\bar{a} = \alpha$, then $f'(a)$ is an R-unit, so for any $x \in m$ we have $f(a + x) = f(a) + f'(a).x + \text{multiple of } x^2 = f'(a).x.(1 + \text{multiple of } x) = x.(\text{R-unit})$. Hence, if $f(a + x) = 0$, then $x = 0$.

12.5.14 LEMMA. Let (R,m) be a local ring and $f(X) \in R[X]$ a monic polynomial with $f(0) \in m$, $f'(0) \notin m$. Put $R[X]/(f) = R[x]$ with $x = X \bmod(f)$, and $n = (m,x)R[x]$. Then n is a maximal ideal of $R[x]$, and the local R-algebra $R' = R[x]_n$ has the following properties:

(1) the natural map $R/m \to R'/nR'$ is a field isomorphism;

(2) $mR' = nR'$, i.e. $x \in mR'$;

(3) for each local morphism $\phi: R \to S$ into a henselian local ring S there is a unique local morphism $\phi': R' \to S$ such that the following diagram commutes:

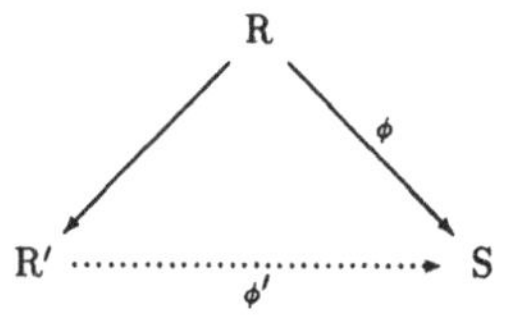

PROOF. Let $f(X) = X^n + \ldots + a_1X + a_0$ with $a_0 = f(0) \in m$ and $a_1 = f'(0) \notin m$. The reduced polynomial $\bar{f}(X) = X^n + \ldots + \bar{a}_1X \in k[X]$ factors as $\bar{f}(X) = X(X^{n-1} + \ldots + \bar{a}_1)$, $\bar{a}_1 \neq 0$, so $\bar{f}(X)$ has a single factor X. Hence, by 12.5.7, $n = (m,x)R[x]$ is indeed a maximal ideal of $R[x]$ and $R/m \simeq R[x]/n$. Localizing $R[x]$ at n yields the local ring R' and does not change the residue field, and we have (1).

The maximal ideal nR' of R' is generated by the images of m and x. Since $f(x) = 0$, i.e. $x(x^{n-1} + \ldots + a_1) = -a_0 \in m$, and since $x^{n-1} + \ldots + a_1 \notin n$, the image of x in R' is in $mR[x]_n = mR'$. This proves (2).

To obtain (3), note that $(\phi f)(X) \in S[X]$ has a unique zero in the maximal ideal $m(S)$ of S (at least one since S is henselian, and at most one by 12.5.13). Therefore ϕ can be extended uniquely to a morphism $R[x] \to S$ sending x to an element of $m(S)$. This morphism then maps $(m,x)R[x] = n$ into $m(S)$, hence it factors via R', i.e., there is a local morphism ϕ' that makes the diagram in (3) commute. The uniqueness of ϕ' comes from the uniqueness property of the morphism $R[x] \to S$.

12.5.15 For our purpose we only need the henselization of integrally closed local domains, and we restrict our attention now to this case. (In exercises at the end of this section we reduce the general case to this special case.)

12.5.16 LEMMA. If m is a maximal ideal of the domain R and S is a local ring with $R \subset S \subset R_m$, then $S = R_m$. (Here we consider R and R_m as subrings of the fraction field K of R.)

PROOF. Let $a \in R\backslash m$. It clearly suffices to show that $a \notin m(S)$. Take $b \in R$ with $ab \equiv 1 \bmod m$, so $ab - 1 \in m$. Since $m \subset mR_m \cap S \subset m(S)$, this implies $ab - 1 \in m(S)$, so $a \notin m(S)$.

12.5.17 LEMMA. Let R be an integrally closed domain with fraction field K, Let $E|K$ be a finite separable extension, and put $R^* =$ integral closure of R in E. Take an $x \in R^*$ with $E = K(x)$. (Note: each field generator of E over K can be multiplied by an element of R to give a new generator inside R^*.) Then the minimum polynomial $f(X)$ of x over K lies in $R[X]$ and $R^* \subset f'(x)^{-1} R[x]$.

PROOF. Let L be a Galois extension of finite degree over K such that $E \subset L$, put $G = \mathrm{Gal}(L|K)$, $H = \mathrm{Gal}(L|E)$, and take a coset decomposition $G = \sigma_1 H \cup \ldots \cup \sigma_k H$, $k = [E : K]$, so $\sigma_1|E, \ldots, \sigma_k|E$ are the distinct K-embeddings of E into L. We also take $\sigma_1 = \mathrm{id}$. Note that $f(X) = \Pi_{i=1}^k (X - \sigma_i(x))$, which shows that f has coefficients integral over R, hence $f(X) \in R[X]$.

Put $g_i(X) = f(X)/(X - \sigma_i(x))$, so $g_i(X) \in L[X]$, and $g_1(X) = X^{k-1} + c_{k-2}X^{k-2} + \ldots + c_0 \in R^*[X]$, since $\sigma_1(x) = x \in R^*$. An easy computation shows that $g_1(x) = f'(x)$, and $g_i(x) = 0$ for $i > 1$. Observe also that $g_i(X) = X^{k-1} + \sigma_i(c_{k-2})X^{k-2} + \ldots + \sigma_i(c_0)$.

Let $b \in R^*$. Then $f'(x).b = g_1(x).b = \Sigma_{i=1}^{k} g_i(x).\sigma_i(b) =$

$\Sigma_{i=1}^{k}(x^{k-1} + \sigma_i(c_{k-2}x)^{k-2} + \ldots + \sigma_i(c_0)).\sigma_i(b) =$

$(\Sigma\, \sigma_i(b)).x^{k-1} + (\Sigma\, \sigma_i(c_{k-2}b))x^{k-2} + \ldots + \Sigma\, \sigma_i(c_0b)$.

The coefficients $\Sigma\, \sigma_i(b)$, $\Sigma\, \sigma_i(c_{k-2}b)$, ... , $\Sigma\, \sigma_i(c_0b)$ are in K since they are traces of elements in E, and they are integral over R, hence they are in R. This shows that $f'(x).b \in R[x]$, i.e., $b \in f'(x)^{-1} R[x]$.

12.5.18 The next lemma is the key fact that gives us direct access to the main properties of the henselization of an integrally closed local domain. To prove and state this lemma we need some decomposition theory, cf. [Lan, Chs. VII-IX], which we now briefly review, see picture:

$$\begin{array}{ccccc} \mathfrak{p} & \subset & S & \subset & L \\ | & & | & & | \\ \mathfrak{p}^{dec} & \subset & S^{dec} & \subset & L^{dec} \\ | & & | & & | \\ \mathfrak{m} & \subset & R & \subset & K \end{array}$$

Meaning of the symbols:
$(R,\mathfrak{m})$: integrally closed local domain,
K: fraction field of R,
S: integral closure of R in the finite Galois extension $L|K$,
G: $Gal(L|K)$.

Note that $S = \sigma(S)$ for all $\sigma \in G$ and that G acts transitively on the finite nonempty set of maximal ideals of S. We fix a maximal ideal $\mathfrak{p}$ of S, so $\mathfrak{p} \cap R = \mathfrak{m}$ i.e. $\mathfrak{p}$ lies over $\mathfrak{m}$ and put:

$G_{\mathfrak{p}} = \{\sigma \in G \mid \sigma(\mathfrak{p}) = \mathfrak{p}\}$ (the decomposition group of $\mathfrak{p}$),

L^{dec} = fixed field of $G_{\mathfrak{p}}$ (the decomposition field of $\mathfrak{p}$),

$S^{dec} = S \cap L^{dec}$ = integral closure of R in L^{dec},

$\mathfrak{p}^{dec} = \mathfrak{p} \cap L^{dec}$, a maximal ideal of S^{dec}.

We also fix a coset decomposition

$$G = \sigma_1 G_{\mathfrak{p}} \cup \ldots \cup \sigma_k G_{\mathfrak{p}}, \ \sigma_1 = \mathrm{id}, \ k = [L^{dec} : K].$$

Notice the following facts:

(i) $\sigma_1 | L^{dec}, \ldots, \sigma_k | L^{dec}$ are the distinct K-embeddings of L^{dec} into L;

(ii) $\sigma_1(\mathfrak{p}), \ldots, \sigma_k(\mathfrak{p})$ are the distinct maximal ideals of S;

(iii) $\mathfrak{p}$ is the only maximal ideal of S containing $\mathfrak{p}^{dec}$; (this is because $G_{\mathfrak{p}}$ acts transitively on the set of maximal ideals of S lying over a given maximal ideal of S^{dec}).

(iv) S^{dec} has only finitely many maximal ideals, and they all lie over $\mathfrak{m}$.

12.5.19 LEMMA. Let $x \in S^{dec}$ belong to $\mathfrak{p}^{dec}$ but not to any other maximal ideal of S^{dec}. (Such an x exists by the Chinese Remainder Theorem.) Then

(1) $K(x) = L^{dec}$;

(2) The minimum polynomial f(X) of x over K lies in R[X], and $f(0) \in \mathfrak{m}$, $f'(0) \notin \mathfrak{m}$;

(3) $(S^{dec})_{\mathfrak{p}^{dec}} = R[x]_{\mathfrak{n}}$, where $\mathfrak{n}$ is the maximal ideal $(\mathfrak{m},x)R[x]$ of R[x].

PROOF. To show $K(x) = L^{dec}$ it is enough to prove that $\mathrm{Gal}(L|K(x)) = G_{\mathfrak{p}}$, the Galoisgroup of L over L^{dec}. Since $K(x) \subset L^{dec}$, we have $\mathrm{Gal}(L|K(x)) \supset G_{\mathfrak{p}}$. Therefore each coset $\sigma_i G_{\mathfrak{p}}$ is either contained in $\mathrm{Gal}(L|K(x))$ (if $\sigma_i(x) = x$), or disjoint from it (if $\sigma_i(x) \neq x$). So it suffices to prove $\sigma_i(x) \neq x$ for $i > 1$ to conclude $K(x) = L^{dec}$. For later use we prove a little bit more:

(*) $\sigma_i(x) \notin \mathfrak{p}$ for $i > 1$.

If $\sigma_i(x) \in \mathfrak{p}$, $i > 1$, then $x \in \sigma_i^{-1}(\mathfrak{p}) \cap S^{dec}$, which is a maximal ideal of S^{dec}, hence $\mathfrak{p}^{dec} = (\sigma_i^{-1}\mathfrak{p}) \cap S^{dec}$, by the assumption on x. So $\sigma_i^{-1}(\mathfrak{p})$ is a maximal ideal of S lying over $\mathfrak{p}^{dec}$, therefore $\sigma_i^{-1}(\mathfrak{p}) = \mathfrak{p}$ (by 12.5.18, fact (iii)), i.e. $\mathfrak{p} = \sigma_i(\mathfrak{p})$, contradicting $\sigma_i \notin G_{\mathfrak{p}}$.

Now that we have shown that x generates L^{dec} over K we know that its minimum polynomial over K is $f(X) = \Pi_{i=1}^{k}(X - \sigma_i(x)) = X^k + a_{k-1}X^{k-1} + \ldots + a_1X + a_0$, with all $a_i \in K$ integral over R, hence all $a_i \in R$. Furthermore $a_0 = \pm\Pi\ \sigma_i(x) \in \mathfrak{p} \cap R = \mathfrak{m}$, since $\sigma_1(x) = x \in \mathfrak{p}$. Note that $\pm\ a_1$ is a sum of k terms, each term a product of k-1 factors $\sigma_i x$, and only one term, namely $\sigma_2(x)\ldots\sigma_k(x)$, misses the factor $\sigma_1(x) = x$, and so this is the only term not in $\mathfrak{p}$. (Here we use (*).) Hence $a_1 \notin \mathfrak{p} \cap R = \mathfrak{m}$. Since $a_0 = f(0)$ and $a_1 = f'(0)$, this proves (2).

Clearly $R[x] \simeq R[X]/(f)$, so by 12.5.14 the ideal $\mathfrak{n} = (\mathfrak{m},x)R[x]$ is a maximal ideal of R[x], and clearly $\mathfrak{n} \cap R = \mathfrak{m}$. Since $a_1 \notin \mathfrak{m}$ and $f'(x) = a_1$ + element of xR[x] this gives $f'(x) \notin \mathfrak{m}$. In combination with 12.5.17 and $\mathfrak{n} \subset \mathfrak{p}^{dec}$ this leads to the inclusions

$$S^{dec} \subset f'(x)^{-1}\ R[x] \subset R[x]_{\mathfrak{n}} \subset (S^{dec})_{\mathfrak{p}^{dec}}.$$

Lemma 12.5.16 then gives $R[x]_{\mathfrak{n}} = (S^{dec})_{\mathfrak{p}^{dec}}$.

12.5.20 The lemma shows that the local R-algebra $(S^{dec})_{\mathfrak{p}^{dec}}$ is isomorphic to the R-algebra R' of 12.5.14, so the conclusions of 12.5.14 apply here:

(1) the local morphism $R \to (S^{dec})_{\mathfrak{p}^{dec}}$ induces an isomorphism between the residue fields;

(2) the maximal ideal of $(S^{dec})_{\mathfrak{p}^{dec}}$ is generated by $\mathfrak{m}$;

(3) each local morphism of R into a henselian local ring H can be extended uniquely to a local morphism of $(S^{dec})_{\mathfrak{p}^{dec}}$ into H.

Since R' is a localization of R[x] and R[x] is a free R-module, R' is a flat R-module. Also $\mathfrak{m}R' \neq R'$, hence:

(4) $(S^{dec})_{\mathfrak{p}^{dec}}$ is a faithfully flat R-module.

These facts will be very useful to us later. Before we continue with "henselization" we first derive a characterization of henselian integrally closed local domains.

12.5.21 COROLLARY. For an integrally closed local domain R with fraction field K the following five conditions are equivalent:

(i) R is henselian;

(ii) The integral closure of R in each finite degree Galois extension L of K is a local domain;

(iii) The integral closure of R in the separable closure K_{sep} of K is a local domain;

(iv) The integral closure of R in the algebraic closure $\tilde{K}$ of K is a local domain;

(v) Each domain $D \supset R$ that is integral over R is local.

PROOF. (i) $\Rightarrow$ (ii): assume R is henselian and let S be the integral closure of R in the Galois extension $L|K$ of finite degree. Then the polynomial f(X) of 12.5.19 (2) has a zero in R and is irreducible in K[X], so its degree $k = [L^{dec} : K]$ must equal 1, i.e. $L^{dec} = K$, $S^{dec} = R$, $\mathfrak{p}^{dec} = \mathfrak{m}$ and by 12.5.18 (iii) $\mathfrak{p}$ is the only maximal ideal of S, so S is local.
(ii) $\Rightarrow$ (iii) is clear since K_{sep} is the inductive union of the fields L such that $K \subset L \subset K_{sep}$ and $L|K$ is a Galois extension of finite degree.
(iii) $\Rightarrow$ (iv): if char K = 0 this is obvious since then $\tilde{K} = K_{sep}$. Suppose char K = p > 0. Let R_{sep} be the integral closure of R in K_{sep} and $\tilde{R}$ the integral closure of R in $\tilde{K}$. By assumption R_{sep} is local. Let $\tilde{\mathfrak{p}}$ be a maximal ideal of $\tilde{R}$, and let $x \in \tilde{R}$. Then $x^{p^e} \in K_{sep} \cap \tilde{R} = R_{sep}$ for some e > 0, and for this e we have: $x \in \tilde{\mathfrak{p}} \Leftrightarrow x^{p^e} \in \tilde{\mathfrak{p}} \cap R_{sep}$ = the unique maximal ideal of R_{sep}; clearly $\tilde{\mathfrak{p}}$ is uniquely determined by this equivalence. So $\tilde{R}$ is local.
(iv) $\Rightarrow$ (v): we may assume $D \subset \tilde{K}$. Then $\tilde{R}$ is the integral closure of R as well as of D in $\tilde{K}$. Therefore the maximal ideals of D are the intersections $D \cap \tilde{\mathfrak{p}}$ with $\tilde{\mathfrak{p}}$ ranging over the maximal ideals $\tilde{\mathfrak{p}}$ of $\tilde{R}$, [Ma 80, (5.E)]. Since $\tilde{R}$ is local by assumption, so is D.
(v) $\Rightarrow$ (i): this is just 12.5.9, (i) $\Rightarrow$ (iv).

12.5.22 We now extend 12.5.18 and 12.5.20 to Galois extensions of infinite degree. Let $(R,\mathfrak{m})$ be, as always, an integrally closed local domain with fraction field K, and let $M|K$ be a Galois extension, not necessarily of finite degree. Let T be the integral closure of R in M and fix a maximal ideal $\mathfrak{q}$ of T; so $\mathfrak{q}$ lies over $\mathfrak{m}$.

$$\begin{array}{ccccc} \mathfrak{q} & \subset & T & \subset & M \\ | & & | & & | \\ \mathfrak{q}^{dec} & \subset & T^{dec} & \subset & M^{dec} \\ | & & | & & | \\ \mathfrak{m} & \subset & R & \subset & K \end{array}$$

Put: $G_{\mathfrak{q}} = \{\sigma \in \mathrm{Gal}(M|K): \sigma(\mathfrak{q}) = \mathfrak{q}\}$ (the decomposition group of $\mathfrak{q}$),

M^{dec} = fixed field of $G_{\mathfrak{q}}$,

$T^{dec} = T \cap M^{dec}$ = integral closure of R in M^{dec},

$\mathfrak{q}^{dec} = \mathfrak{q} \cap M^{dec}$, a maximal ideal of T^{dec}.

Let L range over the fields such that $K \subset L \subset M$ and $L|K$ is a Galois extension of finite degree. Then M is the inductive union of the L's. For each L, put $S(L)$ = integral closure of R in L, $\mathfrak{p}(L) = \mathfrak{q} \cap S(L)$, a maximal ideal of $S(L)$. Then the decomposition group $G_{\mathfrak{p}(L)} =$ $\{\sigma \in \mathrm{Gal}(L|K): \sigma(\mathfrak{p}(L)) = \mathfrak{p}(L)\}$ equals $\{\tau|L: \tau \in G_{\mathfrak{q}}\}$. (To see this, let $\sigma \in G_{\mathfrak{p}(L)}$ extend to some $\tilde{\sigma} \in \mathrm{Gal}(M|K)$; by a standard inverse limit argument show that $\mathrm{Gal}(M|L)$ acts transitively on the set of maximal ideals of T which lie over $\mathfrak{p}(L)$, take $\sigma' \in \mathrm{Gal}(M|L)$ with $\sigma'(\tilde{\sigma}(\mathfrak{q})) = \mathfrak{q}$. Then for $\tau = \sigma'\circ\tilde{\sigma}$ we have $\tau \in G_{\mathfrak{q}}$ and $\tau|L = \sigma$.)

Hence $T = \bigcup_L S(L)$, $\mathfrak{q} = \bigcup_L \mathfrak{p}(L)$, $M^{dec} = \bigcup_L L^{dec}$, $T^{dec} = \bigcup_L S(L)^{dec}$ and $(T^{dec})_{\mathfrak{q}^{dec}} = \bigcup_L (S(L)^{dec})_{\mathfrak{p}(L)^{dec}}$.

Since for the L's considered the results of 12.5.18 and 12.5.20 are available we obtain from these inductive unions:

(1) $\mathfrak{q}$ is the only maximal ideal of T lying over $\mathfrak{q}^{dec}$;

(2) the local morphism $R \to (T^{dec})_{\mathfrak{q}^{dec}}$ induces an isomorphism between the residue fields;

(3) the maximal ideal of $(T^{dec})_{q^{dec}}$ is generated by m;

(4) each local morphism of R into a henselian local ring H can be extended uniquely to a local morphism of $(T^{dec})_{q^{dec}}$ into H;

(5) $(T^{dec})_{q^{dec}}$ is a faithfully flat R-algebra.

Here is the main result of this section.

12.5.23 THEOREM. Let $M = K_{sep}$ in 12.5.22. Then $(T^{dec})_{q^{dec}}$ with inclusion is a henselization of R.

PROOF. We only have to show that $(T^{dec})_{q^{dec}}$ is henselian, since the required universal property has already been established: fact (4) of 12.5.22.

Let U be the multiplicative set $T^{dec}\backslash q^{dec}$. Then $U^{-1}T$ is the integral closure in K_{sep} of $U^{-1}T^{dec} = (T^{dec})_{q^{dec}}$. Any maximal ideal of $U^{-1}T$ must contain the unique maximal ideal $U^{-1}(q^{dec})$ of $U^{-1}T^{dec}$, hence is of the form $U^{-1}(q^*)$ for some prime ideal q^* of T containing q^{dec}. Such a q^* must be a maximal ideal, hence $q^* = q$ by 12.5.22, fact (1). This shows that $U^{-1}T$ is local, and hence, by 12.5.21 (i) ⇒ (iv), $(T^{dec})_{q^{dec}}$ is henselian.

12.5.24 Let R = (R,m) continue to denote an integrally closed local domain. By the construction we just gave for the henzelization of R we may regard R as a local *subring* of its henselization; we shall denote this henselization by R^h and its maximal ideal by m^h. For each e > 0 we have:

1. $m^e.R^h = (m^h)^e$;

2. the natural map $R/m^e \to R^h/(m^h)^e$ is an isomorphism.

Note that 1. follows from the case e = 1, which is property (3) of 12.5.22. The injectivity in 2. follows from 1. and the faithful flatness of R^h over R, cf. property (5) of 12.5.22. To get surjectivity one proves by induction on e that each element of R^h is congruent modulo $m^e.R^h$ to an element of R; the starting point e = 1 is provided by property (2) of 12.5.22.

By fact 2. we may identify the completions $\hat{R} = \varprojlim R/m^e$ and $\hat{R^h} = \varprojlim R^h/(m^h)^e$. Since $\hat{R}$ is a complete local ring, it is henselian, and the natural map $j: R^h \to \hat{R^h} = \hat{R}$ is also the unique local morphism making the diagram

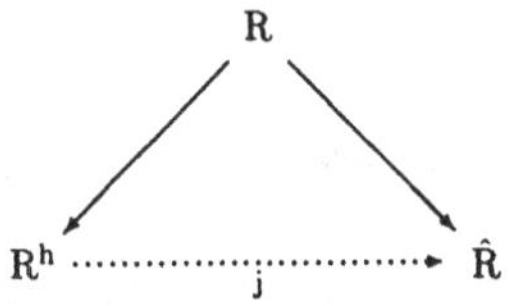

commutative.

12.5.25 PROPOSITION. Let R be an integrally closed local domain and suppose R is noetherian. Then

(i) $\hat{R}$ is a faithfully flat R^h-algebra;

(ii) R^h is noetherian.

REMARK. It follows from (i) that $j: R^h \to \hat{R}$ is injective.

PROOF. Clearly (ii) follows from (i) by [Ma 80, (4.C) (ii)], since $\hat{R}$ is noetherian. Because $m^h.\hat{R} = m\hat{R} \neq \hat{R}$, the proof of (i) reduces to showing that $\hat{R}$ is a flat R^h-module. By [Ma 80, (3.A), Th. 1 (6)] it suffices to show that any solution $(x_1, \ldots x_m) \in (\hat{R})^m$ of a homogeneous linear equation

(*) $a_1X_1 + \ldots + a_mX_m = 0 \qquad (a_i \in R^h)$

is an $\hat{R}$-linear combination of solutions in $(R^h)^m$. Now R^h is the inductive union of subrings $(S(L)^{dec})_{\mathfrak{p}(L)^{dec}}$ as in 12.5.22, and one of these rings, say S, contains the coefficients $a_1, \ldots, a_m$ of (*). The local ring S is clearly noetherian, so $\hat{S}$ is a flat S-module. Moreover, the two maps

$R/m^e \to S/m(S)^e \to R^h/(m^h)^e$ are isomorphisms for each $e > 0$. (Here we use that 12.5.24, 1. and 2. also hold with (R^h, m^h) replaced by $(S, m(S))$, because of 12.5.20, (1) and (2).) Hence $\hat{R} = \hat{S} = \hat{R^h}$, so any solution to (*) which lies in $(\hat{R})^m = (\hat{S})^m$ is an S-linear combination of solutions lying in S^m, and is therefore an R^h-linear combination of solutions lying in $(R^h)^m$.

12.5.26 We have marshalled all the facts that we shall need in the next section on Artin Approximation. However, for completeness' sake we indicate in the following exercises how properties of henselizations of arbitrary local rings can be obtained from the integrally closed case.

12.5.27 EXERCISES.

1. Suppose the local ring (R,m) has a henselization (R^h,i), and let $\alpha \subset m$ be an ideal of R. Then the local ring R/α has the henselization $(R^h/\alpha R^h, i/\alpha)$.

2. Each local ring A is isomorphic to R/α for some integrally closed local domain (R,m) and some ideal $\alpha \subset m$. (Hint: first represent A as the homomorphic image of a polynomial ring over $\mathbb{Z}$ in infinitely many variables.)

3. Each local ring (A,m) has a henselization $((A^h,m^h),i)$, and this henselization has the following properties:

(i) the natural map $A/m \to A^h/m^h$ is an isomorphism;

(ii) $m^h = m.A^h$;

(iii) A^h is a faithfully flat A-algebra, in particular the map $i: A \to A^h$ is injective.

Hint: use 1. and 2., and the corresponding results for integrally closed local domains.

For further information on henselization, see [Na, Ch. VII],

[Ray]. A useful summary is in [Mi, Ch. I, §4].

12.6 AN APPROXIMATION THEOREM OF M. ARTIN

12.6.1 To solve a system of equations over some ring it is often convenient to look first for solutions in a suitable extension ring and then to "specialize" back to solutions in the original ring. This leads to the following notion.

12.6.2 DEFINITION. Let A, B be rings, $A \subset B$. The inclusion $A \to B$ is said to have the specialization property if for each finite set of elements $b_1,\dots,b_N$ of B there is an A-algebra morphism ϕ: $A[b_1,\dots,b_N] \to A$. (Such a ϕ is called a specialization.)

12.6.3 Suppose $A \to B$ has the specialization property. Then each finite system of polynomial equations over A:

$$F_1(Y) = \dots = F_h(Y) = 0,\ F_i(Y) \in A[Y],\ Y = (Y_1,\dots,Y_N),$$

which has a solution in B has a solution in A. Namely, if $b = (b_1,\dots,b_N) \in B^N$ is a solution and ϕ: $A[b] \to A$ a specialization, then $\phi b = (\phi b_1,\dots,\phi b_N) \in A^N$ is also a solution.

The converse is often true:

12.6.4 LEMMA. Suppose A is noetherian or B is a domain, and $A \subset B$. Then $A \to B$ has the specialization property iff each finite system of polynomial equations over A which is soluble in B is soluble in A.

PROOF. We only have to prove one direction. Take first the case that B is a domain, and let $b = (b_1,\dots,b_N) \in B^N$. We have to find a specialization $A[b] \to A$. Let $Q(A)$ be the field of fractions of A and consider the ideal of the polynomial ring $Q(A)[Y]$, $Y = (Y_1,\dots,Y_N)$, consisting of all $F(Y)$ such that $F(b) = 0$. Choose $F_1(Y), \dots, F_h(Y)$ in $A[Y]$ generating this ideal. Now b is a solution of $F_1(Y) = \dots = F_h(Y) = 0$, and if $a \in A^N$ is any solution, then it is clear that we have a specialization $Q(A)[b] \to Q(A)$ sending b to a, and restricting this morphism to $A[b]$ gives a specialization $A[b] \to A$.

The case where A is noetherian is even simpler.

12.6.5 Note that if K is an algebraically closed field and A any K-algebra, $A \neq 0$, then $K \to A$ has the specialization property, by Hilbert's Nullstellensatz.

Artin's approximation theorems state the specialization property for inclusions of certain local rings in their completion. We now study this situation.

12.6.6 In the remainder of this section we fix a pair $(R,\mathfrak{m})$ consisting of a ring R with a finitely generated ideal $\mathfrak{m}$ such that R is separated in its $\mathfrak{m}$-adic topology. We let $\hat{R} = \varprojlim_t R/\mathfrak{m}^t$ denote the $\mathfrak{m}$-adic completion of R, so $R \subset \hat{R}$. The ring R may be too "small" for $R \to \hat{R}$ to have the specialization property. Therefore we look for a ring $\tilde{R}$ with $R \subset \tilde{R} \subset \hat{R}$ which stays "close" to R (e.g. algebraic over R if R is a domain) such that $\tilde{R} \to \hat{R}$ has the specialization property.

Let from now on $\tilde{R}$ be a ring with $R \subset \tilde{R} \subset \hat{R}$.

DEFINITION. $\tilde{R}$ is said to have the approximation property (AP) if $\tilde{R} \to \hat{R}$ has the specialization property.

The next lemma explains this terminology. Before we state it, we recall that, as m is finitely generated, the set R is dense in $\hat{R}$, where $\hat{R}$ is equipped with the $m\hat{R}$-adic topology (with respect to which $\hat{R}$ is separated and complete).

12.6.7 LEMMA. If $\tilde{R}$ has AP, then each solution in $\hat{R}$ of a finite system of polynomial equations over $\tilde{R}$ can be approximated arbitrarily closely, in the $m\hat{R}$-adic sense, by solutions in $\tilde{R}$.

PROOF. Suppose $\hat{y} = (\hat{y}_1,\ldots,\hat{y}_N)$ is a solution in $\hat{R}^N$, and let $k \in \mathbf{N}$. We shall show that there is a solution $y = (y_1,\ldots,y_N) \in \tilde{R}^N$ (of the same system) with $y \equiv \hat{y} \bmod m^k\hat{R}$ (i.e. $y_i \equiv \hat{y}_i \bmod m^k\hat{R}$ for $i = 1,\ldots,N$). Let $m_1,\ldots,m_t$ generate m^k and write

(*) $\hat{y}_i = u_i + \Sigma_{j=1}^t v_{ij}m_j$ with $u_i \in R$ and $v_{ij} \in \hat{R}$ for $i = 1,\ldots,N$, $j = 1,\ldots,t$. Let V be the finite subset of $\hat{R}$ consisting of the v_{ij}. Then, by (*), each specialization $\tilde{R}[V] \to \tilde{R}$ sends $\hat{y}$ to a solution $y = (y_1,\ldots,y_N)$ with $y_i \equiv u_i \bmod m^k\hat{R}$, $i = 1,\ldots,N$, whence $y \equiv \hat{y} \bmod m^k\hat{R}$, again by (*), which ends the proof.

The following technical lemma gives a useful sufficient condition for AP to hold.

12.6.8 LEMMA. Suppose $R \subset \tilde{R} \subset \hat{R}$ are domains and $Q(\tilde{R})$ is algebraic over $Q(R)$. Let $K = Q(R)$, $L = Q(\hat{R})$ be the fraction fields of R and $\hat{R}$ respectively. If furthermore $(f(Z), \alpha_1(Z), \ldots, \alpha_N(Z), \beta(Z), \hat{z})$, $Z = (Z_1,\ldots,Z_t,Z_{t+1})$ is any tuple such that

(i) $f(Z) \in R[Z]$ is separable and irreducible over $K(Z_1,\dots,Z_t)$;

(ii) $\alpha_1(Z), \dots, \alpha_N(Z) \in R[Z]$ and $0 \neq \beta(Z) \in R[Z_1,\dots,Z_t]$;

(iii) $\hat{z} = (\hat{z}_1,\dots,\hat{z}_{t+1}) \in \hat{R}^{t+1}$ and $\hat{z}_1,\dots,\hat{z}_t$ are algebraically independent over K;

(iv) $f(\hat{z}) = 0$ and $\alpha_i(\hat{z}) \equiv 0 \bmod \beta(\hat{z})$ in $\hat{R}$ for $i = 1,\dots,N$; and if the following conditions hold:

(I) $L|K$ is a separable field extension;

(II) For each tuple as above there is $z = (z_1,\dots,z_{t+1}) \in \tilde{R}^{t+1}$ with $f(z) = 0$, $\beta(z) \neq 0$ and $\alpha_i(z) \equiv 0 \bmod \beta(z)$ in $\tilde{R}$ for $i = 1,\dots,N$; then $\tilde{R}$ has AP.

PROOF. We first show:

(*) For each $\hat{y} = (\hat{y}_1,\dots,\hat{y}_N) \in \hat{R}^N$ there is an R-algebra morphism $R[\hat{y}] \to \tilde{R}$. By (I) we can choose $\hat{z}_i \in \hat{R}$, $1 \le i \le t+1$ with $t \in \mathbb{N}$, such that $K(\hat{y}) = K(\hat{z}_1,\dots,\hat{z}_t,\hat{z}_{t+1})$, where $\hat{z}_1,\dots,\hat{z}_t$ is a separating transcendence base over K and $\hat{z}_{t+1}$ is separable algebraic over $K(\hat{z}_1,\dots,\hat{z}_t)$. Of course we may take the $\hat{z}_i$ in $R[\hat{y}] \subset \hat{R}$. Let $f(Z) \in R[Z]$, $Z = (Z_1,\dots,Z_{t+1})$, be irreducible as a polynomial in Z_{t+1} over $K(Z_1,\dots,Z_t)$ such that $f(\hat{z}) = 0$, $\hat{z} = (\hat{z}_1,\dots,\hat{z}_{t+1})$. Then f is in fact separable over $K(Z_1,\dots,Z_t)$.

Write $\hat{y}_i = \alpha_i(\hat{z})/\beta(\hat{z})$, $i = 1,\dots,N$, where $\alpha_i(Z) \in R[Z]$, $i = 1,\dots,N$ and $0 \neq \beta(Z) \in R[Z_1,\dots,Z_t]$. So we have $\alpha_i(\hat{z}) \equiv 0 \bmod \beta(\hat{z})$ in $\hat{R}$, $i = 1,\dots,N$. By (II) there is $z = (z_1,\dots,z_{t+1}) \in \tilde{R}^{t+1}$ with $f(z) = 0$, $\beta(z) \neq 0$ and $\alpha_i(z)/\beta(z)$ in $\tilde{R}$ for $i = 1,\dots,N$.

Without loss of generality we may assume that f is in fact irreducible in $K[Z]$. (If not, remove bad factors.) Then $K[\hat{z}] \simeq K[Z]/(f)$, and as $f(z) = 0$, we obtain a K-algebra morphism $\phi: K[\hat{z}] \to L$ with $\phi(\hat{z}) = z$. Now $\phi(\beta(\hat{z})) = \beta(z) \neq 0$, hence ϕ can be extended to a morphism $\phi: K[\hat{z}, 1/\beta(\hat{z})] \to L$. So $\phi(\hat{y}_i) = \phi(\alpha_i(\hat{z})/\beta(\hat{z})) = \alpha_i(z)/\beta(z) \in \tilde{R}$ for $i = 1,\dots,N$. Hence ϕ restricts to an R-algebra morphism $R[\hat{y}] \to \tilde{R}$, and (*) is proved.

Let $\hat{y} = (\hat{y}_1,\ldots,\hat{y}_N) \in \hat{R}^N$. We are going to construct a specialization $\tilde{R}[\hat{y}] \to \tilde{R}$. Let $F_1(Y),\ldots,F_h(Y) \in \tilde{R}[Y]$, $Y = (Y_1,\ldots,Y_N)$, generate the ideal of all $F(Y)$ in $Q(\tilde{R})[Y]$ such that $F(\hat{y}) = 0$. As the proof of 12.6.4 shows, it suffices to find a solution $y \in \tilde{R}^N$ of

(1) $F_1(Y) = \ldots = F_h(Y) = 0$ (since this gives a specialization $\hat{y} \mapsto y$).

Let $x_1,\ldots,x_n \in \tilde{R}$ be the distinct nonzero coefficients of the F_i's. Take for each $i = 1,\ldots,n$ a nonzero polynomial $p_i(X) \in R[X]$ with

(2) $p_i(x_i) = 0$. (p_i exists because $\tilde{R}$ is algebraic over R.) Take $k \in \mathbf{N}$ large enough so that no other root of p_i in $\tilde{R}$ is congruent to x_i mod $m^k\hat{R}$ for $i = 1,\ldots,n$. Choose $m_1,\ldots,m_s$ in R generating m^k and write:

(3) $x_i = x_{i0} + x_{i1}m_1 + \ldots + x_{is}m_s$, $x_{i0} \in R$, $x_{ij} \in \hat{R}$. By (*) there is an R-algebra morphism

$$\phi\colon R[\hat{y}_1,\ldots,\hat{y}_N,x_1,\ldots,x_n,x_{11},\ldots,x_{ns}] \to \tilde{R}.$$

(2) and (3) imply that $\phi(x_i)$ is a root of $p_i(X)$ which is congruent to x_i mod $m^k\hat{R}$, hence $\phi(x_i) = x_i$ by choice of k. This means that ϕ fixes the coefficients of the F's, hence $y = \phi(\hat{y})$ is a solution of (1), and we are through.

12.6.9 We will apply the lemma in the following situation. Let k be a field, $X = (X_1,\ldots,X_n)$, and put $R = k[X]$, $\hat{R} = k[[X]]$. Let m be the maximal ideal $(X_1,\ldots,X_n)R$ and note that $\hat{R}$ is the completion of the local ring R_m; it is a local ring with maximal ideal $n = m\hat{R}$. Let $\tilde{R}$, with maximal ideal $\tilde{m}$, be the henselization of the integrally closed local domain R_m; since R_m is noetherian it follows from the remark after 12.5.25 that we may consider $\tilde{R}$ as a local subring of $\hat{R}$. So $R \subset \tilde{R} \subset \hat{R}$. In the proof of the main result of this section we shall use the following consequence of the faithful flatness of $\hat{R}$

as an $\tilde{R}$-module: if a, b $\in \tilde{R}$ and a $\in$ b$\hat{R}$, then a $\in$ b$\tilde{R}$.

The main result of this section, due to M. Artin, is that $\tilde{R}$ has AP. For the proof of this we shall check the conditions (I) and (II) of Lemma 12.6.8. Condition (II) will be verified in the proof of Theorem 12.6.11, while the following result 12.6.10 states that (I) also holds. (Readers only interested in the characteristic 0 case, which is all we need for our applications, may skip 12.6.10.)

12.6.10 PROPOSITION. Let k((X)) denote the fraction field of k[[X]]. Then k((X)) is a separable extension of k(X).

PROOF. For standard facts on linear disjointness and separability we refer the reader to [Lan, Ch. X, §§5 & 6].

We may of course assume char k = p > 0. The proof is by induction on n, the essential case being n = 1: Consider the diagram of field inclusions

$$\begin{array}{ccccc}
k((X_1))^p & \text{———} & k((X_1^p)) & \text{———} & k((X_1)) \\
| & & | & & | \\
k(X_1)^p & \text{———} & k(X_1^p) & \text{———} & k(X_1) \\
| & & | & & \\
k^p & \text{———} & k & &
\end{array}$$

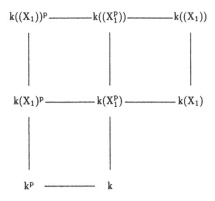

The assertion will follow if we prove that $k((X_1))^p$ and $k(X_1)$ are linearly disjoint over $k(X_1)^p$. First we show:

(*) $k((X_1))^p$ and k are linearly disjoint over k^p.

Let $a_1,\dots,a_n \in k$ satisfy $\Sigma\, b_i{}^p a_i = 0$ where $b_1,\dots,b_n \in k((X_1))$ are not all zero. After multiplication by a suitable element of $k((X_1))$ we may assume that all b_i are in $k[[X_1]]$ and that at least one b_i has a nonzero constant term. (Recall that $k[[X_1]]$ is a valuation ring of $k((X_1))$.)

Then the substitution $X_1 \mapsto 0$ transforms $\Sigma\, b_i{}^p a_i = 0$ into a linear dependence relation of $a_1,\dots,a_n$ over k^p, and (*) is proved. A consequence is that $k((X_1))^p$ and $k(X_1{}^p)$ are linearly disjoint over $k(X_1)^p$.

The basis $1, X_1, \dots, X_1{}^{p-1}$ of $k(X_1)$ over $k(X_1{}^p)$ remains linearly independent over $k((X_1{}^p))$, since $X_1 \notin k((X_1{}^p))$. So $k((X_1{}^p))$ and $k(X_1)$ are linearly disjoint over $k(X_1{}^p)$.

The linear disjointness in the two upper squares of the diagram implies the linear disjointness of $k((X_1))^p$ and $k(X_1)$ over $k(X_1)^p$. The assertion is proved for $n = 1$.

Assume (induction hypothesis) that the statement holds for n. Then $k[[X_1,\dots,X_{n+1}]] \subset k((X_1,\dots,X_n))[[X_{n+1}]] \subset k((X_1,\dots,X_n))((X_{n+1}))$, so

(**) $k((X_1,\dots,X_{n+1})) \subset k((X_1,\dots,X_n))((X_{n+1}))$.

The case $n = 1$ applied to the ground field $k((X_1,\dots,X_n))$ implies:

(***) $k((X_1,\dots,X_n))((X_{n+1}))$ is separable over $k((X_1,\dots,X_n))(X_{n+1})$.

(**) and (***) clearly imply:

(****) $k((X_1,\dots,X_{n+1}))$ is separable over $k((X_1,\dots,X_n))(X_{n+1})$.

The induction hypothesis implies that $k((X_1,\dots,X_n))(X_{n+1})$ is separable over $k(X_1,\dots,X_n)(X_{n+1}) = k(X_1,\dots,X_{n+1})$, and combining this with (****) we get that $k((X_1,\dots,X_{n+1}))$ is separable over $k(X_1,\dots,X_{n+1})$, which establishes the proposition.

12.6.11 THEOREM. (M. Artin) $k[X]^{\sim}$ has AP.

12.6.12 PROOF. Induction on n, the number of X-indeterminates. The case $n = 0$ is trivial.

Suppose $n > 0$ and that the assertion holds if we replace n by n-1 (induction hypothesis). The crucial step in Artin's proof of 12.6.11 is the following lemma. Its proof uses the Weierstrass Theorems to transform a system of equations over $k[X_1,\dots,X_n]$ to one over $k[X_1,\dots,X_{n-1}]$, and then the induction hypothesis is invoked. We use the notations R, $\tilde{R}$, $\hat{R}$, $\mathfrak{m}$, $\tilde{\mathfrak{m}}$, $\mathfrak{n}$ established in 12.6.9. We first prove

12.6.13 DIVISION LEMMA. Let $g_1(Z),\dots,g_m(Z)$, $f(Z) \in R[Z]$, $Z = (Z_1,\dots,Z_M)$, and suppose $\hat{z} = (\hat{z}_1,\dots,\hat{z}_M)$ is a solution of $g_1(Z) = \dots = g_m(Z) = 0$, $f(Z) \neq 0$ in $\hat{R}^M$, and let $\gamma \in \mathbb{N}$. Then there is $z \in \tilde{R}^M$ with $z \equiv \hat{z}$ mod $\mathfrak{n}^\gamma$ and $g_i(z) \equiv 0$ mod $f(z)$ in $\tilde{R}$ $(i = 1,\dots,m)$.

PROOF. If $f(\hat{z})$ is invertible in $\hat{R}$, then $f(z)$ is also invertible in $\tilde{R}$ for all $z \in \tilde{R}^M$ sufficiently close to $\hat{z}$ in the $\mathfrak{n}$-adic topology on $\hat{R}^M$, and this case is trivial. So we may suppose that $f(\hat{z})$ is not invertible in $\hat{R}$. By 12.4.6 we may assume without loss of generality that $f(\hat{z})$ is regular in X_n of order $d > 0$. Note that the automorphisms used in the proof of 12.4.6 map R onto itself, and hence, by the uniqueness properties of henselization, also $\tilde{R}$ onto itself. By enlarging γ if necessary, we may finally assume: $f(z)$ is regular in X_n of order d for all $z \in \hat{R}^M$ with $z \equiv \hat{z}$ mod $\mathfrak{n}^\gamma$.

Let us introduce new variables $A_0, \dots, A_{d-1}$, Z_{vj} $(v = 1,\dots,M,$ $j = 0,\dots,d-1)$ and put:

$$W(X_n) = X_n^d + A_{d-1}X_n^{d-1} + \dots + A_0$$

$$Z_v^* = Z_{v,d-1}X_n^{d-1} + \dots + Z_{v,0} \qquad (v = 1,\dots,M)$$

$$Z^* = (Z_1^*,\dots,Z_M^*).$$

The euclidean division algorithm yields:

$$(\alpha)\quad \begin{cases} f(Z^*) = Q.W(X_n) + F_{d-1}X_n^{d-1} + \dots + F_0 \\ g_i(Z^*) = Q_i.W(X_n) + G_{i,d-1}X_n^{d-1} + \dots + G_{i,0} \end{cases} \quad (i = 1,\dots,m)$$

where Q, Q_i, F_j and G_{ij} are polynomials over k in the variables $X_1, \dots, X_n$, $A_0, \dots, A_{d-1}$, Z_{vj} ($v = 1,\dots,M$, $j = 0,\dots,d-1$) and where the F_j and G_{ij} ($j = 0,\dots,d-1$, $i = 1,\dots,m$) do not involve X_n.

By 12.4.3 and 12.4.4 we can write: $f(\hat{z}) = (\text{unit}) \times \hat{w}(X_n)$, where $\hat{w}(X_n) = X_n^d + \hat{a}_{d-1}X_n^{d-1} + \dots + \hat{a}_0$ with non-units $\hat{a}_{d-1},\dots,\hat{a}_0$ in $k[[X_1,\dots,X_{n-1}]]$. By 12.4.2 we can write:

$$\hat{z}_v = \hat{u}_v.\hat{w}(X_n) + \hat{z}_{v,d-1}X_n^{d-1} + \dots + \hat{z}_{v,0} \text{ with } \hat{u}_v \in \hat{R},$$

$\hat{z}_{vj} \in k[[X_1,\dots,X_{n-1}]]$ ($v = 1,\dots,M$, $j = 0,\dots,d-1$). We define $\hat{z}_v^* = \hat{z}_{v,d-1}X_n^{d-1} + \dots + \hat{z}_{v,0}$ ($v = 1,\dots,M$) and $\hat{z}^* = (\hat{z}_1^*,\dots,\hat{z}_M^*)$. Since $\hat{z}_v^* \equiv \hat{z}_v \bmod \hat{w}(X_n)$ ($v = 1,\dots,M$), we get:

$$f(\hat{z}^*) \equiv f(\hat{z}) \bmod \hat{w}(X_n), \text{ and } g_i(\hat{z}^*)) \equiv g_i(\hat{z}) \bmod \hat{w}(X_n)$$

($i = 1,\dots,m$).

But $f(\hat{z}) \equiv 0 \bmod \hat{w}(X_n)$, and $g_i(\hat{z}) = 0$, hence

$$(\beta)\quad \begin{cases} f(\hat{z}^*) \equiv 0 \bmod \hat{w}(X_n) \\ g_i(\hat{z}^*) \equiv 0 \bmod \hat{w}(X_n) \ (i = 1,\dots,m). \end{cases}$$

Substituting $\hat{a}_0, \dots, \hat{a}_{d-1}, \hat{z}_{vj}$ for $A_0, \dots, A_{d-1}, Z_{vj}$ in (α) we see that (β) implies:

$$(\gamma)\quad \begin{cases} F_j(X_1, \dots, X_{n-1}, \hat{a}_0, \dots, \hat{a}_{d-1}, \hat{z}_{1,0}, \dots, \hat{z}_{M,d-1}) = 0 \\ G_{ij}(X_1, \dots, X_{n-1}, \hat{a}_0, \dots, \hat{a}_{d-1}, \hat{z}_{1,0}, \dots, \hat{z}_{M,d-1}) = 0 \end{cases}$$

($i = 1,\dots,m$, $j = 0,\dots,d-1$).

In other words, $(\hat{a}_0, \dots, \hat{a}_{d-1}, \hat{z}_{1,0}, \dots, \hat{z}_{M,d-1})$ is a solution with components in $k[[X_1,\dots,X_{n-1}]]$ of the following system of polynomial equations with coefficients in $k[X_1,\dots,X_{n-1}]$:

$$(\delta)\quad \begin{cases} F_j(X_1, \dots, X_{n-1}, A_0, \dots, A_{d-1}, Z_{1,0}, \dots, Z_{M,d-1}) = 0 \\ G_{ij}(X_1, \dots, X_{n-1}, A_0, \dots, A_{d-1}, Z_{1,0}, \dots, Z_{M,d-1}) = 0 \\ (i = 1,\dots,m,\ j = 0,\dots,d-1). \end{cases}$$

The induction hypothesis made in 12.6.12 then gives a solution $(a_0, \dots, a_{d-1}, z_{1,0}, \dots, z_{M,d-1})$ of (δ) with components in $k[X_1,\dots,X_{n-1}]^{\sim}$, and such that also:

(ϵ) $\quad a_j \equiv \hat{a}_j \bmod \mathfrak{n}^7,\ z_{vj} \equiv \hat{z}_{vj} \bmod \mathfrak{n}^7\ (j = 0,\dots,d-1,\ v = 1,\dots,M)$.

(For these congruences can be reformulated as extra equations with coefficients in $k[X_1,\dots,X_{n-1}]$.)

Now, for $v = 1,\dots,M$, we choose u_v in $\tilde{R}$ with $u_v \equiv \hat{u}_v \bmod \mathfrak{n}^7$ and put:

$$w(X_n) = X_n^d + a_{d-1}X_n^{d-1} + \dots + a_0 \in \tilde{R}$$

$$z_v^* = z_{v,d-1}X_n^{d-1} + \dots + z_{v,0} \in \tilde{R}\ (v = 1,\dots,M)$$

$$z_v = u_v.w(X_n) + z_v^*$$

$$z^* = (z_1^*,\dots,z_M^*), \text{ so that we get:}$$

$z_v \equiv \hat{z}_v \bmod \mathfrak{n}^7$ (by (ϵ) and the various equations which z_v and $\hat{z}_v$ satisfy).

Because $(a_0, \dots , a_{d-1}, z_{1,0}, \dots , z_{M,d-1})$ is a solution of (δ), we get by (α):

$$(\zeta) \quad \begin{cases} f(z^*) \equiv 0 \bmod w(X_n) \text{ and} \\ g_i(z^*) \equiv 0 \bmod w(X_n)\ (i = 1,\dots,m) \text{ in the ring } \tilde{R}. \end{cases}$$

But $z_v \equiv z_v^* \bmod w(X_n)$ in $\tilde{R}$ for $v = 1,\dots,M$, hence by (ζ):

$$(\eta) \quad \begin{cases} f(z) \equiv 0 \bmod w(X_n) \text{ (where } z = (z_1,\dots,z_M) \in \tilde{R}^M\text{) and} \\ g_i(z) \equiv 0 \bmod w(X_n) \quad (i = 1,\dots,m) \text{ in the ring } \tilde{R}. \end{cases}$$

But $z_v \equiv \hat{z}_v \bmod \mathfrak{n}^7$ implies that $f(z)$ is regular in X_n of order d, hence we get by 12.4.3 and 12.4.4:

$$f(z) = (\text{unit}) \times (X_n^d + b_{d-1}X_n^{d-1} + \dots + b_0),\ b_{d-1},\dots,b_0 \text{ non-units}$$

of $k[[X_1,\dots,X_{n-1}]]$. So (η) implies: $w(X_n) = X_n^d + a_{d-1}X_n^{d-1} + \dots + a_0$ divides $X_n^d + b_{d-1}X_n^{d-1} + \dots + b_0$ in $\hat{R}$, which implies by the uniqueness part of 12.4.3: $a_{d-1} = b_{d-1}, \dots , a_0 = b_0$, so $f(z) = (\text{unit}) \times w(X_n)$ in $\hat{R}$, and because $f(z)$ and $w(X_n)$ are in $\tilde{R}$, also (unit) is in $\tilde{R}$. (Here we use that $\hat{R}$ is a faithfully flat $\tilde{R}$-module as established in 12.5.25 and 12.6.9.) Now (η) implies: $g_i(z) \equiv 0 \bmod f(z)$ in $\tilde{R}$ ($i = 1,\dots,m$) and 12.6.13 is proved.

12.6.14 PROOF OF THEOREM 12.6.11 CONTINUED.

We already established that condition (I) of 12.6.8 is

satisfied in our situation. Therefore, to finish the proof of Artin's Theorem it suffices to show that condition (II) of 12.6.8 is satisfied too.

So consider a tuple $(f(Z), \alpha_1(Z), \ldots, \alpha_N(Z), \beta(Z), \hat{z})$ with the properties (i) - (iv) of 12.5.8.

Take $\hat{y}_i \in \hat{R}$ such that $\alpha_i(\hat{z}) = \hat{y}_i.\beta(\hat{z})$ and put $\hat{y} = (\hat{y}_1,\ldots,\hat{y}_N)$. Define $g(Z) = X_n.\beta(Z).(\frac{\partial f}{\partial Z_{t+1}})^2$ and observe that $g(\hat{z}) \neq 0$ and that the polynomials $f(Z)$ and $\alpha_i(Z) - Y_i.\beta(Z)$ in $R[Y,Z]$, $Y = (Y_1,\ldots,Y_N)$, $i = 1,\ldots,N$, have $(\hat{y},\hat{z})$ as a common zero. So by the Division Lemma 12.6.13 there is $(y,z_1,\ldots,z_t,z'_{t+1})$ in $\tilde{R}^{N+t+1}$ arbitrarily close to $(\hat{y},\hat{z})$ which is, in $\tilde{R}$ modulo $g(z_1,\ldots,z_t,z'_{t+1})$, a common zero of the polynomials $f(Z)$ and $\alpha_i(Z) - Y_i.\beta(Z)$, $i = 1,\ldots,N$. In particular $f(z_1,\ldots,z_t,z'_{t+1}) =$
$dX_n.\beta(z_1,\ldots,z_t).\frac{\partial f}{\partial Z_{t+1}}(z_1,\ldots,z_t,z'_{t+1})^2$ for some $d \in \tilde{R}$, which by 12.5.4 (i) $\Rightarrow$ (ii) gives us a $z_{t+1} \in \tilde{R}$ with

(1) $\quad f(z_1,\ldots,z_{t+1}) = 0$ and $z_{t+1} \equiv z'_{t+1} \mod \beta(z_1,\ldots,z_t)$.

Furthermore by choosing $z_1,\ldots,z_t$ sufficiently close to $\hat{z}_1,\ldots,\hat{z}_t$ we may assume that $\beta(z_1,\ldots,z_t) \neq 0$. Now put $z = (z_1,\ldots,z_{t+1})$. The proof will be finished if we show that $\alpha_i(z) \equiv 0 \mod \beta(z)$ in $\tilde{R}$ for $i = 1,\ldots,N$. Note that $\alpha_i(z_1,\ldots,z_t,z'_{t+1}) - y_i.\beta(z) \equiv 0 \mod g(z_1,\ldots,z_t,z'_{t+1})$ in $\tilde{R}$. Using the definition of $g(Z)$ and the congruence in (1) we find:

(2) $\quad \alpha_i(z_1,\ldots,z_t,z_{t+1}) \equiv 0 \mod \beta(z)$ in $\tilde{R}$ for $i = 1,\ldots,N$,

so we have proved Artin's Theorem.

12.6.15 Very strong limitations are known on rings with AP, some of which are given in the exercise below. Our definition in 12.6.6 of a ring having AP referred to an inclusion of the ring in the completion of a distinguished subring. This is easily remedied by giving a definition which does not have this defect and conforms better with the literature on the subject.

We consider in the following pairs (A,m) consisting of a ring

A with a finitely generated ideal $\mathfrak{m}$ such that A is separated in its $\mathfrak{m}$-adic topology. We write $\hat{A}$ for the $\mathfrak{m}$-adic completion and consider A as a dense subring of $\hat{A}$.

DEFINITION. $(A,\mathfrak{m})$ is called an approximation ring if $A \to \hat{A}$ has the specialization property.

If $(R,\mathfrak{m})$ is an integrally closed noetherian local domain with henselization $(\tilde{R},\tilde{\mathfrak{m}})$, then $R \subset \tilde{R} \subset \hat{R} = \hat{\tilde{R}}$, cf. 12.5.24, and clearly $\tilde{R}$ has AP in the sense of 12.6.6 if and only if $(\tilde{R},\tilde{\mathfrak{m}})$ is an approximation ring.

Nonlocal approximation rings are also important. Some properties of approximation rings are discussed in the following exercise.

12.6.16 EXERCISE. Let $(A,\mathfrak{m})$ be an approximation ring. Prove:

(a) Each solution in $\hat{A}$ of a finite system of polynomial equations over A can be approximated arbitrarily closely, in the $\mathfrak{m}\hat{A}$-adic sense, by solutions in A.

(b) If a finite system of polynomial equations and inequations over A is soluble in $\hat{A}$, then it is soluble in A.

(c) $\mathfrak{m} \subset \operatorname{rad} A$, and $(A,\mathfrak{n})$ is an approximation ring for each finitely generated ideal $\mathfrak{n} \subset \mathfrak{m}$.

(d) $(A,\mathfrak{m})$ is henselian, where Definition 12.5.2 is extended to the nonlocal case.

(e) If A is a domain, so is $\hat{A}$, A is algebraically closed in $\hat{A}$ and $Q(\hat{A})$ is a separable extension of $Q(A)$.

(f) If A is a domain, $\Phi(C)$ is a C-constructible property and $Q(\hat{A}) \models \Phi(c)$ for a tuple c in $\hat{A}$, then there is an A-algebra morphism $\psi\colon A[c] \to A$ with $Q(A) \models \Phi(\psi(c))$.

(g) If A is a domain and $\hat{A}$ is normal, then A is normal.

(h) If $\hat{A}$ is noetherian, so is A.

For the remaining parts we also assume that A is noetherian. Write $\mathfrak{R}(m)$ for the radical of m.

(i) $(A,\mathfrak{R}(m))$ is an approximation ring.

(j) For each ideal α of A the pair $(A/\alpha, m+\alpha/\alpha)$ is an approximation ring.

(k) For each monic polynomial $f(T) \in A[T]$, $f \neq 1$, and $t = T + (f)$ the pair $(A[t], mA[t])$ is an approximation ring.

(l) For each A-algebra B which is module-finite over A the pair (B, mB) is an approximation ring.

Hints: For (a) use the proof of 12.6.7; (b) follows from (a) and the specialization property; for (c), (d), (e), (f), (g) express the desired conclusion in terms of equations and inequations, and apply (b); for (h), take any subset of A and obtain a finite subset generating the same ideal in $\hat{A}$, then show it also generates the same ideal in A. For (i), (j), (k) use properties of completions of noetherian rings like flatness, see section 2.2; (l) follows from (j) and (k).

12.6.17 The parts (e) and (h) of the above exercise have as consequences the well-known facts that $K[X]^{\sim}$ as described in 12.6.9 is noetherian, and algebraically closed in $K[[X]]$.

12.6.18 HISTORICAL REMARKS. Artin's Theorem 12.6.11 has been generalized and strengthened in many directions, e.g. in [El]. Artin's original treatment can be found in [Ar 68] and [Ar 69]. We shall resist the temptation to say much more on this fascinating subject, except to point out that approximation theorems are important at various places in algebraic geometry, and that Artin, in [Ar 69], also proved the mixed characteristic case: 12.6.11 remains true if k is replaced by a so-called excellent discrete valuation ring V (and $k[[X_1,\dots,X_n]]$ by $\hat{V}[[X_1,\dots,X_n]]$); excellent in this case just means that the fraction field of $\hat{V}$ is a separable extension of the fraction field of V. Recently a proof has appeared of the conjecture that all excellent henselian local rings are approximation rings [Po 85], [Po 86]. See also [Rott].

12.7 FROM NOETHERIAN TO FINITELY GENERATED ALGEBRAS

12.7.1 In this section we shall prove Lemma V of 12.0 as a corollary of a more general result which may have some independent interest. Its proof shows a nice interplay of constructible properties with Artin's Approximation Theorem.

12.7.2 PROPOSITION. Let R be a ring finitely generated over a prime field $\mathbf{F}$, and $x_1,\dots,x_n \in R$. If there is a noetherian R-algebra A with $ht(x_1,\dots,x_n)A = n$, then there is an R-algebra A′, finitely generated as a ring over a subfield $K \subset A'$, which satisfies $ht(x_1,\dots,x_n)A' = n$.

PROOF. Localizing A at a prime ideal of height n lying above $(x_1,\dots,x_n)A$ we may assume that A is local of dimension n. Then completing A at its maximal ideal and dividing out by a suitable minimal prime we reduce to the case that A is a complete local noetherian domain containing a field, namely $\mathbf{F}$. By 9.3.3, A is module-finite over its subring $K[[\phi x_1,\dots,\phi x_n]]$

where K is a field of representatives of A and $\phi x_1,\dots,\phi x_n$ are analytically independent over K, ϕ: R → A indicating the structural morphism.

Write $R = \mathbf{F}[y_1,\dots,y_q]$ with $q \ge n$, $y_i = x_i$ for $i = 1,\dots,n$, and let $G_1(Y),\dots,G_m(Y) \in \mathbf{Z}[Y]$, $Y = (Y_1,\dots,Y_q)$, generate the ideal of $\mathbf{Z}[Y]$ consisting of all G(Y) with G(y) = 0, where $y = (y_1,\dots,y_q)$.

The ring A is integral over $K[[\phi x_1,\dots,\phi x_n]]$, so each $\phi(y_i)$ satisfies an equation

$$Y_i^{d_i} + c_{i1}Y_i^{d_i-1} + \dots + c_{id_i} = 0, \quad c_{ij} \in K[[\phi x_1,\dots,\phi x_n]].$$

(For $i = 1,\dots,n$, we take $Y_i - \phi(x_i) = 0$ as our equation.) We now introduce new variables C_{ij} ($1 \le i \le q$, $1 \le j \le d_i$) and put

$$C = (C_{11},\dots,C_{1d_1},\dots,C_{q1},\dots,C_{qd_q}),$$

$$c = (c_{11},\dots,c_{1d_1},\dots,c_{q1},\dots,c_{qd_q}),$$

$G_{m+i}(C,Y) = Y_i^{d_i} + C_{i1}Y_i^{d_i-1} + \dots + C_{id_i} \in \mathbf{Z}[C,Y]$, $1 \le i \le q$. Consider the C-property $\Phi(C|G_1,\dots,G_m,G_{m+1},\dots,G_{m+q};1)$, which is constructible by 12.2.10, so the disjunction of finitely many C-systems. Since $Q(A) \models \Phi(c|G_1,\dots,G_{m+q};1)$, the tuple c is a solution for at least one of these C-systems, say

$$H_1(c) = \dots = H_k(c) = 0, \ H(c) \ne 0.$$

Let B be the ring $K[\phi x_1,\dots,\phi x_n]^{\sim} \subset K[[\phi x_1,\dots,\phi x_n]]$. By Artin's Theorem 12.6.11 and the proof of 12.6.7, there is a specialization ψ: B[c] → B with $\psi(H(c)) = H(\psi(c)) \ne 0$. Because trivially $H_1(\psi(c)) = \dots = H_k(\psi(c)) = 0$ and B is a domain, we have $Q(B) \models \Phi(\psi(c)|G_1,\dots,G_{m+q};1)$. By definition of Φ this means there is a q-tuple $y' = (y'_1,\dots,y'_q)$ in an extension field of Q(B) such that

$$G_1(y') = \dots = G_m(y') = G_{m+1}(\psi(c),y') = \dots =$$
$$= G_{m+q}(\psi(c),y') = 0.$$

The last q equations imply that the y'_i are integral over the subring $K[\phi x_1,\dots,\phi x_n,\psi(c)]$ of B, hence, see 12.5.22, they are integral over a subring B'[1/s] of B, where $B' \supset K[\phi x_1,\dots,\phi x_n]$ is module finite over $K[\phi x_1,\dots,\phi x_n]$, and $s \in B'\setminus$(maximal ideal of B).

Put $A' = B'[1/s, y'_1, \ldots, y'_q]$, and let $\phi'\colon R \to A'$ be the ring homomorphism sending y_i to y'_i (well defined because y' satisfies all relations satisfied by y over $\mathbb{Z}$). So A' is an R-algebra via ϕ' and A' is finitely generated over K (because B' is). Furthermore $\phi'(x_i) = \phi'(y_i) = y'_i = y_i = \phi(x_i)$ for $i = 1, \ldots, n$.

From the diagram of ring extensions

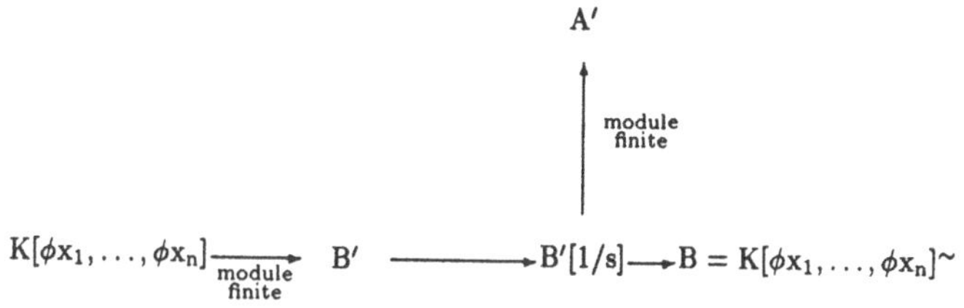

and going up properties of module-finite extensions [Ma 80, (5.E)] one immediately deduces that $ht(x_1, \ldots, x_n)A' = ht(\phi x_1, \ldots, \phi x_n)A' = ht(\phi x_1, \ldots, \phi x_n)K[\phi x_1, \ldots, \phi x_n] = n$.

This finishes the proof of the proposition.

12.7.3 PROOF OF LEMMA V.

We are in the following situation: ξ is a finite system of polynomial equations over $\mathbb{Z}$ in $n+q$ variables $(X,Y) = (X_1, \ldots, X_n, Y_1, \ldots, Y_q)$, and $(x,y) = (x_1, \ldots, x_n, y_1, \ldots, y_q)$ is a solution of ξ in A where A is a noetherian ring containing a field and $ht(x_1, \ldots, x_n)A = n$. Let $\mathbf{F}$ = prime field of A. We have to show that there is a ring A' which is finitely generated over a subfield $K \supset \mathbf{F}$ such that ξ has a solution (x', y') in A' with $ht(x')A' = n$.

In the previous proposition we simply take $R = \mathbf{F}[x,y] \subset A$. Then A', as in the proposition, has the required properties, with (x', y') = image of (x,y) under the structural morphism $R \to A'$.

12.7.4 REMARK. With a bit more effort we can take the R-algebra A' in Proposition 12.7.2 finitely generated, in other words, one can take $K = \mathbf{F}$, see also [Ko 88]. This raises the question whether the following natural generalization of 12.7.2 holds:

Let R be a noetherian ring and $x_1, \ldots, x_n \in R$. If there is a noetherian R-algebra A with $ht(x_1, \ldots, x_n)A = n$, does it follow that there is a finitely generated R-algebra A' with $ht(x_1, \ldots, x_n)A' = n$? This is in general not the case, as shown by an example in [Ko 88].

13 USES OF BIG COHEN MACAULAY MODULES

These are numerous, but in this final chapter we treat only two, of unequal importance. The first is a proof of the New Intersection Theorem 13.1.1. There exist other proofs, [PS 74, Th. 1], [Ho 75b, p. 171], [Ro 76], [Ro 80c, Th. 1], but the present one, due to Foxby [Fo 77b, §1], quickly and elegantly establishes this result in equal characteristic. In a major breakthrough, P. Roberts recently succeeded in deriving the unequal characteristic case from the result in equal characteristic p. His proof, which uses the intersection theory developed in algebraic geometry by Fulton and MacPherson [Fu], we can only reproduce in part, 13.1.2. An important consequence of the New Intersection is the Homological Height Theorem 8.4.3 and its derivates discussed in section 8.4. Two other results which follow are the Auslander Zerodivisor Theorem 13.1.11 and an answer to a question of Bass 13.1.7.

Our second application is more parochial and establishes the nonvanishing of Bass numbers in their allowed range, Theorem 13.2.5. Again there exist other proofs of this result [Ro 76, Th. 2], [Ro 80b, Ch. 2, Prop. 4.3] but ours, which is taken from [FFGR, Th. 1.1], has an amusing feature. A weak kind of Big Cohen-Macaulay module suffices, and this is shown to exist independently from the characteristic, Proposition 13.2.1.

The existence of Big Cohen-Macaulay modules allows one to prove a stronger version of New Intersection [EG, Th. 1.13] which is needed for the proof of the Syzygies Conjecture. The latter therefore remains unproven in mixed characteristic, although at present it is being investigated whether the improved New Intersection would also yield to Roberts' reduction to characteristic p. Since the Syzygies Conjecture is the subject of [EG], a companion volume to the present book, we say no more about it.

In 9.1.3 we saw that the existence of Big Cohen-Macaulay modules also implies the truth of the Monomial Conjecture. Hochster has shown, in a rich and detailed paper [Ho 83], that the latter implies New Intersection, even in its stronger form. But since the Monomial Conjecture and the closely related Direct Summand Conjecture are still open in mixed characteristic, these issues remain undecided, though in this direction a promising theory of tightly closed ideals is being developed by Hochster and Huneke [HH].

13.1 NEW INTERSECTION THEOREMS AND A FEW CONSEQUENCES

13.1.1 THEOREM. (New Intersection) Let (A,m,k) be an equicharacteristic noetherian local ring. Let $L_\bullet = 0 \to L_s \to \ldots L_0 \to 0$ be a finite complex of finitely generated free A-modules. If $\ell(H_i(L_\bullet)) < \infty$ for $i = 0,\ldots,s$ and $L_\bullet$ is not exact, then $\dim A \le s$.

PROOF. Let M be a Big Cohen-Macaulay module over A, i.e. $mM \neq M$ and $\text{E-dp } M = \text{dp } M = \dim A$. Since $\text{E-dp } M \le s$ according to Corollary 6.1.4, the theorem is proved.

This result, known as New Intersection, was proved by Peskine-Szpiro in [PS 74, Th. 1] and, independently, by P. Roberts in [Ro 76] for positive characteristic and in [Ro 80c, Th. 1] for characteristic 0. The present slick proof is due to Foxby [Fo 77b, §1].

In a recent breakthrough, P. Roberts succeeded in extending the result to mixed characteristic. In his note [Ro 78a], he spoke of the Intersection Theorem, but we shall stick to the old terminology and reserve the name Intersection Theorem for its corollary, Theorem 8.4.4. We only give a sketch of his proof.

13.1.2 THEOREM. The above is also true when $(A,\mathfrak{m},k)$ is of mixed characteristic.

PROOF. It is easy to see that we may assume A to be complete in its $\mathfrak{m}$-adic topology with perfect residue class field and that we may suppose A is an integral domain by dividing out a prime $\mathfrak{p}$ with $\dim A/\mathfrak{p} = \dim A$. If $p = \operatorname{char} k$, the element $p = p.1$ is a non zerodivisor of A so that the ring $\bar{A} = A/(p)$ has characteristic p and dimension $d = \dim A - 1$.

Now suppose the theorem is false; clearly there is then an offending $L_\bullet$ of length $s = d < \dim A$ and with $H_0(L_\bullet) \neq 0$. Its reduction $\bar{L}_\bullet = L_\bullet/pL_\bullet$ is a nonexact complex (e.g. by 7.2.14 with $M = A/(p)$) over $\bar{A}$ with finite length homology and $H_0(\bar{L}_\bullet) \neq 0$. We can apply Frobenius, and 11.3.9 tells us that its asymptotic Euler-Poincaré characteristic $\chi_\infty(\bar{L}_\bullet) > 0$. On the other hand, the fact that $\bar{L}_\bullet$ is $L_\bullet$ reduced mod p, allows one to conclude that $\chi_\infty(\bar{L}_\bullet) = 0$. This is a contradiction, so $\dim A \leq s$.

Roberts' proof that $\chi_\infty(\bar{L}_\bullet) = 0$ is based on the properties of local Chern characters and a local Riemann-Roch Theorem. These matters pertain to the new treatment of intersections in algebraic geometry due to

W. Fulton and R.D. MacPherson and are explained in the authoritative treatise [Fu, Ch. 18]. This Riemann-Roch Theorem was partly inspired by the type of problem in local algebra which is discussed in our book. It fell to P. Roberts to bring the power of these methods to bear and prove New Intersection in full generality [Ro 87a], [Ro 89]. Elsewhere [Ro 85], [Ro 87b], he used them to settle a question of Serre, as mentioned in the Preface.

A further remark is in order. Though we announced 13.1.2 as a reduction to the known case of characteristic p, just like its author did in [Ro 87a], our proof is arranged in such a way that it actually furnishes a proof whenever char k = p. One induces on dim A, the case of dimension 0 being trivial. In this integral domain A one just divides out by an element $\neq 0$ (not necessarily p = p.1) to reduce to a ring of characteristic p and lower dimension, and the arguments that $\chi_\infty(\bar{L}_\bullet) > 0$ and $\chi_\infty(\bar{L}_\bullet) = 0$ work as before. Using techniques as in Chapter 12, one can now also prove New Intersection in equal characteristic 0. Roberts, of course fully aware of this, has therefore provided a new proof in all three cases, to set alongside his old one in characteristic p [Ro 76] (and the one by Peskine-Szpiro [PS 74, Th. 1]) as well as his "analytic" proof over the complex numbers [Ro 80c, Th. 1].

At long last we are able to prove the Homological Height Theorem 8.4.3 and hence its consequences in Chapters 8 and 9.

13.1.3 THEOREM. New Intersection implies Theorem 8.4.3.

PROOF. Since we may assume pd M < ∞, take a minimal free resolution of M and let $L_\bullet = 0 \to L_s \to \ldots \to L_0 \to 0$ be the concomitant complex with its homology concentrated in 0 and $H_0(L_\bullet) = M$. If Ann M = $\mathfrak{a}$ and $\mathfrak{B}$ is a prime in B minimally containing $\mathfrak{a}B$, then $L_\bullet \otimes_A B_{\mathfrak{B}}$ is a finite free complex over the ring $B_{\mathfrak{P}}$ with finite length homology. Since $H_0(L.\otimes_A B_{\mathfrak{P}}) = M_{\mathfrak{P}} \neq 0$, this complex is not exact, so ht $\mathfrak{P}$ = dim $B_{\mathfrak{P}} \leq s$ = pd M.

There exist several versions of New Intersection for finite free complexes, some of which are only apparently stronger than 13.1.1, others may be genuinely so. They are most easily proved by plugging a Big Cohen-Macaulay module into the appropriate version of the Acyclicity Lemma [Fo 77b, Th. 1.2], [Ho 83, Th. 2.6]. Whether they are also susceptible to proof in mixed characteristic by reduction to characteristic p, is under investigation. While not dealing with these, we prove a version which features (not necessarily finitely generated) flat modules. Here our Big Cohen-Macaulay module is not inserted into an Acyclicity Lemma of Chapter 6, but rather into the somewhat more general Theorem 7.2.14. We retain the notation f_+^M of that theorem.

13.1.4 THEOREM. Let $F_\bullet = 0 \to F_s \to \ldots \to F_0 \to 0$ be a complex of flat modules over an equicharacteristic noetherian local ring (A,m,k). If $F_\bullet$ is not exact and Supp $H_i(F_\bullet) \subset \{m\}$ for $i = 0,\ldots,s$, then dim $A \leq f_+^k \leq s$.

PROOF. First invoke 7.2.14 (ii) with M = A to ensure that $F_\bullet \otimes_A k$ is not exact. Next let M be a Big Cohen-Macaulay A-module, then E-dp M = dim $A < \infty$; 7.2.14 (i) tells us that the complex $F_\bullet \otimes_A M$ is not exact. Clearly, its homology can only be supported at m. Now 7.2.14 (ii) yields dim A = E-dp M = $f_+^k - f_+^M \leq f_+^k \leq s$.

13.1.5 COROLLARY. Let A be as above, and $I^\bullet = 0 \to I^0 \to \ldots \to I^s \to 0$ be a complex of injective A-modules. If each homology of $I^\bullet$ is finitely generated and $I^\bullet$ is not exact, then dim $A \leq s$.

PROOF. Apply the faithfully exact Matlis dual to $I^{\bullet}$. Since each $(I^i)^{\vee}$ is flat and $H_i(I^{\bullet})^{\vee} \cong H_i((I^{\bullet})^{\vee})$ is artinian by 3.4.11, we may apply the theorem to $(I^{\bullet})^{\vee}$ and obtain the result.

13.1.6 COROLLARY. If a ring A as above possesses a finitely generated module M with $\text{id } M < \infty$, then A is Cohen-Macaulay.

PROOF. A minimal injective resolution has length $s = \text{id } M = \text{dp } A$ by 7.1.5. By the previous corollary also $\dim A \leq s$ so that A is Cohen-Macaulay.

This answers a question first raised by Bass and later known as Bass' Conjecture, and strengthens Proposition 9.1.10. The quick proof above, deriving 13.1.5 from 13.1.4 by way of the Matlis dual, is due to J. Bartijn, but close in spirit to certain considerations in Foxby [Fo 82]. It has the (temporary?) disadvantage of depending on the existence of Big Cohen-Macaulay modules, so that we do not know whether the proof goes through in mixed characteristic. Actually, 13.1.5 is true in all cases since it can be derived from New Intersection using dualizing complexes [Ro 80b, Ch. 3]. However, 13.1.6 was originally proved in Peskine-Szpiro [PS 73, Ch. II, Th. 5.1] using the Intersection Theorem 8.4.4, so that it is now verified in general. We omit their proof, which is somewhat lengthy. Anyway, we state

13.1.7 THEOREM. A noetherian local ring which possesses a finitely generated module of finite injective dimension is Cohen-Macaulay.

13.1.8 COROLLARY. A noetherian local ring is Gorenstein if and only if it possesses a cyclic module of finite injective dimension.

PROOF. A Gorenstein ring is a cyclic module of finite injective dimension over itself, so assume $A/\mathfrak{a}$ has finite injective dimension as an A-module for some ideal $\mathfrak{a}$. Putting $N = A$ and $M = A/\mathfrak{a}$ in the second identity of 7.1.6, we find $\beta(A/\mathfrak{a},T) = \mu(A/\mathfrak{a},T^{-1})\mu(A,T)$. Now $\mu^i(A/\mathfrak{a}) = 0$ for $i > \mathrm{id}\, A/\mathfrak{a} = \mathrm{dp}\, A = s$, say, while $\mu^j(A) = 0$ for $j < \mathrm{dp}\, A$. Since $\beta_0(A/\mathfrak{a}) = \dim_k(A/\mathfrak{a}\otimes_A k) = \dim_k k = 1$, we see that $1 = \mu^s(A/\mathfrak{a})\mu^s(A)$ so that $\mu^s(A/\mathfrak{a}) = \mu^s(A) = 1$. But A is Cohen-Macaulay, hence $s = d = \dim A$. Thus A is Gorenstein by 10.1.8.

This was first proved in [PS 73, Ch. II, Th. 5.5]. A different approach is taken in [Le, Th. 1.9]. Another consequence of 13.1.4 is an extension to modules of finite flat dimension of the Auslander Zerodivisor property mentioned at the beginning of section 8.5.

13.1.9 THEOREM. Let $(A,\mathfrak{m},k)$ be an equicharacteristic noetherian local ring and M an A-module with $\mathrm{fd}\, M < \infty$ and T-codp $M < \infty$. Then a regular sequence on M is regular in A.

PROOF. It is clearly enough to show that a zerodivisor in A is a zerodivisor on M. So let $x \in \mathfrak{m}$ be such a zerodivisor and put $\mathfrak{a} = \mathrm{Ann}(x) \neq 0$. Then $\mathfrak{m} \in \mathrm{Supp}\, \mathfrak{a}$, and since $\mathfrak{m} \in \mathrm{supp}\, M$, we see that $X = \mathrm{Supp}\, \mathfrak{a} \cap \mathrm{supp}\, M \neq \emptyset$; we can therefore localize at a prime $\mathfrak{p}$ which is minimal in X. A little reflection shows that we need only solve our problem for the module $M_{\mathfrak{p}}$ over $A_{\mathfrak{p}}$, i.e. show that the zerodivisor x/1 in $A_{\mathfrak{p}}$ is a zerodivisor on $M_{\mathfrak{p}}$. We therefore shift gears and assume from the outset that $X = \{\mathfrak{m}\}$.

Choose a prime $\mathfrak{p} \in \mathrm{Ass}\, \mathfrak{a}$. Since $A/\mathfrak{p} \subset \mathfrak{a}$, also $\mathrm{Supp}\, A/\mathfrak{p} \cap \mathrm{supp}\, M = \{\mathfrak{m}\}$. Moreover $\mathfrak{p} \in \mathrm{Ass}\, \mathfrak{a} \subset \mathrm{Ass}\, A$, so dp A ≤ $\dim^{-}A/\mathfrak{p}$ ≤ dim $A/\mathfrak{p}$, by 8.3.9.

Let $0 \to F_s \to \ldots \to F_0 \to M \to 0$ be a shortest flat resolution of M and $F_{\bullet}$ the corresponding complex whose 0-th homology is M. Now $\mathfrak{m} \in$ supp M, so $F_{\bullet}\otimes_A k$ is not exact; neither is therefore the complex $F_{\bullet}\otimes_A A/\mathfrak{p}$ and its homology is only supported in $\mathfrak{m}$. Hence 13.1.4 tells us that dim $A/\mathfrak{p} \le f_k^+ =$ $\mathrm{fd}^k M$. On the other hand we find from 7.1.3 with N = A that $\mathrm{fd}^k M =$ dp A - E-dp M. Combined with the fact that dp A ≤ $\mathrm{fd}^k M$ which stems from the first two inequalities, this implies E-dp M = 0. In other words, $\mathfrak{m} \in$ Ass M, and since $x \in \mathfrak{m}$, it is a zerodivisor on M.

We do not know whether this result of Bartijn's [Ba, Ch. IV, Prop. 5.8] is true in mixed characteristic, since we used the "Flat Intersection" 13.1.4, which was proved using Big Cohen-Macaulay modules. However, the original Auslander Zerodivisor Conjecture has now been proved in general. The reader is invited to furnish a proof either as a simplified version of the above, this time using 13.1.1 and 13.1.2, or else to follow Peskine-Szpiro, who derived the result from the Intersection Theorem 8.4.4 [PS 73, Ch. II, Th. 3.1].

13.1.10 THEOREM. Let M be a finitely generated module over a noetherian local ring A. If pd M < ∞, then every M-sequence is an A-sequence.

This result yields a characterization of integral domains, also due to Auslander [PS 73, Ch. II, Cor. 3.3]:

13.1.11 EXERCISE. Let A be a noetherian local ring. Show that A is an integral domain if and only if there exists a prime ideal $\mathfrak{p}$ with pd $A/\mathfrak{p} < \infty$.

13.2 NONVANISHING OF BASS NUMBERS

We shall need a weak derivate of Big Cohen-Macaulay modules, which has the advantage of also having been established in the mixed characteristic case.

13.2.1 PROPOSITION. Let (A,m,k) be a d-dimensional noetherian local ring. We can find a system of parameters $x_1,\ldots,x_d$ and a module $N \neq mN$ such that

(i) $x_dN = 0$;

(ii) $x_1,\ldots,x_{d-1}$ is a regular sequence on N;

(iii) E-dp N = dp N = d-1.

PROOF. If A is equicharacteristic, there exists a Big Cohen-Macaulay module M for every s.o.p. $x = x_1,\ldots,x_{d-1},x_d$. Put $N = M/x_dM$. Then $d-1 \leq \text{dp } N \leq \text{E-dp } N = \text{E-dp } M - 1 \leq d-1$ which proves (iii). (i) and (ii) are obvious.

In mixed characteristic, if char k = p, so is char A/(p) and we distinguish two cases. If dim A/(p) = d, then any s.o.p. x of A is one for A/(p) and we take an x-regular module M and proceed as before. If dim A/(p) = d-1, we form the system of parameters $x_1,\ldots,x_{d-1},p$ of A and notice that $x_1,\ldots,x_{d-1}$ is a s.o.p. for A/(p). There exists a balanced Big Cohen-Macaulay A/(p)-module N. As an A-module, N is easily seen to satisfy (i), (ii) and (iii) if we set $x_d = p$.

The next lemma slightly varies [FFGR, Lemma 1.4].

13.2.2 LEMMA. Let $(A,\mathfrak{m},k)$ be a noetherian local ring, complete in its $\mathfrak{m}$-adic topology, and let $x_1,\ldots,x_n$ be a regular sequence on an A-module $N \neq \mathfrak{m}N$. Put $\mathfrak{a} = (x_1,\ldots,x_n)$. Then $\mathrm{Ext}^i_A(N/\mathfrak{a}N,M) \neq 0$ for every finitely generated nonnull module M of depth 0 for $0 \le i \le n$.

PROOF. We exploit the Matlis dual for the complete ring, first observing that $M^\vee$ is artinian and that $M^{\vee\vee} \simeq M$ by 3.4.12 and that T-codp $M^\vee$ = E-dp M = dp M = 0 by in view of 3.4.14, so that $\mathfrak{m}M^\vee \neq M^\vee$. Moreover $\mathrm{Tor}^A_i(-,M^\vee)^\vee \simeq \mathrm{Ext}^i_A(-,M^{\vee\vee}) \simeq \mathrm{Ext}^i_A(-,M)$ for all i by the same proposition. It therefore suffices to prove that $\mathrm{Tor}^i_A(N/\mathfrak{a}N,M^\vee) \neq 0$ for $0 \le i \le n$.

We induce on n, the case n = 0 being easy: our data tell us that $N/\mathfrak{m}N \neq 0$ so that $N/\mathfrak{m}N \otimes_k M^\vee/\mathfrak{m}M^\vee$ is the tensor product of two nonnull vectorspaces, hence is $\neq 0$. So is therefore $N \otimes_A M^\vee$.

Putting $\mathfrak{b} = (x_1,\ldots,x_{n-1})$, we may assume that $\mathrm{Tor}^A_i(N/\mathfrak{b}N,M^\vee) \neq 0$ for $0 \le i \le n-1$. Since x_n acts injectively on $N/\mathfrak{b}N$, we find exact sequences $\mathrm{Tor}^A_i(N/\mathfrak{a}N,M^\vee) \to \mathrm{Tor}^A_{i-1}(N/\mathfrak{b}N,M^\vee) \xrightarrow{.x_n} \mathrm{Tor}^A_{i-1}(N/\mathfrak{b}N,M^\vee)$ for $1 \le i \le n$. Now $\mathfrak{m}$ is the only prime in the support of the artinian module $M^\vee$, hence of $\mathrm{Tor}^A_{i-1}(N/\mathfrak{b}N,M^\vee)$ so that x_n is a zerodivisor on these nonnull Tor's. The lemma follows, since the case i = 0 is easily taken care of as above.

We need two more lemma's.

13.2.3 LEMMA. Let M be an A-module over a noetherian local ring $(A,\mathfrak{m},k)$. If $\mu^i(M) = 0$ for a certain integer $i \ge 0$, then $\mathrm{Ext}^i_A(T,M) = 0$ for every A-module T with Supp T = $\{\mathfrak{m}\}$.

PROOF. Write $T = \varinjlim_j T_j$ with $T_j \subset T$ finitely generated submodules. Since $\mathrm{Ext}^i_A(T,M) = \varprojlim_j \mathrm{Ext}^i_A(T_j,M)$, we are reduced to proving the result for finitely generated T, still only supported in m. Such a T has finite length, and we conclude by looking at a composition series of T and the corresponding exact sequences of the Ext's, since $\dim_k \mathrm{Ext}^i_A(k,M) = \mu^i(M) = 0$.

13.2.4 LEMMA. Given a noetherian local ring (A,m,k) of depth 0, there exists an exponent $t \geq 0$ such that m^t has a direct summand isomorphic to k.

PROOF. Since $m \in \mathrm{Ass}\, A$, there is an $x \in A$, $x \neq 0$, which is annihilated by m. Choose t such that $x \in m^t \setminus m^{t+1}$. Put $\bar{x} = x \bmod m^{t+1}$, and prolong to a basis $\bar{x},\bar{y}_1,\ldots,\bar{y}_n$ of the k-vectorspace m^t/m^{t+1}. An arbitrary lifting to m^t yields a minimal set of generators $x,y_1,\ldots,y_n$ of this ideal. We need to prove that $(x) \cap (y_1,\ldots,y_n) = (0)$. If $ax \in (y_1,\ldots,y_n)$, then a cannot be a unit, so must be in m. Then $ax = 0$ and we are through.

We can now prove the main result of the section, [FFGR, Th. 1.1]. This nonvanishing of the Bass numbers is more rigid than the behaviour of the related local cohomology functors, which can very well vanish between either extreme, as we mentioned after 10.2.2.

13.2.5 THEOREM. Let (A,m,k) be a noetherian local ring and M a finitely generated A-module. Then $\mu^i(M) \neq 0$ if and only if $\mathrm{dp}\, M \leq i \leq \mathrm{id}\, M$.

PROOF. Since $\hat{A}$ is faithfully flat over A and M is finitely generated, $\hat{A}\otimes_A \mathrm{Ext}^i_A(k,M) \simeq \mathrm{Ext}^i_{\hat{A}}(k,\hat{M})$ and $\mathrm{dp}_{\hat{A}}\, \hat{M} = \mathrm{dp}_A M$. Furthermore, by 8.2.8, $\dim_{\hat{A}} \hat{M} = \dim_A M$, so we may assume A is complete.

If $\mathrm{id}\, A = \infty$, then $\mu^i(M) \neq 0$ for $i \geq d = \dim A$ by 8.5.6. If $\mathrm{id}\, M < \infty$, then $\mu^i(M) \neq 0$ for $i = \mathrm{id}\, M = \mathrm{dp}\, A \leq d$ (7.1.5 and 8.3.8) and

$\mu^i(M) = 0$ for $i > \text{id } M$. Always $\mu^i(M) = 0$ for $i < \text{dp } M$. A little reflexion shows that it is sufficient to prove that $\mu^i(M) \neq 0$ for $\text{dp } M \le i \le d-1$; incidentally we shall then have proven that the possession of a finitely generated nonnull module of finite injective dimension forces the ring A to be either Cohen-Macaulay or almost Cohen-Macaulay, i.e. $\dim A - \text{dp } A = 1$. This is a weak form of 13.1.7 which is of course entirely fitting: we are going to use N in 13.2.1, a poor man's Big Cohen-Macaulay module.

Proceeding as explained in section 8.5, take a largest possible common regular sequence on M and on A, say $x_1,\ldots,x_r$, and call the ideal which these elements generate $\mathfrak{b}$. Putting $\bar{A} = A/\mathfrak{b}$ and $\bar{M} = M/\mathfrak{b}M$, we see that $\text{dp}_A M = \text{dp}_A \bar{M} + r = \text{dp}_{\bar{A}} \bar{M} + r$ and $\dim A = \dim \bar{A} + r$. Also $\mu_A^{i+r}(M) = \mu_{\bar{A}}^i(\bar{M})$ by 3.3.6 while $\text{id}_A M = \text{id}_{\bar{A}} \bar{M} + r$. We may therefore assume that $\text{dp } M = 0$ or that $\text{dp } A = 0$.

In the first case, consider the s.o.p. $x_1,\ldots,x_d$ and the module N guaranteed by Proposition 13.2.1 and with $\mathfrak{a} = (x_1,\ldots,x_{d-1})$. Now $\text{Ext}_A^i(N/\mathfrak{a}N, M) \neq 0$ for $0 \le i \le d-1$ by 13.2.2. Furthermore $\text{Supp } N/\mathfrak{a}N = \{\mathfrak{m}\}$ because $x_d N = 0$, while $x_1,\ldots,x_d$ form a system of parameters. Lemma 13.2.3 then tells us that $\mu^i(M) \neq 0$ for all i such that $0 \le i \le d-1$, which proves our contention in case $\text{dp } M = 0$.

Assume therefore $\text{dp } M > \text{dp } A = 0$. Clearly $\text{id } M = \infty$ and we shall prove that $\mu^i(M) \neq 0$ implies $\mu^{i+1}(M) \neq 0$. Since $\mu^i(M) \neq 0$ for $i = \text{dp } M$, this will suffice. The short exact sequence $0 \to \mathfrak{m}^t \to A \to A/\mathfrak{m}^t \to 0$ gives rise to isomorphisms $\text{Ext}_A^i(\mathfrak{m}^t, M) \simeq \text{Ext}_A^{i+1}(A/\mathfrak{m}^t, M)$ since $i \ge 1$. By 13.2.4 we can choose t such that there is a splitting $\mathfrak{m}^t \simeq k \oplus \mathfrak{c}$ for some ideal $\mathfrak{c}$, so $\text{Ext}_A^i(\mathfrak{m}^t, M) \simeq \text{Ext}_A^i(k,M) \oplus \text{Ext}_A^i(\mathfrak{c},M)$. But then $\mu^{i+1}(M) \neq 0$ by a composition series argument on $A/\mathfrak{m}^t$.

More is known about the behaviour of the Bass numbers, for which we refer to [Fo 71], [Fo 77b] and [Ro 80b].

REFERENCES

[AF] Anderson, F.W. & K.R. Fuller, Rings and Categories of Modules, Grad. Texts Math. 13, Springer-Verlag, New York etc. 1974.

[AG] Aoyama, Y. & S. Goto, Some special cases of a conjecture of Sharp, J. Math. Kyoto Univ. 26 (1986), pp. 613-634.

[AM] Atiyah, M.F. & I.G. Macdonald, Introduction to Commutative Algebra, Addison-Wesley, Reading Mass. etc. 1969.

[Al] Alfonsi, B., Grade non-noethérien, Commun. Algebra 9 (1981) pp. 811-840.

[Ar 68] Artin, M., On the solution of analytic equations, Invent. Math. 5 (1968), pp. 277-291.

[Ar 69] Artin, M., Algebraic approximation of structures over complete local rings, Publ. Math. Inst. Hautes Étud. Sci. 36 (1969), pp. 23-58.

[Av] Avramov, L.L., Modules of finite virtual projective dimension, Invent. Math. 96 (1989), pp. 71-101.

[BE] Buchsbaum. D.A. & D. Eisenbud, What makes a complex exact?, J. Algebra 25 (1973), pp. 259-268.

[BS] Bartijn, J. & J.R. Strooker, Modifications monomiales, Sémin. Dubreil-Malliavin 1982, Lect. Notes Math. 1029, Springer-Verlag, Berlin etc. 1983, pp. 192-217.

[Ba] Bartijn, J., Flatness, completion, regular sequences: un ménage à trois, Thesis, Univ. of Utrecht 1985.

[Barg] Barger, S.F., A theory of grade for commutative rings, Proc. Am. Math. Soc. 36 (1972) pp. 365-368.

[Bas] Bass, H, On the ubiquity of Gorenstein rings, Math. Z. 82 (1963), pp. 8-28.

[Bo 61a] Bourbaki, N., Algèbre commutative, Ch. 1 et 2, Modules plats et Localisations, Actual. Sci. Ind. 1290, Hermann, Paris 1961.

[Bo 61b] Bourbaki, N., Algèbre commutative, Ch. 3 et 4, Graduations, filtrations et topologies et Idéaux premiers associés et décomposition primaire, Actual. Sci. Ind. 1293, Hermann, Paris 1961.

[Bo 62] Bourbaki, N., Algèbre, Ch. 2, Algèbre linéaire, Actual. Sci. Ind. 1236, Hermann, 3ème éd., Paris 1962.

[Bo 80] Bourbaki, N., Algèbre, Ch. 10, Algèbre homologique, Masson, Paris etc. 1980.

[Bo 83] Bourbaki, N., Algèbre commutative, Ch. 8 et 9, Dimensions et Anneaux locaux noethériens complets, Masson, Paris etc. 1983.

[Bor] Borel, A. (et al.), Intersection cohomology, Prog. Math. 50, Birkhäuser, Boston etc. 1984.

[Br] Bruns, W., The Eisenbud-Evans generalized Principal Ideal Theorem and determinantal ideals, Proc. Am. Math. Soc. 83 (1981), pp. 19-24.

[CE] Cartan, H. & S. Eilenberg, Homological Algebra, Princeton Univ. Press, Princeton N.J. 1956.

[Ca] Caruth, A., A short proof of the Principal Ideal Theorem, Q. J. Math. Oxf. II. Ser. 31 (1980), p. 401.

[Co] Cohen, I.S., On the structure and ideal theory of complete local rings, Trans. Am. Math. Soc. 59 (1946), pp. 54-106.

[Cohn] Cohn, P.M., On the free product of associative rings, Math. Z. 71 (1959), pp. 380-398.

[Da] Davis, E.D., Ideals of the principal class, R-sequences and a certain monoidal transformation, Pac. J. Math. 20 (1967), pp. 197-205.

[Da 78] Davis, E.D., Prime elements and prime sequences in polynomial rings, Proc. Am. Math. Soc. 72 (1978), pp. 33-38.

[Dad] Dade, E.C., Localization of injective modules, J. Algebra 69 (1981), pp. 416-425.

[De] Deligne, P., SGA $4\frac{1}{2}$, Lect. Notes Math. 569, Springer-Verlag, Berlin etc. 1977.

[Dr] Dries, L. van den, Algorithms and bounds for polynomial rings, Logic Coll. 78, Stud. Logic Found. Math. 97, North-Holland, Amsterdam etc. 1979, pp. 147-157.

[Du 83] Dutta, S.P., Frobenius and multiplicities, J. Algebra 85 (1983), pp. 424-448.

[Du 89] Dutta, S.P., Syzygies and homological conjectures, Commutative Algebra, MSRI Publ. 15, Springer-Verlag, New York etc. 1989, pp. 139-156.

[EE] Eisenbud, D. & E.G. Evans, A generalized Principal Ideal Theorem, Nagoya Math. J. 62 (1976), pp. 383-402.

[EF] Eagon, J.A. & M.M. Fraser, A note on the Koszul complex, Proc. Am. Math. Soc. 196 (1968), p. 251.

[EG] Evans E.G. & P. Griffith, Syzygies, Lond. Math. Soc. Lect. Note Ser. 106, Cambridge Univ. Press, Cambridge 1985.

[EN] Eagon, J.A. & D.G. Northcott, On the Buchsbaum-Eisenbud theory of finite free resolutions, J. Reine Angew. Math. 262/263 (1973), pp. 205-219.

[El] Elkik, R., Solutions d'équations à coefficients dans un anneau hensélien, Ann. Sci. Éc. Norm. Supér. IV. Sér. 6 (1973), pp. 553-604.

[En] Enochs, E., Flat covers and flat cotorsion modules, Proc. Am. Math. Soc. 92 (1984), pp. 179-183.

[FFGR] Fossum, R. & H.-B. Foxby, P. Griffith,, I. Reiten, Minimal injective resolutions with applications to dualizing modules and Gorenstein modules, Publ. Math. Inst. Hautes Étud. Sci. 45 (1975), pp. 193-215.

[Fo 71] Foxby, H.-B., On the μ^i in a minimal injective resolution, Math. Scand. 29 (1971), pp. 175-186.

[Fo 77a] Foxby, H.-B., Homomorphisms between complexes with applications to the homological theory of modules, Math. Scand. 40 (1977), pp. 5-19.

[Fo 77b] Foxby, H.-B., On the μ^i in a minimal injective resolution II, Math. Scand. 41 (1977), pp. 19-44.

[Fo 79] Foxby, H.-B., Bounded complexes of flat modules, J. Pure Appl. Algebra 15 (1979), pp. 149-172.

[Fo 82] Foxby, H.-B., Complexes of injective modules, Commutative Algebra: Durham 1981, Lond. Math. Soc. Lect. Note Ser. 72, Cambridge Univ. Press, Cambridge 1982.

[GS] Gillet, H. & C. Soulé, Intersection theory using Adams operations, Invent. Math. 90 (1987), pp. 243-277.

[Ga] Gabriel, P., Objets injectifs dans les catégories abéliennes, Sémin. Dubreil, Dubreil-Jacotin, Pisot 1958/59, pp. 17/01-17/32, Paris 1960.

[Go] Golan, J.S., Torsion Theories, Pitman Monogr. Surv. Pure Appl. Math. 29, Longman, Burnt Mill 1986.

[Gr 76] Griffith, P., A representation theorem for complete local rings,

J. Pure Appl. Algebra 7 (1976), pp. 303-315.

[Gr 78] Griffith, P., Maximal Cohen-Macaulay modules and representation theory, J. Pure Appl. Algebra 13 (1978), pp. 321-334.

[Gre] Greuel, G.-M., Dualität in der lokalen Kohomologie isolierter Singularitäten, Math. Ann. 250 (1980), pp. 157-173.

[Gro 61] Grothendieck, A., Éléments de Géométrie Algébrique IV, Publ. Math. Inst. Hautes Étud. Sci. 11, Paris 1961.

[Gro 64] Grothendieck, A., Éléments de Géométrie Algébrique IV, Publ. Math. Inst. Hautes Étud. Sci. 20, Le Bois Marie etc. 1964.

[Gro 67] Grothendieck, A., Local cohomology, Lect. Notes Math. 41, Springer-Verlag Berlin etc. 1967.

[HH] Hochster, M. & C. Huneke, Tight closure, Commutative Algebra, MSRI Publ. 15, Springer-Verlag, New York etc. 1989, pp. 305-324.

[HIO] Herrmann, M. & S. Ikeda, U. Orbanz, Equimultiplicity and blowing up, Springer-Verlag, Berlin etc. 1988.

[HK] Herzog, J. & E. Kunz, Der kanonische Modul eines Cohen-Macaulay-Rings, Lect. Notes Math. 238, Springer-Verlag, Berlin etc. 1971.

[HM] Hochster, M. & J.E. McLaughlin, Splitting theorems for quadratic ring extensions, Ill. J. Math. 27 (1983), pp. 94-103.

[HR 74] Hochster, M. & J.L. Roberts, Rings of invariants of reductive groups acting on regular rings are Cohen-Macaulay, Adv. Math. 13 (1974), pp. 115-175.

[HR 76] Hochster, M. & J.L. Roberts, The purity of the Frobenius and local cohomology, Adv. Math. 21 (1976), pp. 117-172.

[HSV] Herrmann, M. & R. Schmidt, W. Vogel, Theorie der normalen Flachheit, Teubner-Texte Math., Leipzig 1977.

[HSa] Hilton, P.J. & U. Stammbach, A course in homological algebra, Grad. Texts Math. 4, Springer-Verlag, New York etc. 1971.

[HSb] Hartshorne, R. & R. Speiser, Local cohomological dimension in characteristic p, Ann. Math., II. Ser. 105 (1977), pp. 45-79.

[Ha] Hartshorne, R., Residues and duality, Lect. Notes Math. 20, Springer-Verlag, Berlin etc., 1966.

[He] Heitmann, R.C., A negative answer to the prime sequence question, Proc. Am. Math. Soc. 77 (1979), pp. 23-26.

[Her] Herzog, J., Ringe der Charakteristik p und Frobeniusfunktoren, Math. Z. 140 (1974), pp. 67-78.

[Ho 73a] Hochster, M., Cohen-Macaulay modules, Conf. Commutative Algebra Lawrence 1972, Lect. Notes Math. 311, Springer-Verlag, Berlin etc. 1973, pp. 120-153.

[Ho 73b] Hochster, M., Contracted ideals from integral extensions of regular rings, Nagoya Math. J. 51 (1973), pp. 25-43.

[Ho 74] Hochster, M., Grade-sensitive modules and perfect modules, Proc. Lond. Math. Soc., III. Ser. 29 (1974), pp. 55-76.

[Ho 75a] Hochster, M., Topics in the homological theory of modules over commutative rings, Reg. Conf. Ser. Math. 24, Am. Math. Soc., Providence R.I. 1975.

[Ho 75b] Hochster, M., Big Cohen-Macaulay modules and algebras and embeddability in rings of Witt vectors, Conf. Commutative Algebra 1975, Queen's Pap. Pure Appl. Math. 42, Queen's Univ., Kingston Ont. 1975, pp. 110-195.

[Ho 78] Hochster, M., Some applications of the Frobenius in characteristic 0, Bull. Am. Math. Soc., New Ser. 84 (1978), pp. 868-912.

[Ho 79] Hochster, M., Principal Ideal Theorems, Ring Theory Waterloo 1978, Lect. Notes Math. 734, Springer-Verlag, Berlin etc. 1979, pp. 174-206.

[Ho 82] Hochster, M., The local homological conjectures, Commutative

Algebra: Durham 1981, Lond. Math. Soc. Lect. Note Ser. 72, Cambridge Univ. Press, Cambridge 1982, pp. 32-54.

[Ho 83] Hochster, M., Canonical elements in local cohomology modules and the Direct Summand Conjecture, J. Algebra 84 (1983), pp. 503-553.

[Ho 87] Hochster, M., Intersection problems and Cohen-Macaulay modules, Alg. Geometry Bowdoin 1985, Proc. Symp. Pure Math. 46 (2), Am. Math. Soc., Providence R.I. 1987, pp. 491-501.

[Iv] Iversen, B., Cohomology of sheaves, Universitext, Springer-Verlag, Berlin etc. 1986.

[Ja] Jacobson, N., Basic Algebra I, 2nd ed., Freeman, New York 1985.

[KM] Kirby, D. & H.E. Mehran, The homological grade of a module, Mathematika 35 (1988), pp. 114-125.

[Ka 62] Kaplansky, I., R-sequences and homological dimension, Nagoya Math. J. 20 (1962), pp. 195-199.

[Ka 74] Kaplansky, I., Commutative Rings, Univ. of Chicago Press, rev. ed., Chicago 1974.

[Ke] Kelley, J.L., General Topology, The Univ. Ser. Higher Math., Van Nostrand, Princeton etc. 1955.

[Ko 86] Koh, J., Degree p extensions of an unramified regular local ring of characteristic p, J. Algebra 99 (1986), pp. 310-321.

[Ko 88] Koh, J., Super height of an ideal in a noetherian ring, J. Algebra 116 (1988), pp. 1-6.

[Ku 71] Kunz, E., Characterizations of regular local rings of characteristic p, Am. J. Math. 91 (1969), pp. 772-784.

[Ku 86] Kunz, E., Kähler differentials, Adv. Lect. Math., Vieweg, Braunschweig etc. 1986.

[Kuc] Kucera, T.G., Explicit description of injective envelopes: generalizations of a result of Northcott, Commun. Algebra 17 (1989), pp. 2703-2715.

[La] Lazard, D., Suites régulières dans les idéaux déterminentielles, Commun. Algebra 4 (1976), pp. 327-340.

[Lan] Lang, S., Algebra, 7th pr., Addison-Wesley, Reading Mass. etc. 1977.

[Le] Lescot, J., Séries de Bass des modules de syzygie, Algebra, algebraic topology and interactions, Stockholm 1983, Lect. Notes Math. 1183, Springer-Verlag, Berlin etc. 1986, pp. 277-290.

[Li] Lipman, J., Dualizing sheaves, differentials and residues on algebraic varieties, Astérisque 117, Soc. Math. Fr., Paris 1984.

[MS] Macdonald, I.G. & R.Y. Sharp, An elementary proof of the non-vanishing of certain local cohomology modules, Q. J. Math., Oxf. II. Ser. 23 (1972), pp. 197-204.

[Ma 80] Matsumura, H., Commutative Algebra, 2nd ed., Benjamin/Cummings, Reading Mass. 1980.

[Ma 86] Matsumura, H., Commutative ring theory, Camb. Stud. Adv. Math. 8, Cambridge Univ. Press., Cambridge 1986.

[Mac] Macdonald, I.G., A note on local cohomology, J. Lond. Math. Soc., II. Ser. 10 (1975), pp. 263-264.

[Matl 58] Matlis E., Injective modules over noetherian rings, Pac. J. Math. 8 (1958), pp. 511-528.

[Matl 74] Matlis, E., The Koszul complex and duality, Commun. Algebra 1 (1974), pp. 87-144.

[Matl 78] Matlis, E., The higher properties of R-sequences, J. Algebra 50 (1978), pp. 77-112.

[Mi] Milne, J.S., Etale Cohomology, Princeton Math. Ser. 33, Princeton Univ. Press, Princeton N.J. 1980.

[Na] Nagata, M., Local rings, Intersc. Tracts Pure Appl. Math., Wiley, New York etc. 1962.

[No 60] Northcott, D.G., An Introduction to Homological Algebra, Cambridge Univ. Press, Cambridge 1960.

[No 76] Northcott, D.G., Finite free resolutions, Camb. Tracts Math. 71, Cambridge Univ. Press, Cambridge 1976.

[PS 73] Peskine, C. & L. Szpiro, Dimension projective finie et cohomologie locale, Publ. Math. Inst. Hautes Étud. Sci. 42 (1973), pp. 77-119.

[PS 74] Peskine, C. & L. Szpiro, Syzygies et multiplicités, C. R. Acad. Sci. Paris (A) 278 (1974), pp. 1421-1424.

[Po 85] Popescu, D., General Néron desingularization, Nagoya Math. J. 100 (1985), pp. 97-126.

[Po 86] Popescu, D., General Néron desingularization and approximation, Nagoya Math. J. 104 (1986), pp. 85-115.

[Ra] Ratliff, L.J. jr., Chain conjectures in ring theory, Lect. Notes Math. 647, Springer-Verlag, Berlin etc. 1978.

[Ray] Raynaud, M., Anneaux locaux henséliens, Lect. Notes Math. 169, Springer-Verlag, Berlin etc. 1970.

[Re 56] Rees, D., Two classical theorems of ideal theory, Proc. Camb. Philos. Soc. 52 (1956), pp. 155-157.

[Re 57] Rees, D., The grade of an ideal or module, Proc. Camb. Philos. Soc. 53 (1957), pp. 28-42.

[Re 85] Rees, D., A note on asymptotically unmixed ideals, Math. Proc. Camb. Philos. Soc. 98 (1985), pp. 33-35.

[Ro 76] Roberts, P., Two applications of dualizing complexes over local rings, Ann. Sci. Éc. Norm. Supér., IV. Sér. 9 (1976), pp. 103-106.

[Ro 80a] Roberts, P., Abelian extensions of regular local rings, Proc.

Am. Math. Soc. 78 (1980), pp. 307-310.

[Ro 80b] Roberts, P., Homological invariants of modules over commutative rings, Sémin. Math. Supér. 15, Presses Univ. Montréal, Montréal 1980.

[Ro 80c] Roberts, P., Cohen-Macaulay complexes and an analytic proof of the New Intersection Conjecture, J. Algebra 66 (1980), pp. 220-225.

[Ro 83] Roberts, P., Rings of type 1 are Gorenstein, Bull. Lond. Math. Soc. 15 (1983), pp. 48-50.

[Ro 85] Roberts, P., The vanishing of intersection multiplicities of perfect complexes, Bull. Am. Math. Soc., New Ser. 13 (1985), pp. 127-130.

[Ro 87a] Roberts, P., Le théorème d'intersection, C. R. Acad. Sci. Paris, Sér. I 304, (1987), pp. 177-180.

[Ro 87b] Roberts, P., Local Chern characters and intersection multiplicities, Alg. Geometry Bowdoin 1985, Proc. Symp. Pure Math. 46 (2), Am. Math. Soc., Providence R.I. 1987, pp. 389-400.

[Ro 89] Roberts, P., Intersection Theorems, Commutative Algebra, MSRI Publ. 15, Springer-Verlag, New York etc. 1989, pp. 417-436.

[Rot] Rotman, J., An introduction to homological algebra, Pure Appl. Math., Academic Press, New York 1979.

[Rott] Rotthaus, C., On the approximation property of excellent rings, Invent. Math. 88 (1987), pp. 39-63.

[SS] Scheja, G. & U. Storch, Über Spurfunktionen bei vollständigen Durchsnitten, J. Reine Angew. Math. 278-279 (1975), pp. 174-190.

[ST] Siu, Y.-T. & G. Trautmann, Gap-sheaves and extension of coherent analytic subsheaves, Lect. Notes Math. 172, Springer-Verlag,

Berlin etc. 1971.

[SVa] Sharpe, D.W. & P. Vámos, Injective modules, Cambridge Univ. Press, Cambridge 1972.

[SVb] Sharp, R.Y, & P. Vámos, Baire's category theorem and prime avoidance in complete local rings, Arch. Math. 44 (1985), pp. 243-248.

[Sc 79] Schenzel, P., Dualizing complexes and systems of parameters, J. Algebra 58 (1979), pp. 495-501.

[Sc 82a] Schenzel, P., Dualisierende Komplexe in der lokalen Algebra und Buchsbaum-Ringe, Lect. Notes Math. 907, Springer-Verlag, Berlin etc. 1982.

[Sc 82b] Schenzel, P., Cohomological annihilators, Math. Proc. Camb. Philos. Soc. 91 (1982), pp. 345-350.

[Se] Serre, J.-P., Algèbre locale - Multiplicités, Lect. Notes Math. 11, Springer-Verlag Berlin etc. 1965.

[Sei] Seibert, G., Complexes with homology of finite length and Frobenius functors, J. Algebra 125 (1989), pp. 278-287.

[Sh 70] Sharp, R.Y., Local cohomology theory in commutative algebra, Q. J. Math., Oxf. II. Ser. 21 (1970), pp. 425-434.

[Sh 75] Sharp, R.Y., Some results on the vanishing of local cohomology modules, Proc. Lond. Math. Soc., III. Ser. 30 (1975), pp. 177-195.

[Sh 81] Sharp, R.Y., Cohen-Macaulay properties for balanced big Cohen-Macaulay modules, Math. Proc. Camb. Philos. Soc. 90 (1981), pp. 229-238.

[Sharpe] Sharpe, D.W., Grade and the theory of linear equations, Linear Algebra Appl. 18 (1977), pp. 25-32.

[Si] Simon, A.-M., Some homological properties of complete modules, to appear in Math. Proc. Camb. Philos. Soc. 109 (1990).

[St 78] Strooker, J.R., Introduction to categories, homological algebra and sheaf cohomology, Cambridge Univ. Press, Cambridge 1978.

[St 85] Strooker, J.R., Algunos problemas homológicos en algebra local, Cuarto Col. Latinoamericano Algebra, Publ. Cent. Lat. Am. Math. Inf. 15, Buenos Aires 1985, pp. 1-33.

[St 90] Strooker, J.R., A general acyclicity lemma and its uses, Topics in Algebra, Banach Center Publ. 26, Part 2, PWN-Polish Scient. Publishers, Warsaw 1990, pp. 229-233.

[Sz] Szpiro, L., Sur la théorie des complexes parfaits, Commutative Algebra: Durham 1981, Lond. Math. Soc. Lect. Note Ser. 72, Cambridge Univ. Press, Cambridge 1982, pp. 83-90.

[Va] Vasconcelos, W.V., Divisor Theory in Module Categories, North-Holland Math. Stud. 14, North-Holland, Amsterdam etc. 1974.

[Val] Valla, G., Remarks on generalized analytic independence, Math. Proc. Camb. Philos. Soc. 85 (1979), pp. 281-289.

[ZS] Zariski, O. & P. Samuel, Commutative Algebra Vol. I & II, Van Nostrand, New York 1958-1960.

[Za 87] Zarzuela, S., Balanced big Cohen-Macaulay modules and flat extensions of rings, Math. Proc. Camb. Philos. Soc. 102 (1987), pp. 203-209.

[Za 88] Zarzuela, S., Systems of parameter for non-finitely generated modules and big Cohen-Macaulay modules, Mathematika 35 (1988), pp. 207-215.